MW01492387

# Construction Management

## Subcontractor Scopes of Work

Jason G. Smith
Jimmie Hinze

CRC Press
Taylor & Francis Group
Boca Raton   London   New York

CRC Press is an imprint of the
Taylor & Francis Group, an **informa** business

**Cover photos:** Photo of mechanical installation by author and courtesy of Hathaway Dinwiddie Construction Company and California State University Northridge. Photo of curtain wall installation in progress provided by Walters and Wolf, Inc.

CRC Press
Taylor & Francis Group
6000 Broken Sound Parkway NW, Suite 300
Boca Raton, FL 33487-2742

© 2010 by Taylor and Francis Group, LLC
CRC Press is an imprint of Taylor & Francis Group, an Informa business

No claim to original U.S. Government works

Printed in the United States of America on acid-free paper
10 9 8 7 6 5 4 3 2 1

International Standard Book Number: 978-1-4398-0941-9 (Hardback)

---
**Library of Congress Cataloging-in-Publication Data**
---

Smith, Jason G.
    Construction management : subcontractor scopes of work / Jason G. Smith, Jimmie Hinze.
        p. cm.
    Includes index.
    ISBN 978-1-4398-0941-9 (hardcover : alk. paper)
    1. Building--Superintendence. 2. Building--Planning. 3. Construction industry--Subcontracting. I. Hinze, Jimmie. II. Title.

TH438.S537 2010
692'.8--dc22                                                        2009036643
---

**Visit the Taylor & Francis Web site at**
**http://www.taylorandfrancis.com**

**and the CRC Press Web site at**
**http://www.crcpress.com**

*To my grandfather, George Cooke Freund, who began teaching me how to build things before I could walk. He taught me many things, but never simply how to do them. He would always teach me why things are done the way they are, and from that I was able to gain an applied knowledge, not just a memory of how to do a specific task.*

—Jason George Smith

# Contents

## *Module One*

## *Module Two*

## Module Three

## Module Four

## *Module Five*

## *Module Six*

## Module Nine

# Preface

My grandfather was a huge influence in my life. Without him, I would not be where I am today. Perhaps the biggest lessons my grandfather taught me were ones he actually had no idea he was teaching. He would never simply show me how to do things; he would always teach me the reason why things are done the way they are, which is the most important lesson of all. By continually teaching me why things are done the way they are, the lessons I learned from my grandfather were exponential. To illustrate, when he taught me to fasten two boards together with a screw he taught me the reason why we were using a screw instead of a nail, namely that the pullout strength of screws is stronger than nails. When he would teach me to place a $2'' \times 8''$ joist on end rather than flat, he explained why the board is stronger on end than flat and why it would not bend as much. When we were building things together, from go-carts to cabinets to bunk beds to forts, I did not realize he was teaching me lessons that would impact the rest of my life. We were just pals spending time together.

This is the essence of what I hope readers will take away from this book. Take the knowledge and experience we have written here not simply as a checklist of issues and loose ends to be aware of, but as examples. Approach this book not as teachings on how subcontract scopes of work are written, but as teachings in why subcontract scopes are written the way they are. By applying the varied teachings of this book to the unique challenges of your projects, the knowledge and experience you gain will be exponential.

My love of construction began when I was about six months old and my grandfather gave me my first plastic hammer. While I was not quite sure what to do with it at that early age, as I grew up I figured it out and throughout childhood my passion for building grew. Even in my middle and high school years the shop classes were my favorite learning experiences.

Upon arriving at college there was no better fit for me than the construction management program where, again, I found the curriculum to fit me like a glove. The difference this time was that I began my transition from building things with my hands to managing others in building large-scale projects. This was an exciting challenge that I took on with ambition and passion. As graduation neared, I began sending out resumes to companies that appeared to be leaders in the construction industry and quickly landed my first job with a general contractor whose expertise lay in the high-tech sector. Building bio-technology facilities, clean rooms, and other cutting edge facilities excited me to no end.

When I began my career I was somewhat taken aback to realize what a huge business construction was and that a phenomenal amount of management time was spent doing what seemed at the time to be meaningless paperwork. I will never forget my first week on the job when I was asked by my project manager to send out a submittal with a transmittal on it and I wondered, what the heck is a "mittal" and why are there so many different kinds of them? As my first year in the industry progressed I came to realize just how complex and intense a major construction project can be. I began to realize the importance of things like submittals and how much money can be lost if something goes wrong. I began to realize the importance of a well-developed schedule and how much money can be lost if a project runs late. I began to realize how important a thorough review of the documents at the onset of construction can be, and again how much money can be lost if every piece of the project is not clearly included in a subcontractor's scope of work or the general contractor's estimate for self-performed work. The latter realization got me thinking the most, and is the genesis of this book.

It was not long into my first project that I noticed how much effort the estimator had put into the financial planning. This consisted not only of developing the estimate itself, but also in allocating each and every element of the project to either a subcontractor or a line item for self-performed work in our project budget. The level of detail was excruciating and I was quite reluctant to accept the necessity for that much paperwork to build a project. I thought it was a complete waste of time—at first. As the project went on, I saw on a daily basis how the bits, pieces, nooks and crannies of the project all came together in a completed facility. Without the estimator's diligent attention to detail, there would surely have been many change order issues. About midway through the project my doubts about the need for the considerable effort our estimator had put into the project turned into a great deal of respect for the job he had done. This was one of the most important lessons I learned in my career, and the first defining moment in the creation of this book.

After learning a great deal about what a large general contractor does and what it takes to pull together a major construction project, I moved on to my next projects with a much greater focus on the ever-important small and tedious details. One of the best things I did early in my career was to set up and maintain a database that I still use today. This database is in an MS Excel® spreadsheet and consists of separate worksheets for each different subcontracted trade. I use this database to keep track of all the subcontractor scope issues I have come across in my career that are likely to occur on future projects. The items in this database come from a wide variety of sources, including items that have become problems on my projects, items that have become problems on other projects that I have heard about, items that I caught before they became problems, and myriad items that have randomly come to my attention in one way or another. This database has now grown to be a tremendous tool for use in allocating subcontractor scopes of work, not to mention outlining this book.

The construction industry is extremely complex, such that no one person could ever learn everything there is to know within their lifetime. Actually, I do not believe any one person could even learn 5% of the intricacies of this industry in their lifetime. This is why we directly employ so many subcontractors, suppliers, and other individual companies for projects, each of whom have an in-depth knowledge and expertise in their respective trade. Most of these directly employed companies will in turn hire multiple material suppliers, manufacturers, sub-subcontractors, professional services firms, and other companies. Once a project is completed, it is not uncommon for nearly 1000 different companies to have been involved from the project's conception, through design, bidding, construction, and eventually completion. With this in mind, it is extremely important for the general contractor to humbly seek the advice of subcontractors in regard to their trades when it comes to scheduling, allocating the scopes of work, and any other questions that may be best answered by someone with in-depth experience in the subject area. Subcontractors spend their careers concentrating on and learning a single specific trade, whereas a general contractor's personnel will spend their careers learning numerous trades. The general contractor's personnel have only enough time in their lives to learn enough about each trade to effectively coordinate and manage them.

A unique situation in construction is that we are spread out across a region on jobsites. Unlike a business with all of their employees in one building, regular one-on-one contact and coaching from the more experienced executives to the younger people is just not geographically possible. In this industry, a format for knowledge gained by younger people simply by being in the presence of more experienced people is lacking. Secondly, this industry runs at a much faster and more frantic pace than a run-of-the-mill industry, leaving little time in the day for experienced executives to coach, or just sit down and talk to, the younger generation and future leaders of our industry.

Since young project engineers are regularly left to figure things out for themselves, they are forced to repeatedly reinvent the wheel, and make mistakes that with proper coaching could be avoided. With proper instruction, young project engineers or assistant project managers would get a valuable boost to careers. One of the personal missions I have undertaken in the industry is to find ways of bringing this coaching to young construction professionals.

In an effort to further my personal missions of increasing the early project planning efforts throughout the industry and bringing construction experience to the younger generation of builders, I found that textbooks were an excellent vehicle to forge knowledge in the industry. But, just as young project engineers need help in beginning their careers in construction, I needed help beginning my new career as an author. So I sought to find an experienced author with a great deal of educational experience to partner with for the creation of this textbook. After a great deal of research reviewing the work of other published authors in the field I found one experienced author who stood out well above the rest, Dr. Jimmie Hinze, PhD, PE. When speaking with Jimmie we found almost immediately that our missions in the industry were almost identical and after hitting it off, he enthusiastically agreed to partner with me on this project. Henceforth we began the project and are tremendously excited about the knowledge and experience this book brings to the industry.

Since general contractors naturally pair project scope of work issues with which subcontractor will perform the respective scope of work, we have organized this book such that each chapter focuses on a single, specific, subcontracted trade and the work for which that trade is, or is not, responsible. Further, we have grouped the chapters into modules representing the various phases and coordinated systems of a project. This organization not only aids the reader in a classroom setting, but also aids in the use of this manual as a reference book for use throughout a person's career.

This book is quite unique in that it concentrates on the nuts and bolts of a construction project by use of countless real-life examples, rather than on the basic philosophies and concepts of a construction project as most books tend to do. After a great deal of research I found no other book like it on the market and I am excited to bring this unique project management tool to the industry. Whether you are a young project engineer or an experienced vice president, Jimmie and I truly hope this book provides a boost to your career development and wish you the very best in furthering your exciting career in construction.

I hear... I forget
I see... and I remember
I do... and I understand
(Ancient Chinese Proverb)

**Jason G. Smith**

# Acknowledgments

Amcol International
American Hydrotech, Inc.
Anning Johnson Company
Balco USA
The Bilco Company
Burdick Painting
California Institute of Technology
California State University Northridge
Carpenters Union
Cetco Building Materials Group
Chutes International
City of Santa Monica Fire Marshal
Conco Companies
Construction Analysis and Planning, LLC
Construction Specialties, Inc.
Diversified Fire Products
Douglas Lucas
DriTherm International
Floor Seal Technology, Inc.
Forest City Development
George Donnelly Testing and Inspections
Griffin Dewatering Corporation
Hathaway Dinwiddie Construction Company
Herrick Corporation
Hi-Tech Flooring
Intracorp Companies
Iron Works Union
John Gambatese
LJ Interiors
Los Angeles Department of Building and Safety
Malcolm Drilling
Nancy Holland
Otis Elevator
Plant Construction Company
The Plumbing and Drainage Institute
Q Real Estate Partners
Rafael Vinoly Architects, PC
Regional Steel Corporation
Rite Hite Corporation
Rosendin Electric, Inc.
Si Durney

Stan Westfall
TBD Consultants
Texas A&M University
Thyssen Krupp Elevator
Tishman Speyer
University of Florida
University of Southern California
Viking Drillers
Walters and Wolf, Inc.
Wilkinson Hi-Rise

# Authors

With an extensive background as a builder, Jason G. Smith has constructed projects ranging from $10,000 to $850,000,000. During his career with a Top Ten general contractor, Jason rose quickly through the ranks to the position of senior project manager on multiple high profile projects. Known for his expertise as a builder, Jason has been welcomed by architects and owners at the forefront of the design effort, bringing expertise in constructability to the team.

As a true leader, he has taken on additional responsibilities of varying capacities as the superintendent, owner's representative, and various other roles. Through these experiences a true understanding and appreciation for the different perspectives of the various project team members have developed.

Bringing together a superior knowledge of construction means and methods and an understanding and appreciation for the different perspectives of the various project team members, Jason founded Construction Analysis and Planning, LLC, the premiere constructability consulting firm in the nation.

Jimmie Hinze, PhD, PE is a professor at the M. E. Rinker, Sr., School of Building Construction and director of the Fluor Program for Construction Safety at the University of Florida. He received a BS and MS in architectural engineering from the University of Texas and a PhD from Stanford University. He was previously a professor of civil engineering at the University of Washington and also at the University of Missouri-Columbia. For more than 30 years, he has conducted research in a variety of construction-related topics, but primarily in the areas of construction safety. He has authored textbooks on construction safety, construction contracts, and construction scheduling. He has written over 100 articles and conference papers on various construction topics.

# *Module One*

# 1 Demolition*

The scope of work of a demolition subcontractor is generally fairly easy to write. Most demolition work consists of getting rid of anything that is not shown to be there at the end of the project. Despite the simplicity, there still are a few things to keep in mind.

## SCOPE OF WORK ISSUES RELATED TO DEMOLITION

1. Clearly identify the party (the general contractor or the surveying subcontractor) who has responsibility for layout. It is often advisable for the party who is completing the primary site layout (which means establishing and physically marking the building grid lines and elevations for coordinated use by all subcontractors) for construction to also complete the layout work for the demolition subcontractor. Often the architectural documents will indicate the line of demolition to match the building footprint, but this is not usually the full extent of demolition work. The demolition subcontractor is commonly held responsible for additional tasks that may lie outside of the building footprint. It is therefore important that the following information be considered by the general contractor for inclusion in the demolition scope of work:

   (a) Account for hardscape in the way of the shoring piles.
   (b) Account for hardscape in the way of the dewatering wells.
   (c) Account for hardscape in the way of the tower crane and/or material hoist pad (when located outside the building line).

   For renovation projects the demolition layout may be quite elaborate. Being sure the demolition subcontractor knows exactly where to begin and end the destructive operations will be quite important.

   (a) Be sure that clean, properly cut ends of walls are left. When demolition work is scheduled to stop along a drywall partition, do not let the demolition subcontractor destroy an extra foot or more, as this will only result in additional work for the drywall subcontractor who will have to come back, trim, and replace the work that was removed. On a large renovation project there can be hundreds of these conditions, which can eventually add up to a large drywall change order request.
   (b) When demolishing piping, be sure the pipe is cut cleanly. If the demolition subcontractor simply yanks the piping down, this will invariably result in additional damage, for which the plumbing

---

subcontractor will issue a change order request to replace the pipe. If the demolition work is not performed carefully and properly, piping that will be left in place may also sustain damage from the demolition crew's excessive shaking which rattles the piping twenty, thirty, or more feet down the piping run.

2. A building in a dense urban environment will often be constructed on an existing paved lot. In this case it is advisable to complete the site-work demolition in two phases. First, demolish the paving within the building footprint, with an allowance for additional space to accommodate shoring piles, dewatering wells, etc. It is a good construction practice to leave the perimeter hardscape in place through the first two-thirds of the construction effort to act as a solid and clean working surface around the building. Demolish this perimeter hardscape shortly before construction of the permanent hardscape is planned.

3. A clear division between what is removed by the demolition subcontractor and what is removed by the excavation subcontractor needs to be established (Figure 1.1). One way of distinguishing between the two is to focus on the difference between man-made and natural materials. The demolition subcontractor will remove the existing asphaltic-concrete paving, concrete, vapor barrier, and other man-made items. The mass excavation subcontractor will then remove all aggregate bases, sand, and other natural materials. Since it is located along with the natural materials, it is also advisable for the mass excavation subcontractor to remove the underground utility piping, conduit, and wire. The demarcation between the materials removed by the demolition subcontractor and the excavation subcontractor may overlap without careful consideration of the in-place conditions. Therefore, providing a suitable description in the bid instructions is vital for these subcontractors.

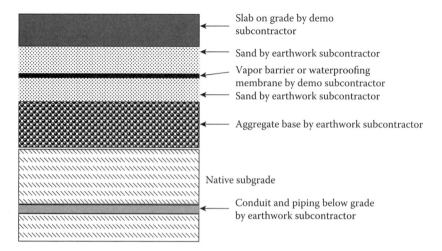

Slab on grade by demo subcontractor

Sand by earthwork subcontractor

Vapor barrier or waterproofing membrane by demo subcontractor

Sand by earthwork subcontractor

Aggregate base by earthwork subcontractor

Native subgrade

Conduit and piping below grade by earthwork subcontractor

**FIGURE 1.1**   Demolition of slab on grade.

4. Include work outside the limits of construction shown on the contract drawings. A common example is trees along the public sidewalk that line the streets. These trees are often in the way of trucking and other general construction work, so it is beneficial to cut them down at the onset of construction and simply replace them at project completion. The sidewalk will also possibly incur significant construction damage and may require significant repair or replacement in the trucking path, if not in its entirety.

5. When a tree is to be removed, be sure the subcontractor removes the entire tree, including the stump and roots. Trees can be removed by either the demolition or excavation subcontractor, though it is commonly most efficient for the excavation subcontractor to perform this work as the removal of the roots of medium to large trees will require significant digging work. This is to ensure that when transplanting a new tree, along with constructing a footing, running utility piping, or any other planned new construction, there is adequate organic-free stable soil below finish grade to accommodate it. Omitting the cost of grinding the stumps and removing the roots is a common mistake. The criteria for root removal will commonly be described in the contract documents as stipulating that all roots above a specified diameter are to be completely removed.

6. The subcontractor responsible for concrete coring and saw-cutting operations (whether it is the demolition subcontractor or another subcontractor performing this work) must maintain control of dust, as well as complete containment and cleanup of the concrete slurry created from the water-lubricated concrete cutting equipment. If not vacuumed up before it dries, this slurry can permanently stain adjacent surfaces. This is a common cause of damage to hardscape and typically requires demolishing and replacing the stained items.

7. Because of the curvature that is typical in concrete saw blades, a clean 90-degree corner cannot be made. Overcutting of a concrete deck is rarely allowed; therefore the demolition subcontractor must include the cost of chipping and smooth grinding of any corners.

8. Protection of adjacent buildings, hardscape, landscape, and other existing features is very important. This protection should ultimately be provided by the general contractor. The reason is that demolition subcontractors are experts at their specific trade, namely demolition. Their crews are trained to smash things and this constitutes most of their work. Protection is not something for which they have received training, and generally it is not what they are asked to do. It is best to have a detail-oriented crew complete the protection work, workers who are trained in caring for building corners, landscaping, etc.

9. In most cases, when a site is being demolished the existing structure(s) is more than 40 years old, an age where asbestos (termed ACM, or asbestos containing materials) and lead paint are common occurrences. It is common to encounter transite pipe, insulation containing asbestos, floor tiles containing asbestos, ceiling tiles containing asbestos, plaster containing asbestos, taping mud containing asbestos, and/or lead paint through the course of demolition. Where

**FIGURE 1.2** Dust barrier to isolate construction work area from occupied spaces on renovation projects. (Photo by author, courtesy of Hathaway Dinwiddie Construction Company and California Institute of Technology.)

these hazardous materials are encountered in a building, they will generally be addressed by the owner at the onset of a project, prior to the general contractor commencing work. This preemptive remediation work does not always result in the removal of 100% of the hazardous materials, as these materials will be concealed in walls and in the ground. It is important to address the acceptable response period for the hazardous materials abatement subcontractor when a surprise construction discovery is made of lead, asbestos, or any other hazardous material. A quick response is needed to prevent significant schedule delay. In most cases, primarily due to liability and insurance concerns, the hazardous materials abatement subcontractor will be under direct contract with the owner of a project rather than the general contractor.

The best way to address concealed contaminated materials discovered during the course of construction is to enlist the services of an abatement subcontractor at the onset of the construction phase with an indefinite scope agreement. If such a subcontractor is identified early, when contaminated materials are found there will be a minimal loss of scheduling time in getting someone to take care of the problem. This proactive approach will often pay off within the first two months of a renovation project.

10. For renovation work in a partially occupied building, ensure that proper and effective negative air containment (Figure 1.2) is provided to prevent dust migration from the construction area into the occupied spaces. Because it is a good construction practice to keep this containment system in operation throughout the construction period, it is most appropriate for the general contractor to provide and maintain the system rather than the demolition subcontractor who will demobilize from the site early in the project.

# 2 Shoring and Underpinning*

The construction of multi-story structures generally is associated with deep excavations to construct parking garages, provide lateral building anchorage, or gain access to suitable foundation materials. The earthen embankments of an excavation must be supported during the excavation work and while the permanent building structure is being constructed. The excavation support (shoring) system will be designed to support and resist the lateral loads generated by the excavation walls and the equipment/materials that might be placed near the edge of the excavation. The support system will be designed and installed by the shoring subcontractor and a different firm, the excavation subcontractor, will dig and remove the material to create the excavation.

Shoring and underpinning, because of its apparent simplicity, is often overlooked as a significant concern when scheduling a project or addressing issues requiring considerable coordination effort. As a result, this work can often become the genesis of substantial change order requests. Shoring and underpinning are among the earlier activities encountered on many projects. Though this scope of work generally requires relatively few crew members, only a handful of different materials, and has comparatively constant means and methods from project to project, the shoring and underpinning work is crucial to a successful project.

Because of the size of the shoring scope, a single mistake in planning or execution, or an exponential growth in the scope of work can cost tens of thousands of dollars. Also, this scope of work typically has a day-for-day impact on the critical path of the project schedule, making a smooth operation imperative to overall project success. Because any delay in the shoring work will directly impact the schedule, it is often advisable to pay the overtime premium for a crew to selectively work 50 to 60-hour workweeks. Since this phase of construction is often limited to a single crew driving the critical path (as opposed to later in the project when eight or ten crews may concurrently drive the schedule) paying the premium cost for only a handful of workers is an inexpensive way to expedite the schedule. The term "selectively" is used for one very important reason-tieback testing. An important part of this scope consists of waiting for grouted tiebacks (Figure 2.1) to sufficiently cure to a point in which the design strength is achieved, so further excavation can continue. This method of schedule compression consists of an iterative process of working overtime such that the schedule is driven to take advantage of weekends for grout cure time (generally four days). If grouting is done on a Friday, the tiebacks will be ready for testing the following Tuesday morning at the latest, meaning that

---

* MasterFormat Specifications Division 31

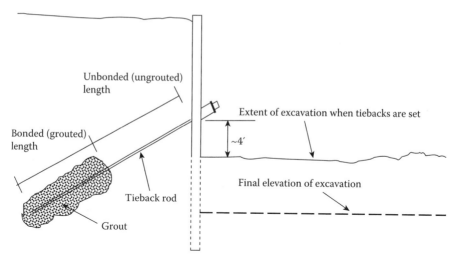

**FIGURE 2.1**   Tieback components.

only one regular work day is lost. If the grouting is done on a Monday, the grout may not be ready for testing until the following Friday, meaning that three regular work days are lost. It is not possible to efficiently and effectively coordinate all of the grouting activities to occur on Fridays, but this scheduling approach should always be considered.

## COORDINATION WITH THE WATERPROOFING SYSTEM

It is of utmost importance that the shoring system be compatible with the below-grade waterproofing system. Experience has shown that shotcrete lagging with a wood float finish is far and away the best substrate for any below-grade waterproofing system. Figure 2.2 this substrate is preferred over backfilled earth, primarily due to the potential for damage to the membrane during the backfilling operation. The traditional wood-lagging systems leave voids behind and between the timbers, not to mention the eventual degradation of the timbers (even 4x treated timbers will degrade over time), whereas a shotcrete lagging system will form a positive, continuous, bond with the earth.

## PROCURING THE SHORING PILES

Setting the shoring piles is one of the very first activities on a project, but these piles often take four to eight weeks, or more, to procure. Waiting for the shoring design to be complete and the permit to be issued before procuring the shoring piles will result in a tremendous delay. To mitigate this delay, it is advisable that the shoring piles be sized by an engineer making rough calculations immediately upon award of the shoring subcontract. Based on conservative calculations, the piles can then be procured in an effort to get them on site as soon as possible. As a contingency factor to ensure that the rough calculations completed by the shoring engineer are not

**FIGURE 2.2**  Shortcrete lagging. (Illustration provided by Malcolm Drilling.)

too light, it is advisable to pay the small increased price for increasing the pile size slightly to add a cushion of safety rather than risking having undersized piles. It will cost a little more, but avoiding the high potential for schedule delays is well worth it. The project completion date will be delayed one extra day for every day those piles are needed, but are not available.

## SCOPE OF WORK ISSUES RELATED TO SHORING AND UNDERPINNING

1. Be sure that all permits are covered, not just the shoring permit itself. Consider the following list of permits that may be required for the project:
   (a)  Shoring permit.
   (b)  Encroachment permits for the tiebacks and any soldier piles that will cross the property line(s). This permit is a staple for dense urban environments.
   (c)  Neighbors' permission for encroachment under their property for the tiebacks and any soldier piles that may cross the property line (this is often a requirement that must be fulfilled prior to municipalities issuing the encroachment permit).
   (d)  If piles are driven (in lieu of vibrating or drilling installation methods), a variance for the local noise ordinance may be an additional step. In residential neighborhoods there may be limitations on the hours of operation.
   (e)  Peripheral work associated with the permits (encroachment in particular) is typically required. For example, the municipality may require the contractor to survey the sanitary and storm sewers adjacent to the site before construction begins. The survey will be repeated after construction to document any damage that might

have been caused by the tieback drilling and/or other construction operations. (This survey is conducted by a specialty firm with the use of a device that consists of a video camera mounted to a remote controlled chassis, similar to a very durable remote controlled car. This camera-equipped vehicle is then driven from manhole to manhole through the city sewers.)

2. There are three different methods of shoring pile installation and it is important to understand which method is to be used by the shoring subcontractor, as they each have their own trade coordination issues.

    (a)　The most common method of shoring pile installation is driving. This method is extremely noisy, which can be quite disruptive to nearby residences and businesses. Because of this disruption many municipalities have prohibited pile driving. If pile driving is allowed, the hours of operation will likely be limited. A benefit that pile driving has over drilling and setting the piles in a pre-drilled hole is that there are no drilling spoils to remove from the site, thus resulting in a savings from the associated loading, trucking, and disposal costs.

    (b)　Drilling and setting the piles in a concrete base is the second most common method of pile installation. A tremendous benefit this method has over driving the piles is quality control. When setting the piles in a pre-drilled hole, the shoring subcontractor maintains excellent control on the bottom of the pile, whereas when driving the piles the bottom tip has a tendency to stray out of alignment.

    (c)　The least common method of setting the shoring piles is vibrating them into place, though this relatively new method of pile installation is rapidly gaining in popularity. This method is nearly identical to the pile driving method. It is just as accurate and even uses a very similar looking rig. This method has one very important benefit over pile driving—it is not nearly as noisy.

3. Verify the line of shoring in relation to the building line:

    (a)　To achieve the desired excavation dimensions, the line of shoring must allow for the thickness of the waterproofing system, including protection boards and drainage mats.

    (b)　For shotcrete lagging, the piles are to be located sufficiently back from the main excavation to allow for the thickness of the shotcrete lagging. For wood-lagging systems the face of the pile will be in line with the face of the lagging.

    (c)　Because of construction tolerances it is suggested to hold the piles back slightly for deep excavations to gain assurance that the piles will not migrate into the structural wall as they approach the bottom of the excavation. It is not a significant problem if the piles move away from the building line, but if they encroach into the building line it can be a very significant problem.

    (d)　A sample calculation of the minimum amount of over excavation will clarify how this is computed. Assuming four inches of shotcrete lagging will be applied in an 80 feet deep excavation, the piles are not held back as a contingency factor, and a ½″ thick bentonite waterproofing system with 1″ thick protection board for which the total assembly thickness would be as follows (Figure 2.3):

4″ shotcrete lagging + 1-1/2″ waterproofing system
+ 0″ contingency = 5-1/2″ from the inside face of pile to the outside face of structural wall.

4. Hand trimming for the lagging behind the face of a pile is a commonly orphaned or overlooked item of work. Both the excavation and shoring sub-contractors will generally exclude hand trimming, but it makes the most sense for the shoring subcontractor to pick up this work item.
5. Confirm the number of tiebacks that will be required, and the necessary length and diameter of the bores (Figure 2.4). Then verify that the excavation bidders have included the cost of removing the resultant spoils. Because shoring is a temporary structure, thus not typically shown in the contract drawings, the excavation bidders will not know what quantity of tieback drilling spoils to pick up unless they are informed about it. Quite often, excavation bidders will simply exclude this work when they are not sure of the quantity, which introduces a strong potential for a change order request.
6. Establish whether the piles will be driven, vibrated, or drilled. If driven or vibrated, there are no additional spoils to worry about. If the piles are

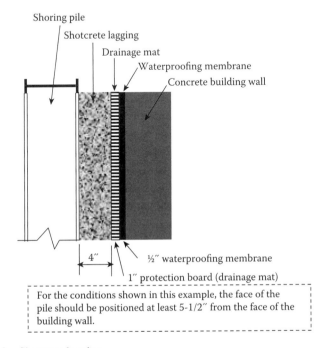

For the conditions shown in this example, the face of the pile should be positioned at least 5-1/2″ from the face of the building wall.

**FIGURE 2.3**   Shotcrete lagging.

**FIGURE 2.4**    Tieback drilling rig. (Illustration provided by Malcolm Drilling.)

drilled, the excavation subcontractor will incur the added expense of hauling off the spoils.

7. Because of equipment availability, the shoring subcontractor will often
exclude the immediate pushing of the tieback drilling spoils out of the way
of the tieback operation. The spoils generated from this drilling operation
will be piled immediately in front of the hole after drilling, so although
the spoils will not likely be immediately hauled off from the site by the
excavation subcontractor, they will still need to be expeditiously pushed
out of the way to allow the shoring subcontractor access to place and grout
the tieback cables. This is a very small amount of work that excavation
subcontractors generally do not mind doing—when they are on the jobsite
anyway. Because these spoils are generated during the long process of drilling tiebacks and the excavation subcontractor will traditionally pull off of a
job for two to five days to allow the shoring subcontractor's tieback operation to catch up, the point in time when the shoring subcontractor needs the
excavation subcontractor to move the spoils out of the way will often be the
time the excavation subcontractor is not on site. The shoring subcontractor will either need express permission to use the excavator's equipment
or will bring in a small skid loader to do the work. To compile their bids,
the subcontractors will refer to the project schedule to determine when the
excavation subcontractor will, or will not, be on site.

8. The trucking ramp and/or excavation perch will cover some of the tiebacks until the very end of the excavation work. Diligent planning is the
only mitigation measure for this issue. Do not fall into the trap of being
nearly complete with the excavation and then lose a great deal of time
by excavating the ramp to the first level of tiebacks. This will involve
demobilizing (and paying for the demobilizing/remobilizing of) the excavation subcontractor, stopping (and paying for) the tieback drilling rig to
be hoisted in and out of the excavation, waiting three days for the tiebacks
to reach strength, waiting a day for tieback testing, then remobilizing the

excavation subcontractor to dig to the next row of tiebacks. This costly process is repeated for each row of tiebacks buried behind the ramp. This will never be a zero cost issue, but there are a few ways to help mitigate the loss of time and money:

(a)    First, the upper level of tiebacks is generally shallow and easy to dig down to, but the trucking ramp will be shut down during this operation. This is an excellent operation to take care of when the excavation subcontractor needs to demobilize for a few days and allow the shoring subcontractor to catch up with their tieback operation. This first row of tiebacks behind the ramp is generally easy to install without losing any time, but when the second row is reached, things get significantly more difficult and expensive due to the extensive excavation and replacement of the ramp soil resulting from their increased depth below the ramp.

(b)    Secondly, the most common method of addressing this problem for lower tieback rows is moving the ramp at least once during the excavation phase, typically when the second tieback level is reached.

9.  The upper two to five feet of the shoring system will be cut off once the building erection passes the ground level. This is necessary to allow room for the landscaping, underground utilities, and/or hardscape to properly tie in with the building. This activity, consisting of digging, torching, and backfilling against the membrane, is a traditional cause for damage to the waterproofing membrane. Damage to the waterproofing membrane may not be fully avoided, but it can be minimized with some of the following approaches:

(a)    For a traditional wood-lagging system: When cutting off the tops of the shoring piles the torch will burn through the waterproofing membrane and cause numerous difficult patches to be made. With the shoring engineer's approval, cut the piles about halfway back from the face of the excavation prior to installing the waterproofing membrane (Figure 2.5). This will enable the torch to be kept about six inches or

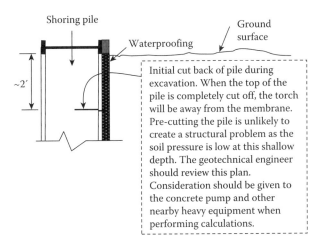

**FIGURE 2.5**   Pre-cutting a shoring pile.

more away from the waterproofing membrane. While the pile will heat up, there is still a potential of causing heat damage to the membrane, but the overall risk of damage is greatly reduced. To accomplish this, the piles will need to be cut when the first excavation bench elevation is reached, commonly at a depth of five feet, so the cutter will have a working platform. When the building is erected out of the excavation the waterproofing membrane will already be covering the piles by the time the erection gets back near ground level.

(b)    For a shotcrete lagging system: The membrane will be protected from the torch by the shotcrete, so the problem of damaging the waterproofing when torch cutting the tops of piles is solved (another good reason for using shotcrete lagging). Unfortunately, the shotcrete lagging will be attached to the piles with studs on the piles, resulting in chipping and demolishing concrete right over the membrane—not a good idea. For this reason the upper portion of a shotcrete lagging system must be lagged with traditional wood timbers, but with this approach the face of the wood lagging will be several inches behind the face of shotcrete lagging. To provide a straight substrate for the waterproofing membrane and proper form for the concrete building wall, the wood-lagging portion must be built out flush with the shotcrete. There are several ways to build up the wood-lagged portion to be flush with the shotcrete below:

   (i)    Tack weld metal decking to the shoring piles to form-up the top of the excavation. This decking can be sized to be flush with the face of the shotcrete lagging. When the tops of the piles are cut off, the metal decking will come right out with them.

  (ii)    Use traditional wood formwork to bring the face of the wood lagging out to the surface of shotcrete lagging. Also, be sure to install the formwork prior to shotcreting so the shotcrete can be screeded to the face of the formwork, thus providing for a smooth transition and acceptable substrate for the waterproofing membrane (Figure 2.6). If the shotcreting is done first, it will not align perfectly with the face of the wood formwork without significant effort.

(iii)    Shotcrete lagging can be used all the way to the top of the excavation by placing a bond breaker around the elevation separating the top (removed) portion of the shotcrete lagging from the abandoned shotcrete lagging below.

10. Be sure to include the excavation and backfill necessary to cut off the tops of the piles in the excavation subcontractor's subcontract. The shoring subcontractors do not often have the equipment or expertise for excavation or backfill, so if this is included in the scope of work they will likely need to subcontract it themselves. For this reason, the most economical route is typically to have the excavation subcontractor include this work in its bid. Regardless, whoever does this excavation and backfill must be trained in the care necessary to work adjacent to a waterproofing membrane.

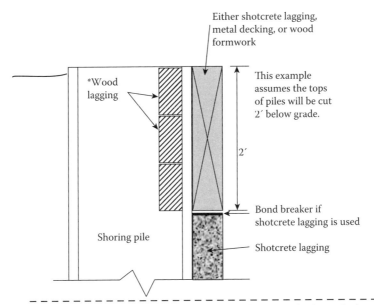

Either shotcrete lagging, metal decking, or wood formwork

*Wood lagging

This example assumes the tops of piles will be cut 2′ below grade.

2′

Shoring pile

Bond breaker if shotcrete lagging is used

Shotcrete lagging

*If wood formwork (most common) or metal decking is used to bring the face of the pile out to the face of shotcrete lagging, wood lagging will be necessary for soil retention behind the wood formwork. This is not necessary in the case of shotcrete lagging.

The wood formwork or metal decking should be installed prior to shotcreting so the shotcrete can be screeded to the formwork. This will ensure that these two substrates are kept in the same plane. In the opposite condition, the shotcrete will inevitably be wavy and the wood formwork will not properly align, thus providing a poor waterproofing substrate.

**FIGURE 2.6**   Forming at the top of shotcrete lagging.

11. Clearly communicate the magnitude of the construction loads that are anticipated adjacent to the excavation (Figure 2.7). Unless specifically told otherwise, a shoring subcontractor will bid and design the shoring system to withstand earthen loads only. Additional construction loads such as the loads attributed to excavators, hauling trucks, concrete trucks, concrete pumps, office trailers, cranes, settling tanks, decontamination systems, and any other heavy loads anticipated adjacent to the excavation, and their specific locations, must be clearly communicated to the shoring subcontractor to ensure they are accounted for in the design.

    (a)   Note that shoring engineers, when accounting for construction loads, will often require loads to be spread by the use of crane mats and sometimes simple trench plates. The shoring subcontractor will not provide these peripheral items, so be sure they are included elsewhere by either the general contractor or the respective subcontractors.

12. Confirm if the tiebacks need to be de-tensioned and clearly relay that information in the bid documents so the shoring subcontractors can properly reflect this issue in their bids. This is a common city requirement and often

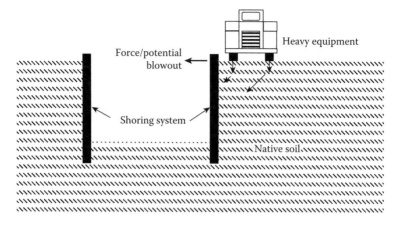

**FIGURE 2.7** Equipment adjacent to shoring.

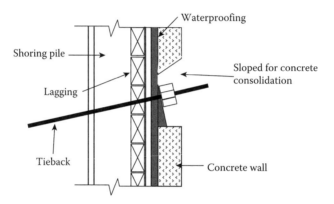

**FIGURE 2.8** Tieback de-tensioning blockout.

an owner of a large campus will also have a similar requirement. (Note, due to the blockouts required for de-tensioning, this will affect the place and finish, rebar, shotcrete, formwork, and waterproofing subcontractors as well) (Figure 2.8).

13. Coordinate the pile layout with the utilities entering the building. This is especially true of the electrical service. Telephone and wet utilities typically have some lateral leeway as to where they enter the building, but the switchgear for the electrical service is usually in a fixed position in the electrical room, with its back to the perimeter wall, and the electrical service will need to enter the building directly behind the switchgear. If a soldier pile is in the way of these conduits it will cause a significant problem in properly routing the conduits into the building and effectively sealing the waterproofing membrane. The best way to handle this coordination issue is in the shoring design phase. The shoring engineer will establish a standard spacing for all piles around the excavation, but the

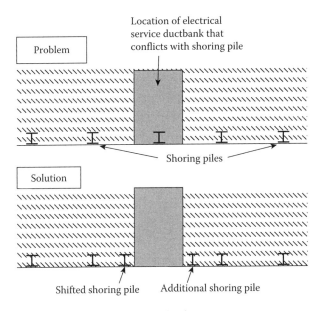

**FIGURE 2.9**   Pile conflict with electrical ductbank.

pile in conflict with the electrical service will need to be shifted one way or the other. Because this shift will increase the spacing between piles opposite from the shifting direction, an additional pile will need to be added (Figure 2.9).

14. To expedite the shoring design and soldier pile procurement, the shoring scope is often bid from very conceptual documents. Because of this, it is very important to clearly relay the exact depth(s) at the bottom of excavation in the bid documents. It is equally important to err on the conservative side when anticipating the bottom of the excavation. It is quite inexpensive to size soldier piles an extra foot or two long, but it is extremely expensive if soldier piles are subsequently found to be a foot or two short. Be sure the shoring bidders are aware of:

   (a) The exact elevation of the bottom of the excavation. Conceptual bid documents often show the elevation of the basement slab on grade, but the shoring subcontractor will need to know the bottom of footing/mat, as well as the depth the excavation will go below the bottom of footing/mat. Be sure to add up the thicknesses of the aggregate base, sand layer(s), mud slab, waterproofing system, and/or protection slab.

   (b) Identify the elevator, escalator, sump, sewage, and any other pit locations and their respective depths.

   (c) The presence and configuration of the subdrainage system.

   (d) Tower crane foundation location and depth.

15. For underpinning conditions (Figure 2.10) ensure that the as-built information for the structure to be underpinned is as accurate as possible (Figure 2.11). differing conditions, especially for older buildings, are a common cause

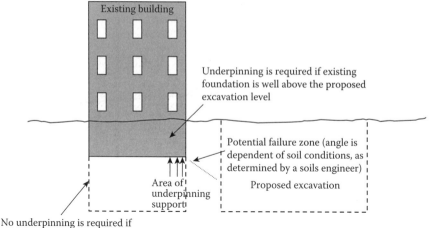

**FIGURE 2.10**   Underpinning.

**FIGURE 2.11**   Underpinning. (Illustration provided by Malcolm Drilling.)

of change order. If there is no basement, a shallow exploratory pit can be dug to verify the foundation type and size. Such an exploratory pit can generally be done quite economically and is highly recommended. If the structure to be underpinned has a basement it may not be economical to dig an exploratory pit to verify the base of the structure. In this case, simply proceed under educated assumptions made by the structural engineer and have the structural engineer review the basement wall periodically as excavation progresses. The structural engineer will establish the frequency of these inspections.

16. Plan for supplemental tiebacks. Because of natural changes in the layers and patterns of the earth, it can only be speculated, based upon soil borings, that

**FIGURE 2.12**    Rakers and corner bracing.

tiebacks will all be drilled into soils typical of those found in the soil borings. However, some weaker, and some denser, spots are bound to be encountered. The dense areas rarely cause concern, but when hitting weak areas, the tiebacks may fail the pull test and supplemental ties will need to be installed. Anticipate this in the project schedule, as well as when planning the number of de-tensioning blockouts and waterproofing boots for the tieback heads.

17. Determine responsibility for the shoring monitoring. Typically a weekly survey to monitor deflection of the piles and settlement of the adjacent grade will be required both by good construction practice and the encroachment permit. It is best for either the general contractor or owner to pick up this surveying work, as it could be construed as a conflict of interest if the shoring subcontractor was required to monitor the shoring. (This surveying work is also required to monitor settlement caused by the dewatering operation.)

18. Whenever possible, avoid internal bracing, including rakers, knee braces, and corner bracing (Figure 2.12). Dealing with internal supports while constructing the building walls, foundation, and below-grade waterproofing around them is extremely difficult. These internal supports should only be used when it is impossible to use tiebacks due to conflicts with adjacent building basements or the inability to attain permission for tiebacks to cross a neighbor's or city property line.

19. The guardrails around the excavation are most commonly furnished and installed by the shoring subcontractor, although this work will often be excluded or it will be shown as an additive alternate to their base bid. Additionally, the ledge of an excavation is very vulnerable to items rolling or being accidentally kicked over the edge, therefore toe-boards of some sort are necessary. Most often, a combination of lagging boards and sandbags are used for this purpose, but regardless of the method, it is best for the shoring subcontractor to pick up this means of protection.

# 3 Mass Excavation and Site Grading*

Whenever mass quantities of excavated earth are involved in a construction project, care must be exercised to control costs. Excavation costs are commonly tracked and benchmarked as a cost per cubic yard of excavation. With this method of quantification and cost measurement any cost-saving ideas that are implemented will result in the savings being accrued for every cubic yard of material that is handled. The saving of just a few pennies per cubic yard can ultimately add up to a substantial sum on a large project. Similarly, any inefficiency, error, or other problem in the operation will result in an escalated cost per cubic yard of material until the problem is corrected (Figure 3.1).

Similar to shoring, the change order and delay potential of mass excavation work is quite often underestimated. This operation has a potential day-for-day impact on the project schedule and, if not run in an efficient manner, a single error can result in a substantial delay to the project schedule and commensurate financial losses.

The various nuances of mass excavation work that must be considered are often overlooked when planning and scheduling a project. This is particularly true in urban areas where a parcel of land may have been subjected to many different uses over the years. Such parcels may harbor a variety of unforeseen conditions that must be anticipated. While it may sound absurd to anticipate the unforeseen, this is a suggested practice where mass excavation work is concerned. Transite (asbestos-containing) pipe, old brick footings, relics, concrete foundations, contaminated soils, unmarked utilities, waste materials, abandoned vehicles, and archeological treasures are all potential discoveries on such parcels. The top 10 feet of excavated materials are likely to yield at least 90% of the unforeseen conditions. While every contingency cannot be accounted for in the work plan, it is imperative, as a first step, that the contractor recognizes the potential for the discovery of unforeseen conditions, also referred to as differing site conditions. It is important to carefully examine all available historical documents for any information that might reveal something about the subsurface conditions. It is also important to carefully walk the site and note any anomalies.

The second step is to research the history of the site. If a paint shop, old laundry, or auto repair shop was previously located on a site, it should be recognized that there is a high probability of encountering contaminated soils. If a gas station was located on a site, the concern includes not only contaminated soils, but also abandoned

---

* MasterFormat Specifications Division 31

**FIGURE 3.1**   Excavation.

underground fuel tanks. Note that the construction documents may contain no information about the prior usage of a site. This is where an independent research effort can provide valuable information. Such due diligence in being prepared for the unexpected will keep the surprises to a minimum.

## SCOPE OF WORK ISSUES RELATED TO MASS EXCAVATION AND SITE GRADING

1. As modifications to the site were made in the past in existing facilities, the mechanical, electrical, plumbing, telephone, and fire sprinkler subcontractors may have capped or rerouted their respective underground utilities. The abandoned pipe, conduit, wire, etc. will need to be removed by the excavation subcontractor in conjunction with the mass excavation work. Construction drawings of the original facility, if available, may be helpful in identifying the potential locations of such abandoned utilities. When the excavated material is to be used for fill material, these utility materials will need to be isolated and segregated as the excavation work progresses. Hauling off and disposing of the underground materials will be completed by the excavation subcontractor; therefore, an estimated quantity of these materials must be made to the best of the project team's abilities.

2. The final elevation of the excavation should be clearly noted in the construction drawings, but on a fast track project the excavation quantities are often estimated on the basis of conceptual drawings which may not even indicate the thickness of the slab. The bottom elevation of the excavation must be clearly established prior to bid time to stave off the potential of a costly change order request from the excavation subcontractor. This elevation must not be confused with the finish floor elevation, which is often the only elevation that is clearly specified. When the

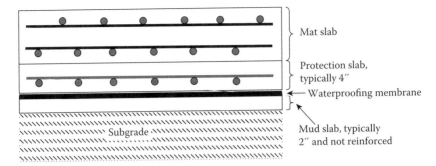

FIGURE 3.2   Base of a mass excavation.

finish floor elevation is given, be sure to indicate the thickness of the underlying components, including the slab, the aggregate base, the sand fill, the mud slab, the protection slab, and the waterproofing membrane (Figure 3.2). An estimate based on an error of a few inches in the excavation elevation may seem to be a minor issue, but on a large site this can add up to be a substantial sum. For example, an estimate based on an error of two inches in a 250' × 400' slab will constitute an error of over 600 cubic yards (250' × 400' × 2/12 = 16,667 cubic feet or 617 cubic yards). Even when bidding fully developed construction documents, the depth added due to the contractor's means and methods (such as mud and protection slabs) will not be illustrated in the drawings and must be conveyed in writing in the bid instructions from the general contractor to the subcontractors.

3. Just as the elevation of the bottom of the excavation must be clearly understood, so too must the dimensions of the sides of the excavation (Figure 3.3). To establish the neat excavation dimensions, it is important to include the thickness of each element of the excavation, including the waterproofing membrane and lagging. It is a safe practice to align the piles slightly back (as much as one or two inches varying with the depth of the excavation) from the original location to account for minor deflection and minor placement variations for driven piles at deep excavations. When this is done, be sure the excavation subcontractor includes the additional volume of excavated materials. Since this variance is generally small, the width of the excavation tends to be easier to estimate with reasonable accuracy on conceptual design drawings than the base materials.

4. The estimate of the materials to be hauled off is often greater than the simple volume computed for the excavation. The excavation subcontractor should be aware of any additional materials that might be encountered from the following additional sources:

   (a)   Drilling for soldier piles (not applicable for driven or vibrated piles).
   (b)   Drilling for dewatering wells.
   (c)   Drilling for tiebacks (and anticipated supplemental tiebacks).
   (d)   Drilling for caissons.

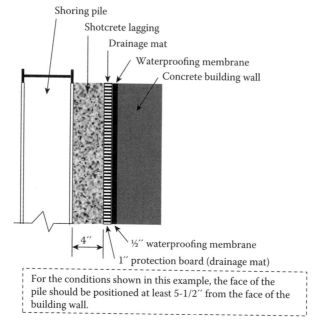

**FIGURE 3.3**　Side of a mass excavation.

    (e)　Underpinning operation, including undermining an adjacent structure, drilling for pile supports and excavating for working pits.

    (f)　Elevator hydraulic ram casings.

  5.　The dewatering system for an excavation will generally be routed above ground, but an exception will be made where it crosses the trucking ramp(s) that provides trucks with access to and egress from the excavation. As the dewatering subcontractor will probably not have the appropriate equipment on site for this work, it is a good idea to include the trenching and backfilling for dewatering pipes across the truck ramps in the scope of the excavation subcontractor. Even then, the dewatering subcontractor should be directed to provide the steel protective sleeves for the dewatering pipes and to lay the piping.

  6.　Although it may seem minor, additional excavation work is commonly required on construction projects to enable the installation of elevators, escalators, sumps, sewage systems, thickened slabs, depressed slabs, etc. Pit dimensions (length, width, and depth) and their locations must be clearly understood. This is generally not a problem when bidding the mass excavation work with a complete set of construction drawings, but significantly greater potential for error is introduced when the estimate of excavated materials is based upon conceptual drawings. Information on the required pits must be quite clear to the bidders.

  7.　On large excavations, it is common to construct trucking ramps, so that trucks and other equipment can enter and exit the excavation. This makes it convenient to haul the materials from the excavation. Before these ramps are constructed (if possible, prior to awarding the subcontracts), coordinate the trucking ramp locations, tieback installation flow of work, and sequencing for any underpinning work between the shoring and excavation subcontractors. This is discussed in greater detail in the chapter on shoring and underpinning.

8. High-rise buildings and other large structures are often constructed with the aid of tower cranes. These cranes take up little space from the project footprint and, when properly cited, can efficiently reach desired locations on the project site. On large projects, multiple tower cranes may be needed. The loads carried by these tower cranes are borne by a seemingly small tower, which will rest on a foundation that distributes the construction loads to the ground. Due to the tremendous size of the cranes themselves and the potential magnitude of the loads, these foundations can be substantial. These foundations must be designed by an engineer and this design may call for a considerable amount of excavation that is not depicted on the contract drawings. This additional excavation item must be included, but responsibility for this work is dependent on the building's foundation design. Commonly, a project consisting of a mass excavation will be designed with a mat slab foundation. With the mat slab it is most appropriate for the excavation subcontractor to excavate for the tower crane foundation because the formwork subcontractor will have no other digging work at the base of the excavation. If the building foundation is designed with spread footings this tower crane foundation will be excavated and materials will be hauled off by the formwork subcontractor, as they are traditionally responsible for footing excavations.

9. Cleanliness outside the borders of the construction site is an important issue during mass excavation, particularly in dense urban environments. Spillage of excavated materials from trucks and dirt tracked out of the site on the truck tires can create a hazard on heavily traveled streets and it can blemish a contractor's reputation (especially in the eyes of the neighborhood). The protocol for cleanliness on public streets will often be quite vague in the contract documents, and this makes it all the more important to have an understanding of the definition of this work prior to entering into any agreements. Street sweepers are a staple during earth moving operations, but they may not be adequate for some projects. Vacuum trucks and periodic steam cleaning may also be necessary. Additionally, pressure washing of truck tires prior to leaving the site, sweeping off the rims of truck beds, rock plates, and other means of reducing dust and dirt debris are highly recommended to ensure that cleanliness does not become a problem adjacent to the site. To reduce dust and spillage in sensitive neighborhoods or environments, it is commonly necessary for the truck beds to be covered prior to leaving the site.

10. It is common to cut off the tops of the piles several feet below grade to prevent conflict with sidewalks, planting beds, and other permanent installations surrounding the building (Figure 3.4). This cutting cannot be done

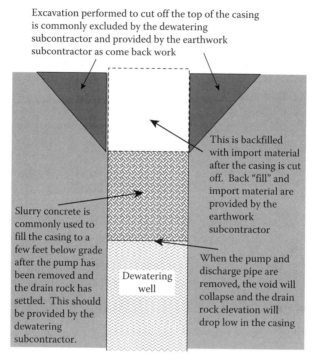

Excavation performed to cut off the top of the casing is commonly excluded by the dewatering subcontractor and provided by the earthwork subcontractor as come back work

This is backfilled with import material after the casing is cut off. Back "fill" and import material are provided by the earthwork subcontractor

Slurry concrete is commonly used to fill the casing to a few feet below grade after the pump has been removed and the drain rock has settled. This should be provided by the dewatering subcontractor.

Dewatering well

When the pump and discharge pipe are removed, the void will collapse and the drain rock elevation will drop low in the casing

**FIGURE 3.4**  Dewatering well capped below grade.

without some excavation and backfilling work around the tops of the piles. It is very important that the excavation work and particularly the back-filling work are carefully performed to avoid damaging the waterproofing membrane. Due to the importance of maintaining the integrity of the water-proofing membrane, it is advisable to have continuous inspection of this work by a waterproofing consultant. This is especially important during the backfilling phase, where damage could result and, if this is not detected, will be covered up within minutes.

11. Substantial construction loads will be generated by various large pieces of equipment that might be used on a project site. In some instances, it will be important to distribute the loads over a broad area. This is especially true of loads that will be applied adjacent to the shoring or on weak soils. Examples include mobile crane mats, trench plates, or other means of spreading equipment loads adjacent to the shoring, as deemed appropriate or as required by the shoring engineer of record. (This issue is discussed in further detail in the chapter on shoring and underpinning.)

12. Dust control is an ongoing concern with most mass excavation operations, but the excavation subcontractor will only plan for dust control while on site. The excavation subcontractor may demobilize while other operations (such as tieback installation work) are catching up. The excavating sub-contractor may also be absent as underpinning work is performed. Dust

**FIGURE 3.5** Dust control. (Photo by author, courtesy of Hathaway Dinwiddie Construction Company and Tishman Speyer.)

**FIGURE 3.6** Site grading. (Photo by author, courtesy of Hathaway Dinwiddie Construction Company and Tishman Speyer.)

control must be performed during these periods, but because the excavation subcontractor may not be on site the general contractor will be required to fill in for them. The general contractor is also the most appropriate source for dust control after the excavation subcontractor has completed their work and demobilized from the project (Figure 3.5).

13. The time of year, soil conditions, and historical local rainfall (or snowfall) will be quite important in influencing the conditions of access roads for both the mass excavation trucking operation and general construction use throughout the project. A good determination and layout of the necessary roads, frequency of maintenance, and cost of removal is quite important. It is often advisable to have the excavation subcontractor include this work in

the excavation bid. Rather than the general contractor simply including an allowance in their estimate for this work, a competitive bid will generally prove to be the most cost-effective and expedient way to address this potentially costly work item.

14. Loose soils and highly plastic soils such as clay, which experience considerable differential expansion and contraction with changes in moisture content, can be extremely difficult to maintain during even light rainfall. To help stabilize the soil, lime can be added to chemically alter it. The time of year, soil conditions, and historical local climate also play a significant role in determining if lime treatment will be beneficial below the slab on grade, the mat slab, the site access roads, around the building perimeter, etc. Lime treatment is expensive, but it can be cost-effective if crew efficiency is increased or if soil conditions are improved. It is important to remember that most vegetation and plants will die in lime-treated soils, so be sure to keep the treatment under the paved areas. Also, be sure to avoid lime-treating of soils that are hardscape, but which may very well receive planting in the future (Figure 3.6).

# 4 Dewatering*

Dewatering can consist of managing surface water, subsurface water, or both. The primary focus of this chapter will be on the more difficult task of controlling subsurface water. With proper planning, most problems related to excess surface water can be addressed with careful design of site sloping, terracing, and project layout. Even with these controls, unseasonably high rainfalls can result in damage, delays and/or rework that may only be compensated if the contract specifies such relief for the contractor. Subsurface water holds a significantly higher risk and poses a greater potential cost, largely because subsurface water flows through subsurface conditions for which the different soil and rock gradients cannot be understood with total accuracy based upon the only tool available to aid in predicting subsurface water flow—the soil borings. A project will have only a few soil borings performed, primarily due to their high cost.

Due to the vast uncertainty associated with subsurface water, the design of a dewatering system (Figures 4.1 and 4.2) involves a considerable amount of intuition and guesswork. Determining the depth to drill the dewatering wells, based upon the determination of the depth of the water table relative to the depth of the excavation, is relatively easy and will likely take care of 98% of the water. Unfortunately, 98% is not satisfactory and that remaining 2% needs to be addressed. The majority of the sources for this rogue water are difficult to identify as the soil borings may not disclose a layer of free-flowing sand trapped above a clay or impermeable rock layer. Rain water that reaches the sand layer will run right through the sand over the top of the clay or rock, through the lagging, and into the excavation.

A variety of suggestions will be offered concerning dewatering. Dewatering can be a costly operation and is particularly burdensome since this is not a cost that is readily apparent from a cursory examination of the construction drawings. It is important to make a judgment about including a contingency amount for spot dewatering (dewatering of footings, pits, and other surface spots around the project) at the onset of a project. Predicting the weather is not a reasonable expectation, but including this contingency is more prudent than ignoring the potential for water intrusion. Naturally, this judgment will vary greatly in different regions based upon the historical local climate.

## SCOPE OF WORK ISSUES RELATED TO DEWATERING

1. Route dewatering piping to avoid congested walkways around the rim of the excavation. The best place to route the dewatering piping is generally

---

* MasterFormat Specifications Division 31

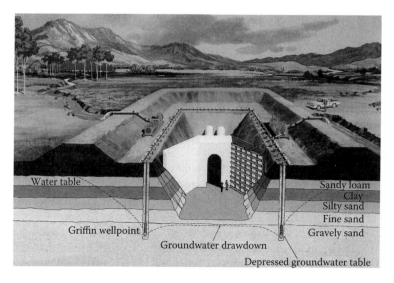

**FIGURE 4.1** Dewatering system. (Illustration provided by Griffin Dewatering Corporation.)

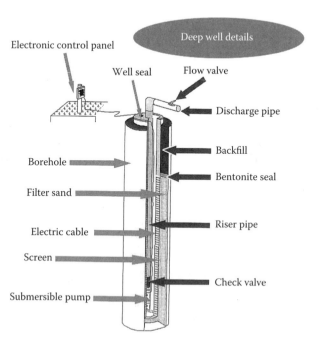

**FIGURE 4.2** Dewatering well. (Illustration provided by Griffin Dewatering Corporation.)

on the guardrails around the excavation, but this needs to be coordinated (before the shoring piles are drilled) because the dewatering system is often in place and functional before the guardrails are scheduled to be in place.

2. Include steel sleeves under the trucking ramps, across construction roads, and at any other locations where dewatering piping must be placed underground. As noted in the chapter on mass excavation, the excavation subcontractor is generally best equipped to perform the trenching and backfill work for piping due to the simplicity of the work and the fact that the excavation equipment will usually already be available on the project site. The steel sleeve(s) and all piping will still be provided by the dewatering subcontractor.

3. Backup the dewatering system with a generator. If the local utility company provided electrical power fails for even a few hours, the water table may rise into the bottom of the excavation, and almost definitely into the bottom of any deep pits. This can cause significant damage to the subgrade, aggregate base compaction, and waterproofing. If the rebar is already placed in the bottom of the excavation and recompaction of the subgrade is required, this will be extremely expensive. Dewatering subcontractors are commonly capable of providing these generators and do so regularly.

   Severe damage is more likely if the power failure lasts for 12–24 hours or more. In this case the water table will probably rise high enough to undermine the shoring system. Shoring systems are designed for dry environments, assuming the dewatering system will run continuously without interruption. For shotcrete lagging in sandy soils, the water table can rise quickly when power fails, and water can be trapped outside the excavation by the shotcrete and exert pressure on the back of the lagging. This excess water can impose lateral pressures that far exceed the design loads of the shotcrete. Ultimately, the risk is that the shotcrete may suffer structural failure and collapse.

   Many municipalities require a backup generator system to be put in place for dewatering systems, but this requirement is not universal. Until such backup systems become mandatory for all projects, it is important to recognize the need for this backup power, just as if it was required by a federal regulation—i.e., plan on providing backup power on all projects.

4. Backup generators must remain ready for operation at all times. To ensure this, it is important to keep the generator fueled. The responsibility for fueling the generator should be clearly established between the general contractor and the dewatering subcontractor, because the generator system will need to be exercised, or tested, on a regular basis. In these regular tests, fuel will be expended, and a reliable party must be responsible for calling a fuel truck each time the generators are run and having the fuel level in the tank maintained at a full level. The generators should be set to activate automatically via a timer on a designated morning. Subsequently, a fuel

truck should be scheduled to make a routine weekly stop early that same afternoon.

5. On some occasions, especially during and after rainstorms, it may become necessary to do spot dewatering. This type of intermittent dewatering usually utilizes the building pits for construction sumps. Even when a dewatering subcontractor is employed on the site, it is generally more efficient for the general contractor to perform this work. The dewatering subcontractors commit most of their effort to the installation and removal of dewatering systems. A small crew will be employed to perform routine maintenance on their dewatering systems located in the region, but the dewatering subcontractor will not maintain a large pool of laborers for general spot dewatering and it really is not economical or efficient for them to do so.

6. Dewatering wells are to be kept a significant distance back from the edge of the excavation. When this is not possible, wells should be made with slotted casings set in the bored holes. If casings are not used and the well is close to the excavation, there is a high probability that the wells (typically filled with drain rock to keep the sides from eroding) (Figure 4.3) will cave-in and compromise the stability of the excavation wall. Keep in mind that this cave-in will not occur until significant pressure is developed in the well—meaning the cave-in will not likely occur until the excavation is between 20 and 60 feet down (depending on soil types and how far the well is from the edge of excavation), which raises a safety concern.

7. Dewatering systems are temporary, constructed as inexpensively as possible, run continuously (24/7) for months and consist of a great deal of flimsy PVC piping. When damage to the piping occurs, it needs to be repaired quickly. It is advisable to keep a supply of extra pipe, fittings, primer, and glue, on hand. This pipe replacement is an expected occurrence due to trucks running over pipes, forklifts setting pallets on them, or other unexpected accidental damage. This pipe replacement is just part of the construction

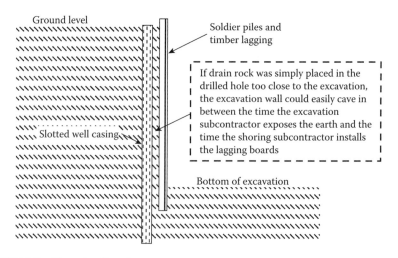

**FIGURE 4.3**   Slotted well casing.

process. Although the pipe is relatively inexpensive, a small budget should be set aside for this added expenditure. Due to the urgency associated with this minor pipe repair, it is advisable for the general contractor to perform this work. The general contractor can perform the work quickly and at a low cost.

8. Since the dewatering system must operate continuously, some allowance must be made for contingencies. For example, the failure of a dewatering pump can cause costly damage on a project if a replacement pump cannot be obtained quickly. To address this potential problem, it is suggested that the dewatering subcontractors always keep one or two spare pumps on the site for quick replacement.

9. The job superintendent, or a responsible designee, must check the wells routinely to assure the pumps are running. This is a task that should be done as a regular routine each morning before work begins and at the end of each workday. This task entails a quick trip to each pump where it is as simple as looking down each well to verify it is not filling up with water.

10. Dewatering is a very specialized and very important trade, and consequently there are few subcontractors that do this work. As a result, there are generally few choices when dewatering subcontractors are being selected. Before using a small and/or new company it is important to make a judgment about the credibility of the firm. If the project is in the middle of a dense urban setting surrounded by skyscrapers (Lower Manhattan, for example), do not take chances with the selection of an unknown dewatering subcontractor. The selection must be based on experience and industry credibility. If the project is in the middle of a parcel with no adjacent buildings and the excavation is not very deep, then it might be acceptable to consider the selection of a smaller/newer company. When life safety and major property damage issues are at stake, the selection criteria must be based on proven ability and not on baseline price.

11. Once the dewatering operation sufficiently drops the water table, the excavation work can proceed. Once the excavation work has begun, it is imperative that the dewatering operation functions and operates continuously. For this reason, especially for remote sites or sites that are a considerable distance from the home base of the subcontractor, confirm that the dewatering subcontractor has a round-the-clock emergency call center and that a repair technician can be on site within a reasonable response time.

12. A settlement tank will be required in nearly all situations, especially during the initial drawdown period, to collect the silts before they enter the city storm system. Dewatering subcontractors will commonly exclude this settlement tank by qualifying in their bids that the system provided will terminate at the connection to the settlement tank. This tank can be furnished by either the dewatering subcontractor or the general contractor. The important thing is to ensure that the tank is covered, but not double covered.

13. Estimate the time required to lower the water table to the desired elevation and verify that this is consistent with the time that is in the project schedule. If the excavation subcontractor digs faster than the water table subsides,

they will either stop working for a while and incur demobilization and remobilization costs or they will proceed by excavating damp soils and incur additional costs due to heavier loads. With free-draining sandy soil this will not likely become a problem, but in dense clays this can prove to be a real concern and place the dewatering operation drawdown on the critical path.

14. If possible, tie the dewatering system into the permanent storm drain connection; otherwise a temporary connection will be necessary. The dewatering subcontractor will not be equipped to trench into the streets or perform underground utility work of any kind, so be sure that work is covered by the site utilities subcontractor. If it is possible to discharge water into an existing catch basin adjacent to the site, it would be an optimal choice.

15. Generally it is most appropriate for the general contractor to procure and pay for the city tie-in permits, discharge permits, and usage fees associated with a dewatering system.

16. Be sure the dewatering subcontractor is aware of the pit locations and their respective depths, as this additional depth will need to be accounted for when bidding this work off of conceptual drawings. The dewatering work is commonly bid based upon conceptual drawings, and the pits are generally not identified on them.

17. When the project is completed and the dewatering system is shut down, it will be most economical to leave some components in the ground. It should be clearly communicated to the dewatering subcontractor how this is to be done. First, confirm how deep the wells must be demolished below grade and properly address this in the bid instructions to prospective subcontractors. For example, the requirements may be to remove the pump and PVC sleeves, cut off the casings five feet below the finish grade, top off the wells with slurry to five feet below grade, and backfill with imported soil for the top five feet (Figure 4.4).

18. For sites with contaminated soils (very common in dense urban environments) an in-line decontamination system will be required between the dewatering wells and the discharge location into the city system. For sites with excessive contamination, the dewatering system may need to tie into the city sanitary sewer in lieu of the storm drain system. This is required so that the discharge water is run through the treatment plant before it is discharged to a stream, ocean, or wherever the local water goes. Storm drains (that are separate from the sanitary sewer system) often discharge directly to streams, rivers, oceans, etc. A few issues to be cognizant of for a decontamination system are as follows:

    (a) First and foremost, the specific contaminants in the soil must be identified; this will determine the composition of the decontamination system, as different contaminants can have very different treatment methods.

    (b) Determine if the dewatering subcontractor, general contractor, or decontamination subcontractor will provide the settlement tank, as this item is often excluded by all parties. It really does not matter who includes this, so long as it is covered, but not double covered.

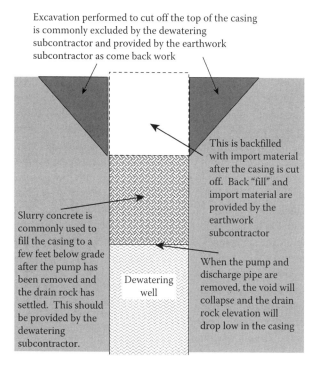

Excavation performed to cut off the top of the casing is commonly excluded by the dewatering subcontractor and provided by the earthwork subcontractor as come back work

This is backfilled with import material after the casing is cut off. Back "fill" and import material are provided by the earthwork subcontractor

Slurry concrete is commonly used to fill the casing to a few feet below grade after the pump has been removed and the drain rock has settled. This should be provided by the dewatering subcontractor.

Dewatering well

When the pump and discharge pipe are removed, the void will collapse and the drain rock elevation will drop low in the casing

**FIGURE 4.4**   Dewatering well capped below grade.

(c)  Since the water will need to be tested by a laboratory on a regular basis, the decontamination subcontractor should take care of this testing, along with any associated lab fees. Note, as discussed in later chapters, that in most circumstances—such as concrete, steel, and waterproofing systems—it is highly advisable for the inspections to be completed by an independent third party employed by the owner, in order to avoid a potential conflict of interest. Due to the exemplary reputation of decontamination subcontractors as a whole in the industry, decontamination systems have become a widely accepted exception to this rule.

(d)  If it is necessary to discharge to the sanitary sewer system instead of the storm system, a separate sanitary sewer discharge permit will need to be pulled. The general contractor should handle this permit submission and all related fees. Note that this permit will dictate the frequency of the mandated laboratory testing and the periodic test results will need to be provided to the city sewer department. By taking care of this portion of the work, the general contractor can establish early in the process the various requirements associated with the decontamination of the ground water.

(e)  When discharging ground water into the storm system, usage fees are generally not charged. When the ground water is discharged into the sanitary sewer system (or a combination sanitary/storm system as is common is some large and old cities, such as New York City), a meter will typically be provided by the same city agency that issues the permit. This meter should be installed by the decontamination subcontractor. The discharge to the city sewer is monitored and recorded, and discharge fees are paid, on a monthly basis. Be sure these discharge fees are included in the estimate as they can amount to tens of thousands of dollars every month. The actual discharge quantity will be a variable, but the soils report will be helpful in providing an expected discharge rate in the form of an estimated range. The soils engineer will ideally provide two estimated ranges, namely the initial drawdown rate and a lower range for once the drawdown is complete and the system is simply maintaining the water level. Since this is a variable, it is unfair for the general contractor or any of the subcontractors to be held accountable for the actual water discharge amount. Therefore, it is suggested that an allowance be created in the general contractor's budget to pay for these fees with a change order being issued once the system is finally shut down. The change order will return any remaining or unused funds to the owner, or if the allowance is overrun, an additive change order request will be submitted by the contractor to the owner.

(f)    Depending on the nature of the contaminants, carbon filters may very well be required and these must be changed regularly. These filters are very expensive, as is the labor for each trip to the site to change them. The rate at which these filters need to be changed will be a variable and the frequency will decrease as the project progresses. During the initial drawdown, the concentration of contaminants in the discharge water can be high, but as the system runs for multiple months and the ground water is constantly purged, the contaminant level will decrease. Since the cost of changing these filters is a variable, it is recommended that an allowance for them be included in the estimate, similar to the discharge fees, with the ultimate financial risk being borne by the owner. Be sure a unit cost per filter change is agreed upon with the decontamination subcontractor.

(g)    If there is no decontamination system, the dewatering subcontractor will generally tie directly into a city lateral line, but with a decontamination system the dewatering subcontractor will simply tie into the decontamination system (or settlement tank, depending on how the general contractor has allocated the scope of work). Decontamination subcontractors regularly exclude the costs of making the tie-in to the lateral lines from their bids. Thus, the line from the downstream end of the decontamination system to the lateral is often not included. It will be necessary to specifically allocate this segment of the system, which is most commonly placed into the site utilities subcontractor's scope of work.

(h)   All periodic or regular decontamination system maintenance (such as cleaning the sand filters) should remain the decontamination subcontractor's responsibility.

(i)   The decontamination bids will be based upon monthly rental rates, which is only fair for the decontamination subcontractors because they have no direct control over the project schedule. The subcontract issued to the decontamination subcontractor will be based on the duration indicated in the project schedule with the application of the monthly (usually defined in 30-day increments) rental rate. Depending on conditions, it may be advisable to include a contingency amount to cover added costs, such as any project delays over which the decontamination subcontractor will have no control.

# Questions—Module One (Chapters 1–4)

1. What type of lagging is the optimal shoring substrate for a below-grade waterproofing membrane, especially for a waterproofing membrane which acts under pressure, such as a bentonite system? Explain why.

2. Which activity (pile driving, tieback drilling, tieback grout curing, or lagging installation) requires no direct labor, but constitutes a timely process that must be accounted for in the project schedule? Utilizing weekends for this process is optimal. Explain why.

3. Discuss the statement that wood-lagging timbers that are pressure treated with a wood preservative will not degrade over time.

4. Which of the following—encroachment permit, shoring permit, site utilities permit, neighbor's written permission for tiebacks crossing the property lines, or a variance from the local noise ordinance—is not required for a shoring operation? Explain why.

5. Which method of shoring pile installation requires work to be performed by the excavation subcontractor? Explain why.

6. Which method of shoring pile installation is the most costly, but provides for the greatest quality control with regards to maintaining the plumbness of the piles? Explain why.

7. Which subcontractor should perform the trimming excavation work behind the face of the pile to make way for lagging board installation?

8. Which party (or parties) is responsible for pushing the tieback spoils out of the way immediately after drilling to make way for the tieback installation? Explain why.

9. What aspect(s) of cutting off the tops of the shoring piles (digging, torching, or backfilling) is a potential cause of damage to the below-grade water-proofing membrane?

10. When shotcrete lagging is used, the upper portion of the shoring system (equivalent to the depth to which the piles will be removed) will be wood lagging. In relation to the shotcrete lagging installation, when should the formwork in front of the wood lagging be installed?

11. Which subcontractor should ideally perform the excavation and backfill for cutting off the tops of the shoring piles? Explain why.

12. Which subcontractor (excavation, formwork, shoring, or waterproofing) is least likely to be impacted by tieback de-tensioning? Explain why.

13. Since soil borings only provide general information concerning the different soil layers and patterns, how will this impact the estimation of the number of tiebacks?

14. For which operation(s) (dewatering, shoring, or excavation) is it important to monitor shoring deflection and settlement of the adjacent grade? Explain why.

15. Which subcontractor bears the responsibility of the guardrails around an excavation, including furnishing, installing, and subsequently removing and disposing of them?

16. Describe property that would be prone to contain unforeseen conditions.

17. Describe the slabs that are commonly deemed a contractor's means and methods of construction, and thus will not be reflected on the contract drawings and need to be defined in the bid instructions prepared by the general contractor.

18. Name six sources of additional spoils that the excavation subcontractor will be responsible for hauling from the site, but which are not shown on the contract drawings.

19. For a project with a mat slab foundation, which subcontractor is ideally suited to excavate for the tower crane foundation? Which subcontractor is ideally suited for the tower crane foundation excavation work for a building designed with a spread footing type foundation?

20. In city centers where cleanliness outside a construction site is extremely important, what additional measures beyond common street sweepers may be necessary?

21. Discuss the merits of having continuous inspection by the waterproofing inspector during backfilling of basement walls.

22. During periods in which the excavation subcontractor is off site and after the excavation subcontractor has completed their work, which party is ideally suited for performing dust control?

23. For which types of soils (landscaped areas, sidewalks, building slab, and footings) is lime treatment not to be used to stabilize soils?

24. Is the design of a dewatering system better characterized as an exact science or as a process that involves a considerable amount of intuition and guesswork?

25. Which party should be held responsible for providing backup generators for the dewatering system? Explain why.

26. Which party is ideally suited to perform spot dewatering on a project? Explain why.

27. If a dewatering well is two feet from the edge of an excavation, describe the requirements of whether the well must be stabilized with a slotted casing.

28. Should broken dewatering pipe be scheduled for repair by the dewatering subcontractor? Explain why.

29. Which party commonly provides the settlement tank for the dewatering system? Explain why.

30. Discuss the merits of the owner's third-party special inspector performing the laboratory testing for the decontamination system discharge water.

31. Because the discharge quantity of groundwater is a variable and the cost of discharging the groundwater to a municipality's sanitary sewer system is quite costly, which party should bear the financial risk and responsibility for this variable cost? Explain why.

32. Which party should be held responsible for layout of the demolition work? Explain why.

33. Discuss the merits of viewing demolition work as a destructive activity, where quality control is not important and the general contractor can simply turn the demolition subcontractor loose in the building while focusing their full efforts on the imminent construction activities that will follow the demolition work.

34. Which party should be responsible for removing from the site any underground utility conduit, piping, and other buried items that are unearthed? Explain why.

35. Hazardous material abatement subcontractors are most commonly under direct contract with which party? Explain why.

# Module Two

# 5 Below-Grade Waterproofing*

High-rise and other types of specialty buildings typically are built with deep excavations for several reasons, including reaching good foundation materials. The excavated space is commonly used to create parking garages or other functional spaces. To achieve the deep excavation work, it is common for considerable dewatering effort to be required to lower the water table. When the dewatering efforts end, the water table will rise to its prior level which often is well above the bottom of the building foundation. Since water intrusion is detrimental to the use of the building spaces, an effort must be made to prevent water from penetrating into the building. This waterproof membrane must remain intact for the useful life of the structure.

There are various types of below-grade waterproofing systems. These systems can differ considerably in terms of materials and in price. When choosing a below-grade waterproofing membrane many factors must be considered, including the following:

a. Maximum expected height of the water table.
b. The intended use of the below-grade space. For example, some minor moisture intrusion in a parking garage or mechanical area may not be particularly damaging or detrimental to the use of the space. On the other hand, if the below-grade space is used for the storage of rare books, as a clean room, or for residences, it may be prudent to commit sufficient funds to install a particularly reliable waterproofing system.
c. Local climate conditions—for example, the arid Mohave Desert or rainy Seattle.
d. The potential need for a dual purpose membrane that will prevent the intrusion of water and some other liquids or gases, such as naturally occurring methane gas or hydrogen sulfide.
e. The level of soil contamination must be considered as some membranes (such as bentonite membranes) can have an adverse chemical reaction to certain common contaminates.

The waterproofing membrane is commonly specified in the contract documents. The particular system specified was hopefully chosen by the architect and/or owner as

---

* MasterFormat Specifications Division 7

the best membrane for the project conditions. The contractor must then ensure that the system is properly bid and constructed.

Third-party waterproofing inspectors are commonly employed on projects to verify the quality of the waterproofing membrane installation, by comparing the installed work to the architectural drawings and manufacturers installation instructions. If such an inspector is not specified, it is advisable that one be brought on board by the owner. It is especially important, even vital, not to permanently cover any membrane without first having it signed-off by the inspector. If an inspection for electrical wiring in a wall has been missed, it is expensive, but it is always still possible to cut a hole in the wall to make the necessary inspection. Similarly, a missed inspection of a steel connection can still be made by demolishing and patching an opening in the ceiling. Unfortunately, a missed inspection of a below-grade waterproofing section is virtually impossible to inspect without incurring tremendous demolition and reconstruction expenditures.

## SCOPE OF WORK ISSUES RELATED TO BELOW-GRADE WATERPROOFING

1. The waterproofing subcontractor should be fully apprised of the number of tiebacks (including supplemental tiebacks) and whether or not they need to be de-tensioned. Tiebacks are not de-tensioned until the project structure has been completed to a point in which the perimeter walls will independently support the earthen pressure, thus rendering the temporary shoring system obsolete. Therefore, to maintain clear access to the tieback heads a blockout is constructed in the basement wall by the formwork, rebar, and shotcrete (or place and finish) subcontractors. The waterproofing subcontractor will need to return to the project to patch each tieback blockout after the blockout forms are removed, because the waterproofing boot that is used to cover the tieback heads would otherwise be in the way of the blockout formwork. (Many city regulations require the tiebacks to be de-tensioned as a safety precaution. Tensioned strands pose a danger to future trenching/excavation operations as they can snap or break explosively when hit with a backhoe bucket.)

2. If the general contractor has planned for shotcrete to be applied over the membrane of the perimeter basement walls, this must be approved by the waterproofing system manufacturer. There is an increased risk of damage to the membrane from the intense shotcrete nozzle blast and, consequently, few manufacturers will still warrant waterproofing systems behind shotcrete walls.

3. The number of cold joints must be established, as the waterproofing subcontractor will commonly need to place waterstop at all of them (Figure 5.1). In shotcrete situations the waterproofing subcontractor will need to be present during all shotcrete operations, which can be a significant added expense. (In one day of a shotcrete operation, the work will typically consist of first placing a 4′ lift, waiting an hour or two for that lift to set up, and then placing a second 4′ lift above the first. The waterproofing subcontractor will

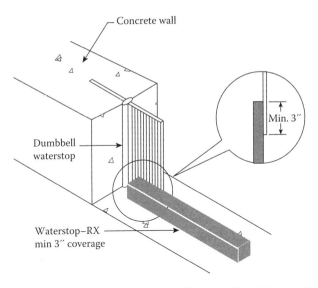

Concrete wall

Dumbbell
waterstop

Min. 3″

Waterstop–RX
min 3″ coverage

**FIGURE 5.1**   Waterstop. (Illustration provided by Cetco Building Materials Group.)

commonly need to place waterstop between these two lifts. Although this work can generally be done in about an hour, the waterproofing subcontractor will probably need to pay for at least four hours of wages for the employee to be on site to perform this work.)

4. Patching of inboard dewatering wells is extremely difficult; primarily because the building must be constructed around these wells until such time when the dewatering system can be shut down (typically once the first or second-level deck above ground level is placed). All waterproofing manufacturers have an approved method of patching these wells, but the methods generally require planning and a considerable amount of work. Be sure the waterproofing subcontractor (as well as the concrete subcontractors) knows prior to submitting their bid about the incorporation of any inboard wells for dewatering (Figure 5.2).

5. Determine if and at what frequency penetrations will be made through the membrane to hold up the rebar mats and inside face wall forms. It is highly recommended, particularly for work below the water table, that rebar mats and inside wall forms be internally braced to avoid penetrations through the membrane. Though an additional expense is incurred with the internal bracing, this is becoming a common and preferred practice. When penetrations are made through the membrane for temporary support of the rebar and forms, the penetrations are commonly spaced 4′ on center, both vertically and horizontally. These penetrations essentially turn the perimeter membrane into a sieve, to be patched by the waterproofing subcontractor.

   (a)  Another practice increasing in popularity consists of welding reinforcing steel bars horizontally from one tieback head to an adjacent

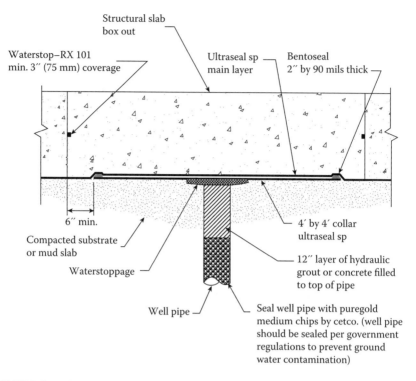

**FIGURE 5.2** Inboard dewatering well cap. (Illustration provided by Cetco Building Materials Group.)

tieback head, to provide support for the perimeter rebar mats. This approach is most effective for projects in which the shoring tiebacks are scheduled for de-tensioning, as these projects will have block-outs at each tieback head. In this condition, after the perimeter walls have cured, these support bars are simply cut inside the blockouts to clear the way for waterproofing boots to be installed over the tieback heads.

6. Determine the number of utilities that will penetrate the horizontal water-proofing membrane.

(a) The branch electrical circuits may be run in the slab, in a sand layer between the slab and waterproofing membrane, or below the waterproofing membrane. Running the electrical lines below the membrane dramatically increases the penetrations through the membrane and is not recommended. Any breach in the underground conduit below the water table will cause a leak through the conduit which provides a direct avenue for water to enter into the building (Figure 5.3).

(b) Determine if the sanitary sewer and storm drain will be run below the membrane. In mat slab situations it is recommended that the sanitary sewer and storm drain run through the mat slab. While this may result

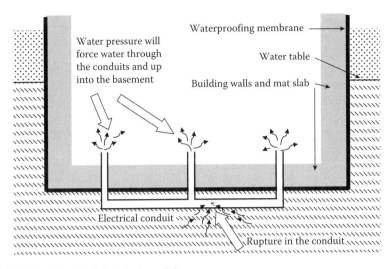

**FIGURE 5.3** Breached electrical conduit.

in the need for additional rebar around the piping and potentially a thickened slab at the low ends of the piping, this is preferred to the alternative of penetrating the membrane. The structural engineer will need to review and give advice on this condition, as additional steel may be required to compensate for the voids created in the concrete mat by the piping.

7. If the sanitary sewer and storm drain are run through the mat slab, it is important that the pipe installation be performed without penetrating the membrane with any pipe supports. If not carefully coordinated, the plumbers will be tempted to drive form stakes into the subgrade or to set expansion anchors in the mud slab to anchor their piping. There are two viable alternatives:

   (a) Support all the piping with the rebar cage. Note that, because the piping needs an accurate slope, additional diagonal reinforcing will probably be required to hold the rebar mat steady enough during the concrete pour to ensure that the piping does not move within the shaky rebar cage. Be sure to coordinate this work with the rebar subcontractor.

   (b) Support the piping with 2″ to 3″ embedments in a 4″ thick protection slab. Drilling into the protection slab will require significant quality control and should be performed under continuous inspection to ensure over-drilling does not occur. Note that there are some drawbacks (described later in this chapter) to using protection slabs between the membrane and the bottom of the mat, but this is a common practice in the industry.

8. Determine if mud slabs (placed below the membrane) or protection slabs (placed above the membrane) will be used as part of the general contractor's means and methods. It will also be important to know the manufacturer

of the waterproofing system. Some waterproofing manufacturers will not allow the use of mud slabs and/or protection slabs while other manufacturers will require them.

It is recommended that protection slabs only be used for bentonite or other systems capable of expansion and self-healing. It is presumed that when a building is completed (for example, a high-rise building with a basement) the weight of the completed building will often shatter the protection slab. It is theorized that if the membrane is bound to this slab, the membrane will crack right along with the protection slab.

9. Confirm the nature of the vertical substrate and be sure the waterproofing subcontractor has performed all necessary preparatory work. The required work will depend on various factors, the extent of which will be dependent on the specific type of waterproofing system scheduled for use. Some considerations are as follows:

   (a)   For shotcrete lagging, very little preparatory work will be required.
   (b)   For sloped excavations with earthen backfill, no preparatory work will be necessary.
   (c)   For wood lagging there are several elements that can be detrimental to a below-grade waterproofing system. A few common problems are as follows:
       (i)   Large gaps between the lagging boards will need to be filled, especially for a bentonite system.
       (ii)  Nails used for positioning the lagging boards that are bent over the soldier piles will need to be covered. Drainage mat materials are generally used for this purpose, but are never shown on the contract documents specifically for this purpose. Typically, for installations above the water table, a drainage mat will be part of the waterproofing system, covers the nails, and in effect becomes a dual purpose component. For installations below the water table, a drainage mat will not be a part of the designed waterproofing system.
       (iii) Voids behind the lagging boards cause them to be loose, which can be particularly detrimental with a bentonite waterproofing system. Filling the void space behind the lagging boards with grout is a common solution to this problem, but this decision is often left to the contractor's means and methods of providing a waterproof system. If grouting behind the lagging boards is required it will be completed by the shoring subcontractor.

10. Include an allowance to patch the membrane where damage is caused when the tops of the shoring piles are cut off. Methods of minimizing this type of damage are discussed in the chapter on shoring and underpinning. Since the amount of damage is difficult to estimate at bid time, it is advisable to estimate the number of expected patches and include this quantity in the subcontract agreement, with a unit cost for additive or deductive quantities of patches.

11. Include an allowance for patching minor damage that might occur at anytime throughout construction. Minor damage can be minimized with good quality control, but accidents happen and some type of damage to the membrane is very likely to occur. Be sure to address the waterproofing integrity of not only the bottom and sides of the excavation, but also the top rim of the membrane at grade. The membrane at grade will be exposed to potential damage for many months and in this time will probably incur significant nicks and tears.

Further, because of the importance of an impenetrable below-grade membrane, it is advisable to institute a protocol for the project such that absolutely no backcharges or reprimands of any kind will be made for accidentally damaging the membrane. One way of achieving this is by asking the party that causes the damage to circle the damaged area with marking paint. At the appropriate time, the general contractor will pay the waterproofing subcontractor to take care of the necessary repairs via a contingency amount, similar to the amount recommended for damage to the membrane caused while cutting off the tops of shoring piles. Since damage to a membrane is often impossible to see with the naked eye, it is important that the damage be clearly marked before it is forgotten. The following tips can be useful in ensuring that this is taken care of with minimal effort:

(a)  Provide several cans of orange upside-down marking paint for everyone's use at various locations in the excavation.

(b)  Make sure the subcontractors' office personnel, foremen, and especially the crews know that all they need to do is circle any damage, and reinforce the fact that there will be no reprimands for any accidental damage that is caused. All that is required is for the damaged area to be circled with the orange paint. They do not need to tell the job superintendent, their foreman, or anyone else. The orange paint will be quite visible and the damage will be found later when the repairs are being made.

(c)  To stress the use of the orange marking paint, include clear language in the subcontracts about marking damaged areas for those trades that will be working around the below-grade waterproofing. A similar approach is also advisable for the roofing work.

(d)  Use the term accidental damage, so that if a subcontractor acts carelessly or intentionally vandalizes the waterproofing membrane, they are not immune to backcharges (or termination for that matter). A definition of "accidental damage" needs to be established for contractual purposes, but it should be an extremely lenient definition that should apply in most instances except when the intent of the vandalizing party is clear and obvious.

12. Explore the means of attaining a sole-source warranty for the waterproofing system. Such warranties are somewhat rare, but can often be negotiated. Traditionally, the waterproofing subcontractor will provide a labor warranty (workmanship) for the system and the manufacturer will provide

a separate materials warranty. This often results in a dilemma in the event of a failure, because the manufacturer will insist the damage was the result of a labor error and the subcontractor will insist the materials were faulty. The only way to resolve this with certainty is to lift up the building and look underneath—not a feasible task. A sole-source warranty avoids this complication.

Negotiating and attaining a sole-source warranty is not free, i.e., there will be a cost associated with it. Sole-source warranties are best negotiated with the manufacturers. Such a warranty will probably include additional compensation for a full-time inspector from the manufacturer to witness the membrane installation and monitor the protection of the membrane throughout the rebar and concrete activities until the membrane is properly and permanently concealed. The added cost of an inspector is often justifiable when due consideration is given to the fact that this ensures the delivery of a quality membrane for a spectacular new building, where the basement will stay dry.

Regardless of whether separate warranties or a sole-source warranty are provided, be sure to specify the duration allowed for investigating leaks and the duration allowed for making repairs. In particular, be sure the investigation period does not drag out for weeks, months, or even a year while the building is taking in water. Further, be sure that the warranty provider is not only responsible for repairing the work, but also for all consequential or collateral damages and remedial measures necessary to mitigate the water intrusion. Damaged drywall, ruined carpeting, stained ceilings, equipment damage, and other losses often cost several times the amount required to mitigate the waterproofing breach itself. The warranty holder will often try to exclude these consequential damages from the warranty policies. Make sure that this issue is adequately addressed before entering into a binding agreement.

13. Determine with the shoring engineer the number of tiebacks that will need to be sealed (including estimated supplemental tiebacks). The difference of a few tiebacks is not a major concern, but if the waterproofing subcontractor expects three rows of tiebacks and the shoring engineer designs a four-row system, a sizable change order request will be imminent. On the other hand, if the number of tiebacks is less than originally estimated, a deductive change order should be issued. Whether used for additive or deductive quantities, a unit price should be clearly established while still in a competitive bidding environment. This unit price should be the same for additive or deductive quantities, though some subcontractors have a tendency to provide a higher unit price for additive work than they will for deductive work. It is still better to avoid change order altogether with proper planning at the onset of the project, but for hard-bid public works projects with a design-build shoring scope of work, the variables in the forthcoming shoring design often make iron-clad pre-bid planning impossible.

14. Determine if the tiebacks need to be de-tensioned, as this will mean that the waterproofing subcontractor needs to return to the project to cover and seal the tieback heads. This will be less efficient, thus more expensive, than if the heads could be covered at one time with the rest of the below-grade membrane work.

15. For contaminated soil conditions, be sure the waterproofing membrane manufacturer has received a copy of the contaminated soils report and has signed off in writing that the known contaminants will not affect the membrane. There are various common contaminants that degrade different waterproofing materials.

16. The baseline schedule typically establishes the number of times that the waterproofing subcontractor will need to mobilize for the project. Minor items such as pouring the pit bases and the tower crane foundation are sometimes omitted from the baseline schedule (though they should be included) and this information needs to be conveyed to the waterproofing subcontractor as they will incur additional mobilizations.

# 6 Reinforcing Steel*

Work associated with reinforcing steel (commonly referred to as rebar) consumes relatively little project time, but it requires close coordination with the formwork subcontractor to ensure an efficient operation. The sequence of work begins with deep foundations and footings toward the end of the earthwork phase, and the flow of work continues throughout the structural phase with the repetitious activities of single siding wall formwork, followed immediately by rebar, and followed by the installment of the second side of forming. The concrete can then be placed in the walls, which makes way for the equally repetitive activities of deck formwork, deck rebar, and placing the deck concrete. This sequence of work is followed repeatedly throughout many concrete structures (Figure 6.1).

With the association of formwork, reinforcing steel, and concrete placement, it should be evident that regular coordination meetings are imperative to ensure a smooth flow of work among these various subcontractors. The coordination effort includes the careful allocation of space and the efficient distribution of manpower on the project site. In addition to the weekly site subcontractor coordination meeting, it is essential that the subcontractor foremen involved in the structural concrete work meet separately on a regular basis, at least weekly. While it may be difficult to get these subcontractors to commit an hour per week for this additional meeting, this time spent planning and coordinating will be repaid exponentially through improved productivity.

The issues relating to the scope of work for rebar and concrete formwork are obviously related in many instances, so many of the comments that pertain to rebar subcontractors will also apply to the formwork subcontractors.

## SCOPE OF WORK ISSUES RELATED TO REINFORCING STEEL

1. The mechanical, electrical, and plumbing (MEP) penetrations through decks, shear walls, and beams are among the most common and most expensive exclusions regularly found in a rebar bid quotation. Architects, MEP engineers, and structural engineers usually do a good job of coordinating the blockouts for major duct runs and pipe chases, but the miscellaneous branch ducting and individual pipe penetrations typically remain the responsibility of the general contractor and subcontractors to locate during the MEP coordination effort. The determination of the MEP systems routing, whether around or through concrete walls and precisely where to penetrate the concrete decks, is generally left to the contractor team for coordination as a means and methods issue. Since the rebar subcontract will be executed

---

* MasterFormat Specifications Division 3

**FIGURE 6.1** Reinforcing steel. (Photo by author, courtesy of Hathaway Dinwiddie Construction Company and Tishman Speyer.)

considerably before the time the MEP coordination effort is complete, it is difficult to accurately estimate the number of penetrations to include in the rebar subcontract and in subcontracts with other concrete-related trades. Nevertheless, time spent examining the MEP documents will help to generate a good educated estimate of the number of penetrations. Some general contractors simply include an allowance for additional penetrations in their own budget, and then pay the rebar and concrete placement subcontractors via change order. This is not recommended as change order pricing is invariably higher than the pricing received in a competitive bid environment. At a minimum, it is suggested that the rebar subcontractor, and other concrete trades, include the cost of a stipulated number of additional pipe penetrations and a stipulated number of duct blockouts. With a thorough review of the MEP drawings, the estimated number will be very close to the final quantities.

2. Include rebar for the equipment curbs and pads (also termed house keeping pads). These curbs and pads are used to keep MEP equipment raised a few inches above the finish floor so dirt and debris is not swept, kicked, or otherwise collected underneath the equipment. Similar to MEP penetrations, the designers generally do a good job of locating the curbs and pads for the large major equipment, but there will commonly be more housekeeping (incidental but necessary) pads needed throughout the project for electrical motor control centers, controls air compressors, hydronic system expansion tanks, and various other components. Also, similar to MEP penetrations, the full extent and size of the various pads and curbs will be left up to the MEP coordination effort. Another educated estimate will be necessary to quantify the extent of miscellaneous curbs and pads throughout the project prior to executing the rebar and concrete-related subcontracts (Figure 6.2).

**FIGURE 6.2**   Housekeeping (equipment) pad. (Photo by author, courtesy of The University of Southern California.)

3. For large steel structures in dense urban environments, both the tower crane and trucking lane will be tied up by the structural steel subcontractor during steel erection. To accommodate this situation, it is often prudent that the deck rebar, as well as other subcontractor's materials, be hoisted during off hours. It is important to establish this in the bid instructions to ensure that the rebar subcontractor and other subcontractors as deemed necessary by the general contractor, incorporate the respective premium time costs.

4. Be sure the responsibility for the shotcrete test panels is clearly addressed. This is often missed in rebar and formwork bids, simply because these panels are a means and methods issue that is not reflected on the contract drawings. Negotiate an agreement with the shotcrete subcontractor as to how many test panels will be required and coordinate that quantity with both the rebar and formwork subcontractors. (Anticipate that the test panels will be constructed to reflect the most congested conditions on the project and that each shotcrete operator, known as a nozzleman, must shoot a complete test panel. This is a significant amount of work.)

5. Be sure the masonry steel is included in the rebar quotation, but that it is not double covered. The rebar subcontractor should always furnish and install the dowels for the concrete masonry unit (CMU) work. It is common for the rebar subcontractor to only furnish the wall steel for the CMU work, while the masonry subcontractor will install it. This has essentially become a standard practice in the industry, but it is still suggested that the specific arrangements be clarified with all subcontractors impacted by this work.

6. Be sure the detailing for the CMU wall steel is covered (included in one of the subcontractors' quotations). It is often recommended that the masonry subcontractor detail this steel and provide cut lists to the rebar subcontractor. Since this steel is generally small in quantity and very straightforward, this arrangement seems to work well. It is recognized that a less collaborative

rebar subcontractor may not like providing a lump sum bid for this work, and then having another subcontractor dictate the quantities and shapes. Due to the simplicity of this work it really makes little difference who details the steel as long as a subcontractor is identified to perform this work.

7. Be sure any on-site welding of the rebar is covered by one of the subcontractors. Rebar subcontractors rarely have a certified welder on their payroll. Since on-site welding of rebar is a rarity, they tend to exclude welding from their quotations.

Since the rebar subcontractor will probably not have anyone employed on site to perform this work, it is recommended that the rebar subcontractor furnish the bars, but that the welding should be performed by the structural steel subcontractor. This should be carefully planned and discussed during the bidding phase with the steel subcontractor, as welders need a special certification to weld rebar (relatively few welders have this certification). Most large steel companies only have a few employees that are certified to weld rebar, while some do not have any. If the structural steel subcontractor has no certified welders for this task, additional arrangements will need to be made to obtain them. Furnishing and welding rebar to embed plates is very common on projects, but will be performed in the shop by the subcontractor providing the embed plates and this rebar welding is rarely an issue.

8. For wall steel, even high wall steel, rebar crews will typically climb the mats in lieu of using scaffolding to access their work. This work approach is safe if the rebar is properly tied and workers use positioning devices (with harnesses and lanyards). This is generally an efficient, safe, and cost-effective way of performing this work. Be sure to ask the subcontractor prior to awarding the subcontract if there are any unique situations on the site (such as high beams or confined shafts) that will require scaffolding or other forms of access provided by others. On a complex project there will likely be a few locations that need special attention.

9. Safety must be considered when rebar is installed. During construction, the exposed ends of rebar can be hazardous to workers who can be injured by abrasions or by impalements on the rebar ends. These types of injuries will be prevented or reduced with the use of rebar protection caps (Figure 6.3). To ensure that this OSHA-mandated safety measure is addressed, verify that the rebar subcontractor has included the cost of rebar protection caps in the quotation, as they commonly exclude them. Of course, another way to make sure this is covered is for the general contractor to provide the rebar protection caps. Regardless of who provides the caps, a single worker must have the assignment of walking the jobsite every day and replacing any missing caps, as they are frequently knocked off while performing normal construction activities. Note that all subcontractors who have employees at risk of harm from exposed rebar are responsible for replacing any caps that they remove or accidentally knock off. An alternative to protection caps that is accepted in some jurisdictions is to hook (bend over) the tops/ends of all exposed bars (Figure 6.4). This is quite costly, but a good and safe alternative.

**FIGURE 6.3** Rebar protective cap. (Photo by author, courtesy of Hathaway Dinwiddie Construction Company and Tishman Speyer.)

**FIGURE 6.4** Rebar bender. (Photo by author, courtesy of Regional Steel Corporation.)

10. Wire mesh in concrete-filled metal stair pans is often omitted from both the miscellaneous metals and the rebar subcontractor quotations. Verify whether wire mesh is required by the contract documents and, if so, it is most appropriate for the rebar subcontractor to include it in their price quotation.

    At first glance it may appear prudent for the miscellaneous metals subcontractor to simply install this mesh in the shop and tack weld it into the stair pans. The metal stairs are routinely used for construction purposes by placing a 2x timber board in the pan. This board will break loose the wire mesh, rendering it useless, which is why the rebar subcontractor should place this mesh on site shortly before placing the concrete at these stairs.

11. Unless they are unusually small, the shoring tiebacks will require additional steel bars (termed trim steel) above, below, and on each side of the tieback heads to structurally compensate the wall for the bars interrupted by the tieback heads. The estimated number of tiebacks (including supplemental tiebacks) must be communicated to the rebar subcontractor. The shoring design is most often the responsibility of the design-build shoring subcontractor and the final number of tiebacks will be determined by their shoring engineer. An educated estimate of the total quantity will be provided by the shoring subcontractor in their bid proposal, as shoring subcontractors run rough calculations for shoring designs for bidding purposes.

12. For post-tensioned decks the rebar subcontractor needs to include all sleeves, tendons, hardware, and tensioning work. The post-tensioning tendon work should be provided turnkey by the rebar subcontractor.

13. The placing and finishing subcontractors have been known to complain about rebar tie-wire clippings strewn across the decks. The concrete placing and finishing subcontractors are responsible for giving the decks a final blow down to ensure they are clean, prior to concrete placement. Note that they cannot simply blow off the tie-wire clippings with a compressed air hose because when airborne the clippings create an unsafe condition. Just as formwork subcontractors are responsible to ensure that all their nails are picked up off the deck, the rebar subcontractor needs to be held responsible for cleaning up the tie-wire clippings prior to each concrete pour. This task is as simple, but time consuming, as lowering a magnet in each and every tiny grid of reinforcing bars.

14. Be sure to plan for the rebar foreman and (depending on the size of the project) one or two additional workers for pour watch. When pumping concrete for a pour, the pump hose and general construction traffic may knock loose rebar supports, break tie wires, and jar bars out of position. During a pour there is often less than 15 minutes to make the necessary repairs, so be sure the rebar subcontractor is properly and diligently represented during the pour to keep the concrete placement activity progressing smoothly. Rebar and formwork subcontractors commonly claim that their pour watch personnel will be working elsewhere on the job, but available. Since it can take more time to find the rebar foreman and then get a worker to come over to repair the damage than is allowed in a hectic concrete pour, it is advisable that this pour watch actually be performed by personnel standing-by.

15. Rebar templates (2″ × 4″ boards used to both layout and temporarily support wall dowels in a slab pour) are best supplied and installed by the formwork subcontractor, but this item is typically excluded by both the rebar and formwork bidders (Figure 6.5).

16. For work below grade it is important to confirm the allowable means of supporting the rebar mats, both horizontally and vertically, to avoid damage to the waterproofing membrane. In addition to the obvious causes of damage to the horizontal membrane (such as foot traffic, dropping tools, and general construction activities) are the bundles of bars that are lowered by a crane into the excavation. As the bars are lowered, the ends of the bars droop

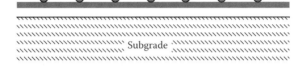

A 2 × 4 is tied off to the rebar by the formwork subcontractor. The rebar subcontractor then ties off the wall dowels to the 2 × 4. During the concrete pour, after placement, but prior to the concrete beginning to harden, the dowels will be able to stand on their own in the concrete and the place and finish subcontractor will remove the 2 × 4.

Rebar wall dowels

2 × 4 rebar template

Slab rebar

Subgrade

**FIGURE 6.5** Rebar template.

downward and are the first part of the bars to hit bottom. Though wood blocks will be spaced along the length of the bundled bars, the ends of the bundles need to be lowered onto plywood or some other type of protective board to ensure that the ends of the bars do not puncture the membrane.

The most common cause of damage to the vertical membrane is the punctures made to tie the rebar mats back to the shoring. Methods to avoid this are discussed in the chapter on below-grade waterproofing.

When developing bid instructions be cautious of special requirements for one trade's materials or installation methods dictated by the manufacturer's instructions for the materials of another trade, as this coordination is commonly the responsibility of the general contractor. For instance, rebar mats for slabs on grade are traditionally supported by concrete chairs, commonly termed dobies. These small concrete blocks are considered harmful to some waterproofing and vapor barrier membranes by the respective manufacturers. The general contractor needs to review the installation instructions for these membranes to verify if other supporting devices, such as plastic chairs, will be necessary, and if so, relay this information to the rebar bidders via the bid instructions.

17. Be sure the bidders include in their quotations the foundations necessary for general construction operations that are not on the design documents, such as the tower crane and manlift foundations.

18. The rebar subcontractor should fabricate and furnish the cages for the caissons free on board (FOB) (which has two commonly recognized definitions of freight on board or the more common FOB) jobsite, but (because it is a very simple operation with the drilling rig already in place) the subcontractor drilling the caissons should set them.

# 7 Formwork*

For projects with concrete structures, the formwork subcontractor, undertaking the most labor-intensive work related to concrete, will be the determining factor as to whether the schedule milestones are met. The rebar, placing and finishing, and shot-crete subcontractors will all do their part as well, but these subcontractors follow the formwork operation. Since their work is comparatively less extensive than the work of the formwork subcontractor, there is generally little difficulty in keeping up with the formwork (Figure 7.1).

To develop accurate schedules, it is important to prepare detailed estimates of the quantity of formwork needed on the project. Due to the impact of formwork materials on the schedule, it is advisable to perform "material loading" of the project schedule to allocate the form usage on the project. This is an effective method to ensure that schedule activities do not overlap, such that forms are scheduled to be in two places at once. All the formwork should be reflected in the schedule. It is quite common in the industry to reflect the allocation of decking materials in the project schedule, but allocating the wall and column formwork can be equally important.

## SCOPE OF WORK ISSUES RELATED TO FORMWORK

1. Elevated deck formwork can pose a serious fall hazard to construction workers. To address this, safety railings must be installed at the perimeter of the decks and at any floor openings in the deck. This worker protection can be nearly as costly as the formwork, but it is most important. For a concrete structure, the railings are often moved on a daily basis in conjunction with the formwork operation, so it is appropriate that the formwork subcontractor furnish, install, maintain, and remove these railings. This is commonplace in the industry and the formwork subcontractor should have this work in their base bid quotation.

   After the concrete work is complete, there will be a continuing need for safety rails. Installing, maintaining, and removing the safety railings from the point that the concrete work is complete to the point that permanent rails or walls are constructed is typically not included in the quotation received from the formwork subcontractor. In fairness, this should not be expected, because they will not consistently be on site after completion of the concrete structure. The following are some suggestions of how this can be addressed:

---

* MasterFormat Specifications Division 3

**FIGURE 7.1**   Integration of safety railings with the formwork.

(a)   Use slab grabbers or other prefabricated railing systems (Figure 7.2) once the concrete work is complete. This work can readily be performed by the general contractor. These railing systems are preferred for the following reasons:

    (i)   These rails are easy to install, so the risk of a faulty installation is significantly reduced.

    (ii)   Wood railings suffer considerable abuse through use and reuse on a project. As they are removed or replaced, and become exposed to general construction traffic, they become flimsy and unsafe. Despite this, there tends to be reluctance (due to both the cost and effort involved) to take deteriorated railings out of service. The prefabricated railing systems, on the other hand, are very durable and maintain their integrity throughout a project.

(b)   Have the formwork subcontractor furnish and leave in place wood railings that are installed at all deck edges and around interior floor openings that are too large to be safely covered with 1-1/8″ plywood. The formwork subcontractor must provide this fall protection through the course of their work anyway, so leaving these safety measures in place is not a problem if properly planned. Thus, the railings should be constructed such that they can remain in place when the forms are stripped—for example, attached to the concrete deck, not to the formwork. After the formwork subcontractor completes their work and demobilizes from the site (or from an area of the site on large scale projects) the general contractor should assume responsibility for the maintenance and eventual removal/disposal of the rails.

(c)   For a steel structure the railings at the perimeter of the decks should consist of cables installed by the structural steel subcontractor. Interior

**FIGURE 7.2** Slab grabber. (Photo by author, courtesy of Hathaway Dinwiddie Construction Company and The University of Southern California.)

railings will be required around openings for mechanical, electrical, and plumbing (MEP) chases, stairs, and so on. For a steel structure it is generally best for the structural steel subcontractor to provide steel posts with cable rails at openings that are too large to be safely covered with 1-1/8″ plywood. These types of railings are extremely durable and will not require any significant maintenance.

2. The formwork subcontractor must include setting anchor slots in cast-in-place concrete or shotcrete walls for the stone and/or brick facing. The slot materials will be furnished by the masonry subcontractor.

3. Each day when shotcreting is performed, each nozzleman must shoot a small flat test panel, typically about 18″ × 18″ × 4″. Be sure the formwork subcontractor fabricates the forms for these panels. It is best if these forms are mass produced. This will ensure that there will be a stack of them, and that there will be no shortage. After they have been used, the shotcrete filled forms can simply be tossed into the dumpster.

4. The formwork subcontractor needs to include setting all steel embeds in the structural concrete that will be furnished by others. Even with design-build trades (for the exterior skin in particular) deriving an educated estimate for this work is not overly complicated. For every large heavy panel to be set, regardless of the material, a connection will be necessary at all

four corners, so counting the corners that will occur over a concrete wall or at the edge of a concrete deck will give a reasonably good estimate of the number of required embeds.

When estimating the cost to set embeds, the degree of difficulty of setting each particular embed must be considered. Formwork subcontractors set embeds routinely on every project, they are very familiar with how they are designed, and can effectively judge the size and complexity of embeds by reviewing the architectural drawings. Therefore, the fact that embeds for design-build work have not yet been designed prior to bid time should not be cause for change order concern, as the architectural documents will indicate which building elements are supported by embeds. Nevertheless, this is a very important issue to clarify with the formwork subcontractor during the bidding phase.

(a)   An exception to the rule: Though the formwork subcontractor is responsible for setting the embeds for all trades falling within MasterFormat Specifications Division 1 through 14, industry standards dictate that embedded items for MEP work are set in place by the MEP subcontractors. This includes conduit, junction boxes, pipe sleeves, hanger supports, and any other items that are specifically and solely for use by the MEP trades.

5. Traditionally, the excavation subcontractor will leave a flat and level dirt pad at the elevation (specified in feet above sea level) indicated on the contract drawings, or the elevation dictated by the general contractor to account for the mud slab, protection slab, etc. The excavation subcontractor will not dig any footings, utility trenches, pits, or other below-grade items, unless specifically directed otherwise. (A common exception is that it is often advisable for the excavation subcontractor to dig deep pits due to the excessive excavation and material to be hauled off, but shallow pits will commonly be completed by the formwork subcontractor. Either of these subcontractors is capable of performing this work, so this is a judgment call to be made by the general contractor when preparing the bid instructions.) The footings will be dug by the formwork subcontractor. The formwork subcontractor will also remove their own spoils from the site.

6. The formwork subcontractor has the tedious responsibility of reviewing the contract documents to pick up all of the miscellaneous depressions, embeds, anchor bolts, blockouts, and other features required for the project. Some of the miscellaneous items to look for are:

(a)   Recessed floor mats

(b)   Floor-recessed door closers

(c)   Bollard anchor bolts

(d)   Depressions for thickset ceramic tile

(e)   Sump pits at elevators

(f)   Stair edge angles

(g)   Embeds for railings

(h)   Grating supports for telecommunications/data troughs

    (i)    Stair nosings

    (j)    Elevator sill angles

7. The formwork subcontractor should maintain responsibility for all equipment pads and curbs within the building (outside the building these items will be provided by the site concrete subcontractor). As discussed in further detail in the chapter on reinforcing steel, the full extent of the curbs and pads will not be fully realized until significant progress has been made in the mechanical–electrical–plumbing (MEP) coordination effort. Nevertheless, a good educated estimate can be made with a proper review of the drawings during the bidding phase.

8. The formwork subcontractor should set all anchor bolts within the building footprint, though the bolts and all necessary hardware (nuts, washers, and square plates) should be furnished by the specific subcontractor for whom the anchor bolts will be set. The site concrete subcontractor should set all anchor bolts outside the building footprint.

    For large and complicated structures (such as high trusses spanning an arena), the structural steel subcontractor may require additional anchor bolts and footings for temporary columns and guy cables needed to erect high trusses or other large complicated members. Any additional anchor bolts and footings necessary for the structural steel or other work must be clearly indicated in the general contractor's bid instructions for inclusion by the formwork (and rebar) subcontractor in their bid quotations.

9. The formwork subcontractors typically exclude rebar templates from their bid quotations. Nonetheless, they are the appropriate subcontractor to provide these templates, as they will be responsible for laying out the walls. Further, in addition to the wall dowel layout at the slab level provided via the rebar templates, be sure the formwork subcontractor provides a three-dimensional layout for the entire concrete structure for use by the rebar and other related subcontractors, including the top of wall layout, plumbness of wall, edge of wall, etc.

10. Formwork subcontractors may include wire mesh for small equipment pads, metal stair pans, and topping slabs, but these items are more appropriately allocated to the rebar subcontractor. Most union jurisdictions also recognize these items as being the responsibility of the rebar subcontractor. Since this work is quite simple, it does not really matter who performs it, as long as this work is covered by one of the subcontractors, but not double covered by both.

11. The formwork subcontractor will remove and/or accidentally knock off rebar end caps regularly throughout their operations. Regardless of who provides the protection end caps, it must be clearly communicated to the formwork subcontractor (specifically included in their subcontract agreement) that they are responsible for replacing any rebar end caps they remove or accidentally knock off during the performance of their work. This is typical for all subcontractors, but it is used as an example in this chapter because the formwork subcontractor will regularly remove and replace these caps throughout the course of their work, much more so than any other trade except the rebar subcontractor.

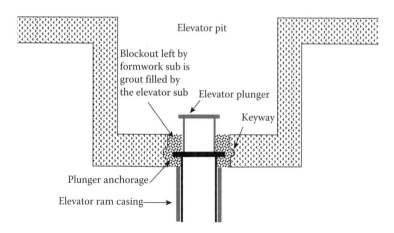

**FIGURE 7.3**    Elevator plunger blockout.

12. Though steel casings will be set for the hydraulic elevator plungers well ahead of placing the pit concrete, the elevator plunger will generally be delivered at a much later date. To accommodate this late delivery, the formwork subcontractor will commonly need to leave a blockout for setting the plunger (Figure 7.3). Rather than having the place and finish subcontractor return to the project to fill in the blockout, it is suggested that the elevator subcontractor grout it in. This is work that the elevator subcontractor is capable of performing, and it is a small amount of work. It is more efficient to have the work completed by a single subcontractor rather than coordinating this minimal work between two subcontractors in the small elevator pit.

13. For shored earth conditions using tieback anchorages, confirm whether the tiebacks need to be de-tensioned. If de-tensioning is required, blockouts will need to be created at each tieback head. After the de-tensioning is complete, it will be necessary to return to the project to fill in the blockouts. The general contractor will need to verify the number of tiebacks planned for the project with the shoring design-builder and convey the quantity to the formwork subcontractor. This work is discussed in greater detail in the chapter on below-grade waterproofing.

    (a)  When constructing blockouts, formwork subcontractors commonly create a square box with 90-degree corners. These function quite well for a shotcrete wall, but, for a cast-in-place (CIP) perimeter wall it is important to slope the top of the blockout to assure proper concrete consolidation (Figure 7.4). With a flat top in a 12 to 24-inch wall, consolidating the concrete completely is nearly impossible, although this is a common practice and regularly passes inspection. Regardless of current industry practice, it is suggested that the tops of these blockouts be sloped.

14. The formwork subcontractor will need to include blockouts for all duct shafts, pipe chases and other large penetrations through the building decks

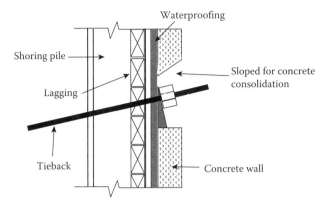

Waterproofing

Shoring pile

Sloped for concrete consolidation

Lagging

Tieback

Concrete wall

**FIGURE 7.4** Tieback blockout.

and walls. The MEP subcontractors will furnish and install their own sleeves for individual conduits and piping, but blockouts for ductwork and multiple pipes or conduits will be completed by the formwork subcontractor. Though this division of work seems ambiguous at first glance, it is actually an industry standard that has developed and is well known by all subcontractors involved. The design team generally does a good job of coordinating and identifying the larger openings on the drawings, but the full extent of the penetrations will not be realized until the MEP coordination effort has made significant progress. Since the general contractor, not the design team, controls the means and methods for the project, this is commonly deemed the responsibility of the general contractor. This work needs to be quantified prior to bid time, and a thorough review of the MEP documents will provide a good educated estimate.

15. Just as a foreman and potentially one or two field employees of the rebar subcontractor are necessary stand-by personnel to ensure that there is an efficient placing and finishing operation, a foreman and (depending on the size of the pour) a small crew from the formwork subcontractor are equally important. Embeds, anchor bolts, and hung forms are highly susceptible to damage from foot traffic, dragging of the concrete pump hose, and the general hectic activity that accompanies a concrete pour. Keeping just one set of structural steel column anchor bolts from being knocked out of position will justify the wages paid for the pour watch personnel.

Keep in mind when scheduling the formwork subcontractor for pour watch that they may very well already have a small crew planned for the pour to wet set embeds (installing embeds loosely in fresh, wet, concrete) for items such as bollards and other weld plates that may appear in the middle of the slab. This crew will have a full day's work in wet setting these embeds, so if anything happens during the pour they will not have time to break away from their duties to fix it. In short, the pour watch personnel need to be relatively free of any other responsibilities during the pour so they will be ready at a moment's notice if anything goes wrong.

16. For formed decks, seepage through the decking is very common. It is important to verify that the formwork subcontractor has personnel under the formed deck during the concrete pour with garden hoses to dilute (hose down) any seepage that makes its way through the deck. If seepage drips through and is permitted to harden on the floor below, it will eventually need to be ground off, which is very laborious. If the seepage is hosed down and diluted, it simply turns to dust when dry and major complications are avoided. Formwork subcontractors may exclude this remedial work with the expectation that the general contractor will assign laborers to the task (Figure 7.5). This issue needs to be resolved in the scope of work discussions.

17. For the below-grade perimeter walls an acceptable means of supporting the inside face wall forms must be made. The simplest and easiest approach for the formwork subcontractor is to tie the forms to the shoring, but this creates a tremendous number of penetrations through the waterproofing membrane that will need to be patched. As discussed in greater detail in the chapter on below-grade waterproofing, patching the membrane is not only costly, but also degrades the quality of the membrane. Though it is expensive, it is recommended that the below-grade perimeter wall forms, and rebar mats, be supported solely off the interior slab (or ground if the slab has not yet been placed) with kickers (diagonal bracing). Though more effort will be spent on the rebar and formwork operations, this additional cost will be partially offset by avoiding the necessity of performing extensive waterproofing patching.

18. For mat slabs in particular, it is recommended that the pit bases be poured as separate elements very early in the project. If the pit bases are poured in a single monolithic pour with the rest of the foundation, either large forms hung from the rebar will be necessary or supports for the pit walls will need to be driven through the waterproofing membrane. Large hung forms

**FIGURE 7.5**   Concrete seepage from concrete placement above.

are difficult to keep plumb and in place during a pour, so this is not a good method of construction. In addition to the obvious problem of penetrating the waterproofing membrane, driving supports through the waterproofing membrane in the middle of a pour means the supports need to be left in place so the membrane will seal to the sides of the form stakes. If the form stake is removed, it is like removing a nail from a tire. A nail in a tire causes only a small leak, but when the nail is removed the tire will be flat within seconds. Although the waterproofing manufacturers will have approved details for sealing the protrusion, this is still not a good waterproofing solution. There is a high probability that this is where a leak will eventually develop even with membranes, such as bentonite, that are engineered to maintain their integrity when penetrated in manners such as this.

19. Resolve the method of penetrating the formwork for piping and conduits at the onset of the project, to avoid conflicts with electricians and plumbers drilling holes all over the formwork subcontractor's forms (Figure 7.6). If not suitably addressed, after three or four uses the lumber begins to resemble Swiss cheese rather than gang forms. For every reuse, the holes drilled by the plumber and electrician need to be plugged, and particularly if steel forms are used, this could be a big problem. As for the plumbing penetrations, sleeves will almost always be used. For larger pipes, through footings in particular, large heavy sleeves will be necessary.

    As for electrical conduit, there is a high potential for the scope of work to expand beyond expectations. If the electrician plans on drilling holes through the formwork at will, but the formwork subcontractor expects the electrician to use form savers (conduit couplers designed such that they screw to the face of the form, but do not penetrate it), one of these subcontractors will assume that they have a legitimate change order request. Either

**FIGURE 7.6**  Conduits penetrating formwork. (Photo by author, courtesy of Hathaway Dinwiddie Construction Company and Tishman Speyer.)

the electrician will request compensation for the expense of the couplers or the formwork subcontractor will request compensation for patching the forms.

20. Just as the rebar subcontractor needs to clean up tie-wire pieces left in the form prior to a pour, the formwork subcontractor needs to clean up all nails, cuttings, and other debris that may have been dropped or left in the forms. The placing and finishing subcontractor will give the decks a final blow off prior to the concrete pour, but will expect the decks to be in the equivalent of a broom-swept condition.

21. Make sure that fire-treated lumber is used for any formwork in dead spaces that become trapped and will not be removed after the concrete is poured. Otherwise the fire marshal may (particularly in California) require sprinklers to be installed in the dead space.

22. Include the tower crane foundation as a separate mobilization. Also, be sure the manlift pad, temporary power pad, and any other concrete items necessary for construction operations are clearly conveyed to the formwork subcontractor. This work for temporary structures is not detailed on the contract drawings. Be sure to include setting the tower crane anchor bolts and embeds at the decks for tower crane lateral support, as well as any other bolts, embeds, and blockouts that are required for construction use.

23. For a concrete slab placed on metal decking, the metal decking subcontractor should include in their bid quotation the edge forms at the perimeter of the floors and all the deck penetrations, but the hung forms for slab depressions will need to be constructed by the formwork subcontractor. These include forms for floor depressions at restrooms with a thickset ceramic tile, balconies, stone flooring, stepped slabs, and topping slabs.

24. Anchor bolts furnished by the structural steel and miscellaneous metals subcontractors should always come with steel templates from the respective subcontractor. Verify that one template is provided for each set of bolts, and do not plan to reuse templates because they will become damaged after a single use. For any miscellaneous sets of anchor bolts not furnished by one of these two subcontractors, the templates will most commonly need to be fabricated by the formwork sub.

25. The formwork subcontractor, not the place and finish subcontractor, needs to patch the holes in the concrete walls created by the form ties. This is standard protocol that is rarely excluded by the formwork subcontractor or double covered by the place and finish subcontractor.

26. Formwork subcontractors are generally the best source for sand, vapor barrier, aggregate base, and any other layers (except waterproofing membrane) between subgrade and the bottom of the slab. Medium to large formwork subcontractors are always capable of performing this work, have the necessary equipment, and do it on a regular basis.

# 8 Concrete Placing and Finishing*

The concrete placing and finishing subcontractor will generally be on site sporadically for only short periods, commonly only one day at a time, although there can be extremely long days. Despite the short, and intermittent, time on the site this subcontractor performs a very important operation. The adage that nothing is a problem until it is set in concrete emphasizes this importance.

The general contractor should not perform any direct or indirect labor for the concrete pours when they have employed a concrete place and finish subcontractor. On concrete placement days, the general contractor's superintendent occasionally will allocate several laborers for traffic control, scraping concrete chutes, and other general activities. This is done solely due to the importance of a successful concrete pour, but is not a financially responsible practice and should be avoided. This can be a significant commitment of labor resources which, unless the general contractor had anticipated these when the job costs were first estimated, will be a needless reduction of the project contingency. It is important to hold the place and finish subcontractor fully responsible for all work associated with a concrete placing operation and if the general contractor in fact needs to allocate additional employees to supplement areas in which the place and finish subcontractor is in any way deficient, the associated costs should be issued as a backcharge to the subcontractor. This assistance from the general contractor is often necessary, but must not be taken lightly, as the costs associated with it are significant.

## SCOPE OF WORK ISSUES RELATED TO CONCRETE PLACING AND FINISHING

1. The general division of work between the placing and finishing subcontractor and the site concrete subcontractor lies at the perimeter of the building between the inside and outside of the building footprint. There are some general contractors that will allocate this work alternatively by making a division between structural and nonstructural concrete. The main difference between these two approaches lies with the interior nonstructural concrete, which includes stair pans, bollards, equipment pads and other miscellaneous work within the building that is not part of the building structure. Either approach is acceptable as long as it is clear to all parties before bids are received. For the purpose of discussion in this chapter, the former approach

---

* MasterFormat Specifications Division 3

will be assumed in which the place and finish subcontractor maintains responsibility for all concrete work within the building footprint.

   (a)   Note that the division of concrete work discussed above is typical for the formwork and rebar subcontractors as well. Contrary to structural concrete work, site concrete subcontractors, in most occasions, perform the forming, rebar, and place and finish work themselves.

2. The general contractor should have a standard, city-approved, lane closure permit for general operations when project activities extend into public roadways. During major concrete pours, there will be a significant increase in activity around the site due to the number of concrete trucks and concrete pumps. This is when a more substantial lane closure permit may be required. The placing and finishing subcontractor should bear the responsibility for drafting this more elaborate lane closure plan, presenting it to the city, paying the fees, and pulling this permit. Further, all work detailed by the lane closure permit must be by the placing and finishing subcontractor. This includes arrow boards, flaggers, cones, and other resources that will provide a turnkey operation.

3. The placing and finishing subcontractor should include the concrete fill at miscellaneous metals items such as metal stair pans designated for concrete fill and interior bollards (the site concrete subcontractor will fill the exterior bollards).

4. The placing and finishing subcontractor should maintain full responsibility for the construction, maintenance, tear down, and disposal of the concrete truck washout area. If the site is too congested to allow for a concrete washout, the subcontractor must include the necessary costs for washing out the drum and disposing of the concrete remnants in a designated drum at the batch plant or some other location.

5. The placing and finishing subcontract should include all equipment pads, curbs, topping slabs, and any other cast-in-place concrete items within the building footprint. Similar to the rebar and formwork scope of work, a good review of the MEP documents will be required to develop a detailed estimate of the quantity and size of the pads. This will ensure that reliable bids are received.

6. Expansion joint cover assemblies located between elevated decks at a building seismic separation commonly require grouting. This work is not typically completed by the expansion joint subcontractor. The placing and finishing subcontractor is generally the best candidate for this work, unless the general contractor elects to self-perform it (Figure 8.1).

7. Grouting of structural steel baseplates is most often completed by the placing and finishing subcontractor, although if preferred, this work might also be performed by the general contractor. Conversely for miscellaneous metals work the miscellaneous metals subcontractors should be required to grout their own work.

8. The placing and finishing subcontractor should cover all concrete finishing, whether it is a wet or dry finish. Placing and finishing subcontractors will

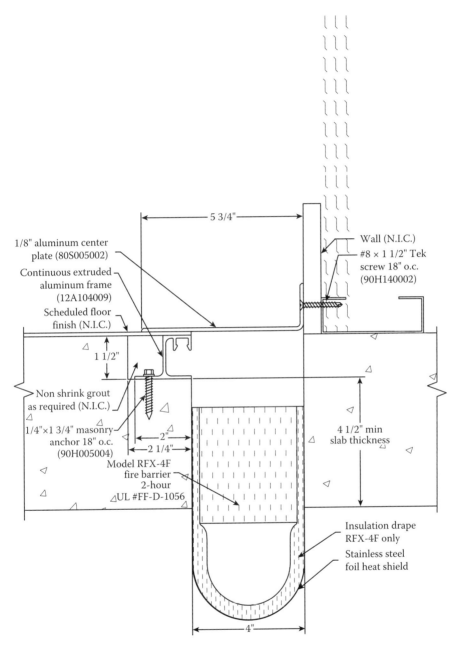

1/8" aluminum center
plate (80S005002)

Continuous extruded
aluminum frame
(12A104009)

Scheduled floor
finish (N.I.C.)

1 1/2"

Non shrink grout
as required (N.I.C.)

1/4"×1 3/4" masonry
anchor 18" o.c.
(90H005004)

2"

2 1/4"

Model RFX-4F
fire barrier
2-hour
UL #FF-D-1056

5 3/4"

Wall (N.I.C.)

#8 × 1 1/2" Tek
screw 18" o.c.
(90H140002)

4 1/2" min
slab thickness

Insulation drape
RFX-4F only

Stainless steel
foil heat shield

4"

**FIGURE 8.1**  Expansion joint. (Illustration provided by Construction Specialties, Inc.)

often exclude sandblasting (which occurs after concrete has cured, so
termed a dry finish), but this should rightfully be allocated to them. Most
large placing and finishing subcontractors do their own sandblasting, while
many small to medium companies have a business partner they rely on

as a lower-tier subcontractor to perform this work. The place and finish subcontractor must also be held responsible for protection of adjacent finishes that will be damaged if exposed to a sandblaster, such as red-brick banding running through or around the concrete surfaces.

9. Colored concrete is a difficult process, as there is no second chance to get the color or consistency that is desired. A high level of diligence is required to achieve good quality control of colored concrete. There are three different methods of coloring concrete, and these will be described in order of difficulty:

   (a)  The easiest method of coloring concrete is to mix a dye into the wet concrete mix, and this will result in an integral color. An integral color will be furnished as part of the concrete purchase and provides for a very consistent finished product. An added benefit that integral colors have over topically applied colors is durability. Over time, chips and dings occur in concrete. With integrally colored concrete, minor flaws are often not noticeable. With a topically applied color, deep chips and major dings will go through the color and make the grey concrete visible when the surface becomes damaged. The added expense of integrally colored concrete for such uses as sidewalks is often a worthwhile expenditure.

   (b)  Powder-based shake-on colors are a means of applying color during the concrete finishing phase. This treatment must be performed by the placing and finishing subcontractor. It is imperative that an experienced crew be assigned to this task, to ensure that there is an even distribution of the colored powder and that it is evenly troweled into the concrete surface. This task requires a conscientious dedication to the work.

   (c)  The most difficult color application on concrete is a spray-applied stain. These stains are acid-based, but are not simply sprayed on like paint as many first-time applicators or general contractors new to the process may assume. The acid-based color must be sprayed uniformly and scrubbed evenly into the concrete surface. The acid must then be neutralized, sprayed off, and cleaned. This is not a simple activity, as it requires not only a crew experienced in this application procedure, but also one that is well-acquainted with the safety precautions that must be taken when working with acidic compounds. The use of respirators, smocks, gloves, and protection for footwear are standard protocol for this operation.

10. Concrete curing compounds and sealers are items commonly either missed or double covered during the bidding phase of a project. Concrete curing compounds should always be designated as being the responsibility of the placing and finishing subcontractor. Some general contractors prefer to self-perform work associated with concrete sealers, though the painting subcontractor is another valid source for this work. Further complicating this issue is the use of single products which both promote the curing process and seal the concrete. If a combination

curing compound/sealer is used, the placing and finishing subcontractor should take care of it. The use of this combination product is not always allowed by the design team and/or the general contractor for the following reasons: First, the sealer component hampers the natural concrete curing process by locking in moisture. Secondly, the sealer component of a combination cure/sealer is not of as high quality as a separate sealer. Lastly, topically applied compounds can be detrimental to the adhesion of other products that will be applied to the concrete surface—for example, flooring. This compatibility must be verified by the general contractor prior to application of any compound. It is commonly preferred that separate concrete curing and concrete sealer products are used.

11. Water repellants on vertical walls or other applications specifically designed for waterproofing protection should be applied by a qualified company, most often the painting subcontractor. Placing and finishing subcontractors are not specialists in waterproofing integrity and should not be expected to satisfactorily perform this work.

12. Similar to the formwork and rebar subcontractors, the place and finish subcontractor must include all temporary concrete structures necessary for construction use, but not necessarily depicted on the contract documents. Examples include the tower crane foundation, manlift pad, mud slab, and protection slab.

13. The number of concrete pours for a project is a very important aspect to quantify prior to formwork, rebar, and place and finish subcontractors bidding for a project. This applies to slabs on grade, elevated decks, shear walls, stair pans, topping slabs, equipment pads, and all other concrete work. This quantity will dictate how many slab construction joints and wall bulkheads the formwork subcontractor will need to construct. It will also dictate the number of splices that the rebar subcontractor will need to provide and, with regards to this chapter in particular, it will dictate the number of concrete placing days allocated for the place and finish subcontractor. The setup and cleanup for a concrete placing operation is very expensive, therefore larger pours provide a significant economy of scale.

14. If the shoring tiebacks need to be de-tensioned, then be sure the placing and finishing subcontractor includes a return trip to infill the blockouts after the de-tensioning has occurred. Naturally, this will not be required of this subcontractor if the below-grade walls are shotcrete, as the shotcrete subcontractor should infill blockouts of that nature. Also, be sure to communicate information about the stage of construction when the blockouts will be ready to be infilled. Quite often there are many obstacles (building walls, parking garages full of subcontractor laydown areas, etc.) by the time the blockouts are ready to be filled. This causes problems when running the pump hoses, especially when the pump hoses are crossing the designated paths of travel for material movement or snaked through finished rooms and corridors.

Be sure the number of tiebacks and their elevations are established. Blockouts a few feet off the floor can be filled while standing on the floor, but blockouts six feet or more above the floor will require the subcontractor to incur the cost of scaffolding at each location.

15. The formwork subcontractor is responsible for all excavation associated with the building footings. With stiff soils, the footings can be dug neat and the concrete cast against the earthen sides of the trench. This is the most economical means for the formwork subcontractor, but because this neat cut leaves sides that are not as true as a board-formed footing, extra concrete will be required to fill the trench, therefore the place and finish subcontractor will be responsible for providing this additional concrete. On the other hand, with loose soil conditions the sides of the unshored trench will not stand up properly if neat cut, as it will cave-in and require a significant sloping of the sides of the trench. This will prevent the footing concrete from being contained within a reasonable width. In this case the sides of the footing will need to be formed and subsequently stripped, which is a greater cost for the formwork subcontractor and a lesser cost for the place and finish subcontractor. Naturally, most project soil conditions will fall somewhere between these two extremes. The general contractor must coordinate and establish whether the footings will be neat cut or formed to ensure that proper bids are received. Otherwise the subcontractors will each figure the least expensive method for their own purposes. The formwork subcontractor will figure the footings are neat cut and the place and finish subcontractor will figure all of the footings are to be formed, thus creating a scope gap.

16. The placing and finishing subcontractor traditionally performs the final cleaning (blow off) of the decks prior to concrete placement. This is after all rebar, tie-wire clippings, nails, and other debris have been removed by the formwork and rebar subcontractors. The general contractor could also blow off the forms, but traditionally the placing and finishing subcontractor will perform this work.

17. Concrete seepage control for elevated concrete deck pours can be confusing, as there are three different situations when seepage can occur, each with a different delegation of responsibility. The formwork subcontractor will be responsible for hosing down the wet concrete seepage through the wood deck forms because they are ultimately responsible for the integrity of their formwork. Secondly, for unvented metal decks, the metal decking subcontractor must maintain responsibility for the integrity of their deck and edge forms and include hosing down any related concrete seepage. Lastly, the seepage through vented metal decking is imminent due to the vent perforations in the bottom of the decking. Since this seepage is not caused by a deficiency in the metal decking subcontractors work, it will most commonly be hosed down by the general contractor (though they may at times opt to allocate this work to the metal decking or placing and finishing subcontractor). If properly constructed, there should theoretically not be any seepage through a non-vented metal deck or through a wood formed deck, though in reality seepage will occur. This is why the subcontractor

responsible for constructing the forms (formwork or metal decking sub-contractor) must attempt to contain the concrete seepage and must be held responsible for any breaches in their own systems. Since seepage is a natu-ral and expected occurrence for vented decks, any one of the three parties mentioned above would be appropriate to handle this work. The main point, as always, is that this work needs to be covered, but not double covered.

18. Typically, the placing and finishing subcontractor will purchase the concrete. There are many general contractors who actually prefer to purchase the concrete themselves to save the markup on materials. In a competitive bid environment the subcontractors generally include little markup on the con-crete purchase. This is done because there is little risk in concrete pricing and this helps to keep their competitive bids low. If interested in purchasing the concrete directly as a general contractor, it is suggested that bidders still include the cost in their base bids, but also provide a quotation for a deductive alternate in their bids if the concrete purchase is taken out of their scope. Note that large place and finish subcontractors may even receive better pricing than the general contractor from concrete batch plants due to their significantly larger volume of business.

19. Inevitably, on large projects the concrete finishing activities will occa-sionally extend beyond the daylight hours (particularly during the shorter winter days). For large pours it is advisable to start placing concrete at or before dawn. This will ensure full use of the sun during the hectic workday. Nevertheless, large pours often extend beyond the daylight hours, requiring the use of multiple light towers to properly illuminate the work area. There is only one opportunity to put a nice finish on a concrete slab, so be sure the proper number of light towers are provided. The placing and finishing subcontractor should be held responsible for the concrete placing operation. Along with this comes the responsibility for all the equipment required to perform the job, including the light stands. There should be a specific inclusion in the subcontract with specific illumination requirements stated in foot-candles at the slab level. Be sure to discuss this lighting during the preconstruction meeting and ensure not only that the place and finish sub-contractor furnishes the lighting, but also that there is no compromise in meeting the lighting requirements. This will assist in the assurance of a quality finish on the concrete, as poor lighting will inevitably result in a poorly finished slab.

20. Rock pockets are an inevitable problem, but are under the control of the place and finish subcontractor and so as such must be patched by them. Similarly, patching the form-tie holes should be the responsibility of the formwork subcontractor. These two patching activities should be addressed in the subcontract agreements such that they are not confused with each other.

# 9 Shotcrete*

Shotcrete is an excellent method of concrete placement, being both less expensive and faster than the traditional method where forms are required to define the shapes of all sides. The only drawbacks of using shotcrete are the limitations imposed by some manufacturers on the below-grade waterproofing systems (as discussed in the chapter on below-grade waterproofing) and the diminished quality of the finish. The quality of the finish is not an issue for walls which will be concealed or out of public view, such as when scheduled to be covered with gypsum board; or for walls located in MEP equipment rooms.

When the use of shotcrete will not affect the finished product of the building as described above, it is considered a means and method of construction, and as such will be used only at the general contractor's discretion. A differentiation between which walls are shotcreted and which are cast with double-sided formwork is not commonly indicated on the construction drawings.

Shotcrete can be a very appealing option because of the anticipated reductions in both cost and time, but there are issues of concern. A few potential problem areas will be described:

A. Due to the intense impact of the shotcrete spray, there are few below-grade waterproofing membrane manufacturers that allow shotcrete to be used with their products. The use of shotcrete will negate the warranty of the manufacturer on a vast majority of membranes due to the anticipated damage to the membrane from the shotcrete blast, most notably when the shotcrete blast blows open the seams between sheets. Be sure the below-grade membrane is compatible with the shotcrete application prior to confirming that shotcrete will be used.

B. Since shotcrete is a very quick operation and requires quite expensive equipment, the overhead for this business is extremely high. Due to of this increased business risk, there are comparatively few shotcrete subcontractors in the market. If using shotcrete, be sure to have a specific and firm clause in the subcontract regarding response time. No more than three days' notice should be required for the shotcrete subcontractor to mobilize. That is, if they are called to begin shotcreting operations, they should be on site completing preparatory work (setting scaffolding and screed wires) within three days, followed immediately by shotcreting on the fourth day. If a subcontractor requires two to four weeks' notice (as is somewhat common in busy times), opting for shotcrete in lieu of double-sided forms could extend the schedule.

---

* MasterFormat Specifications Division 3

C. Since the project progress will be driven by the shotcrete subcontractor's schedule and because there is a scarcity of shotcrete subcontractors, the shotcrete subcontractor is in a position of control. To address this, it is suggested that a delay clause with a fixed cost per workday or liquidated damages be included in the shotcrete subcontract. Time is money, and for every day that the schedule is delayed the general contractor and other subcontractors on site incur added overhead costs. These overhead costs continue whether work is performed or not. As a result, it is only fair that the shotcrete subcontractor be held accountable for delays that are caused by the shotcrete operations. If liquidated damages are in effect, the shotcrete subcontractor will be motivated to act promptly on the project.

D. The above suggestions are primarily focused on mitigating the general contractor's liability. It is only fair to the shotcrete subcontractor that the risks borne by the general contractor be addressed in the subcontract as well.

(a)  The shotcrete bidders will assume a well-run, efficient project. Therefore, they intend to mobilize a finite number of times. The shotcrete setup and teardown is a significant amount of work; so the subcontractor will incur additional costs caused by additional mobilizations beyond what could have been anticipated at bid time if the general contractor calls them to the project more frequently and for smaller amounts of work than initially planned. The subcontractor needs to be protected from these added costs resulting from actions beyond their control. To do this, one or both of the following criteria should be incorporated into the subcontract:

(i)  Specify the number of assumed move-ins in the base subcontract agreement, and then maintain the ability to adjust this number with a unit cost amount for each additional, or deducted, move-in. The project schedule issued for bidding purposes is the appropriate document used to establish this baseline.

(ii) Specify a minimum amount of work (quantified in either cubic yards or square feet) per mobilization.

(b)  Identify a reasonable work postponement notice time for shotcrete work. If told the afternoon prior to mobilization that their work must be delayed, the shotcrete subcontractor will probably not be able to find somewhere else for their crews to work the next day. This will mean that they will lose a day's work, the cost of which is termed an opportunity cost. In other words, because the shotcrete subcontractor is out a day's work they are also out a day's worth of profit. The general contractor must be responsible for providing timely notifications of schedule changes. Since the actual cost of overhead and lost profits is extremely difficult and cumbersome to compute, it is reasonable to establish a lump sum value for postponement announcements that occur within 24 hours of the scheduled mobilization. This amount may be derived by various means, including the following:

(i)  The subcontractor's average daily proceeds realized from the shotcrete crew less their average daily cost with respect to that crew.

(ii)  The overtime cost of the subcontractor's crew for making up each lost day's work on a Saturday. This essentially replaces the day in which the subcontractor lost their opportunity to work.

## SCOPE OF WORK ISSUES RELATED TO SHOTCRETE

1. For small and/or non-critical embeds, either simply tying them off to the rebar mat or wet setting them in the face of the shotcrete wall may be the most economical approach. It is not recommended that wet setting of embeds be considered for large embeds or embeds with small tolerances. On some projects there will be many embeds that are large and/or have small tolerances, and for these, wet setting is not a viable option.

2. The shotcrete subcontractor should always provide their own scaffolding. The primary reason for this is because the scaffolding will become covered with shotcrete splatter (commonly termed rebound). Due to this mess the shotcrete subcontractors own their scaffolding, as the cleaning and damage costs to rented scaffolding make renting costs prohibitive.

3. Establish in the subcontract the number of shotcrete test panels that will be required. These test panels involve a significant amount of work, and because they are indirect work they will not be shown on the contract drawings. Due to this, they are commonly omitted by the formwork, rebar, and demolition bidders. The quantity of test panels must be clearly established and coordinated among the shotcrete, formwork, rebar, and demolition subcontractors. Each nozzleman must be individually certified for each project and the number of nozzlemen will not be established until a shotcrete subcontractor is identified after bidding, as only the shotcrete bidder can accurately anticipate how many nozzlemen will be utilized on the project and each bidder may plan for a different number of nozzlemen (Figure 9.1).

   (a)  The general contractor may consider self-performing the demolition and hauling off of the test panels after completion. It is not critical who demolishes and removes the test panels, just to be sure someone has this demolition covered.

4. If shoring tiebacks are to be de-tensioned at a shotcrete wall, the shotcrete subcontractor will need to come back and fill the blockouts after the de-tensioning has occurred. This costly activity is discussed in further detail in the chapter on concrete placing and finishing.

5. The most common waterproofing material used for below-grade shotcrete walls is bentonite. Bentonite is an expansive waterproofing product that, like a gasket in a car, works under pressure and activates when moisture is introduced. This activation cannot occur until the permanent retaining wall has been shotcreted against it, thus containing the bentonite and creating the above-noted pressure. The shotcrete rebound poses a problem. The

**FIGURE 9.1**   Shotcrete test panel. (Photo by author, courtesy of Hathaway Dinwiddie Construction Company and Tishman Speyer.)

rebound splatter will spot-activate the bentonite product above and to the sides of each day's shotcrete work. Therefore, visqueen (or another sheet product) must be used to protect the membrane above and to the sides of each day's effort. It is best for the shotcrete subcontractor to be held responsible for this protection. Thus, the control lies with one firm and there is a reduced risk of finger-pointing in the event of damage.

6. The current standard of practice in the industry is for the general contractor to clean out, usually with compressed air, any construction joints and keyways at the base of the walls ahead of the shotcrete operation.

7. Often, the shotcrete personnel are not skilled at layout, and they rely on others to provide the face of wall layout. The shotcrete subcontractor will then set the screed wires in accordance with the face of wall laid out on the deck. To ensure proper quality control it is recommended to have this face of wall layout done by the formwork subcontractor, because they will be doing the entire layout for the formwork, as well as the entire layout for the rebar templates. Surveying and layout work is critical; therefore, it is best to keep this work under the control of a single party.

8. Shotcrete is an extremely messy operation and can be disruptive to the other trades. This warrants a stringent clean-up clause in the subcontract that requires clean-up following the shotcrete operation to be completed no later than noon on the following work day.

9. The shotcrete subcontractor must maintain full responsibility for the protection of any adjacent work. During the structural phase this protection will be fairly simple, basically consisting of laying out visqueen to catch the rebound. If/when returning to fill in the tieback blockouts, protection must be provided for not only the adjacent finishes, but also for the path of travel of the shotcrete hoses and the path of travel of the scaffolding. Shotcrete crews are trained to perform this work quickly, and while this

aggressive approach is commended during the structural phase of a project, it can be detrimental to door jambs, corridor walls, and especially gypsum board corners during the finishes phase of a project. This should be clearly addressed in the subcontract agreement and discussed thoroughly during the preconstruction meeting as a quality control effort. Thousands of dollars worth of nicks and dings to the interior finishes can result if proper care is not taken.

10. Keyways at the tops of walls are often specifically called out by structural engineers to be formed with provisions in the specifications that do not allow the shotcrete subcontractor to simply dig out the keyway in the top of the wall while the shotcrete is still wet. Engineers specify this qualification because quite often the keyways dug out by the shotcrete subcontractors are done half-heartedly and thus are smaller than specified by the design. It is generally considered acceptable by the structural engineer to have the keyways dug out by the shotcrete subcontractor if the special inspector employed by the owner is specifically charged with the task of monitoring this work. This inspector will perform continuous (not periodic) inspection during a shotcrete operation and must sign off that the keyways are constructed in accordance with the contract documents. This system of checks and balances which allows the shotcrete subcontractor to dig out the keyways is the most cost-effective approach. If the structural engineer will still require forming of these keyways, this formwork must be completed by the formwork subcontractor.

11. Shotcrete subcontractors regularly exclude the removal from the site of rebound concrete. Some subcontractors do this because they expect to simply put it in the site dumpster, but the rebound from a shotcrete operation can be a tremendous quantity and can quickly fill a dumpster. Consideration should be given to having the shotcrete subcontractor haul off the rebound debris to a recycling facility. This apparently simple issue entails a costly disposal fee and as such must be resolved prior to finalizing the subcontract agreement.

12. The tower crane will probably not be in operation when the lower level basement walls are shotcreted. This means the shotcrete subcontractor will need to include material hoisting for the period of time leading up to the tower crane becoming operational. The project schedule should reflect the milestone date that the tower crane is to be put into service. The scaffolding and other light materials will most likely be hoisted in and out of the excavation by rope over the edge of the excavation. Getting the heavy rebound material out of the excavation will require a crane. The shotcrete subcontractor should bear the responsibility for removing the rebound concrete from the site, including any crane costs that are incurred prior to commissioning the tower crane.

13. When a shaft, such as a stair shaft, is to be shotcreted, the general contractor must establish whether the shaft walls will be formed on the shaft side of the wall or the floor deck side. Since walls are formed from one side, then rebar and shotcrete installed from the other side, the general contractor

must establish whether the formwork subcontractor will be able to perform their work without the floor decks, and whether the rebar and shotcrete subcontractors will have use of the floor decks. The subcontractor(s) who must perform their work from the shaft side will have significantly greater difficulty, and respective added costs, in accessing their work than subcontractors that are given preference for use of the floor decks.

14. For projects with a below-grade parking garage, there is a common cost saving approach that addresses the disposal of shotcrete rebound. This consists of placing the rebound material in the dead space below the lowest level basement parking ramp. This seems appropriate as it saves the cost of a crane and hauling to a recycling facility. It also eliminates the cost of forming and/or backfilling this void space. If placed under the ramp in lieu of compacted fill, it must be placed while still wet and in a plastic state so that it can be properly consolidated. This placement must also take place under the observation of the special inspector.

   Clean-up is often postponed until after the shotcrete operation is complete, simply because the shotcreting pace is so hectic and requires the full attention of the entire crew. This approach results in hardened rebound concrete that will not be an acceptable fill. The shotcrete subcontractor must maintain sufficient numbers of laborers to clean up the rebound, then place and compact it below the ramp while the shotcreting is going on. The special inspector must also be prepared to observe this work during the shotcreting operation.

# Questions—Module Two (Chapters 5–9)

1. What are four primary factors that will impact the selection of a below-grade waterproofing membrane?
2. Which party generally selects the below-grade waterproofing membrane? Explain why.
3. Discuss the suggestion that the general contractor should employ the third-party waterproofing inspector because third-party below-grade waterproofing inspections are not commonly required by code.
4. What additional work is required of the formwork, rebar, and shotcrete (or placing and finishing) subcontractors if a tieback is to be de-tensioned at a later date during the construction phase?
5. What can be done to the perimeter wall rebar mats and interior face wall forms to avoid the multitude of penetrations through the waterproofing membrane that would otherwise be needed for rebar and form support?
6. For a project in which the basement is below the water table, where should the electrical conduits be routed relative to the waterproofing membrane?
7. Where is the mud slab located relative to the waterproofing membrane and the protection slab?
8. What is the recommended approach to address situations when subcontractors accidently damage the waterproofing membrane?
9. Which parties are primarily involved in the construction of equipment pads inside buildings?
10. For the masonry work, which party is typically responsible for installing the wall rebar and which party is responsible for installing the dowels? Which party is responsible for furnishing the rebar and dowels?
11. Which party is responsible for welding rebar to embed plates in the shop?
12. What type of safety precautions must rebar crews exercise when they climb high and all over the wall rebar that has been properly tied?
13. Which party should provide and install the wire mesh to be placed in the concrete-filled metal stair pans?
14. Is it acceptable for tiny rebar tie-wire clippings to be left on the deck form prior to concrete placement?
15. Which parties should provide stand-by personnel for pour watch during major concrete placement activities?
16. Which subcontractor typically provides rebar templates?
17. Which party typically furnishes the rebar cages for caissons and which party installs them?
18. Which subcontractor (place and finish, formwork, rebar, or shotcrete) will have the greatest impact on the project schedule?
19. Which party furnishes the anchor slots in concrete walls for masonry facades and which subcontractor installs them?

20. Which type of temporary safety railing (steel post and cable, wood railings, or slab grabbers) requires the least maintenance?

21. Which subcontractor is responsible for the grout fill in the blockout in the bottom of the elevator pit after the elevator subcontractor has set the hydraulic elevator plunger?

22. Which subcontractor will furnish and install the sleeves to be embedded in the structural concrete work for storm drain piping?

23. Which subcontractor should wet set embeds during a concrete pour so the workers can have something productive to do while they are on pour watch?

24. What is the best construction practice for supporting the pit wall forms?

25. For a concrete slab on metal decking, which subcontractor should provide the edge forms at the perimeter of the deck and at all deck penetrations, and which subcontractor should provide all hung forms for slab depressions?

26. Describe the suggested material for anchor bolt templates and the suggested number of uses?

27. Which subcontractor is responsible for patching form-tie holes, and which subcontractor is responsible for patching all rock pockets?

28. Discuss the merits of the general contractor directly assisting in the performance of the place and finish subcontractors work?

29. When the general contractor pulls a lane closure permit for a concrete placing and finishing operation, how should this be viewed in terms of the responsibility for all work associated with the lane closure, including flaggers, arrow boards, cones, etc.?

30. Which subcontractor(s) should install the equipment pads within the building?

31. Which party has the primary responsibility for grouting baseplates for miscellaneous metals work?

32. Describe the three different methods of producing a colored concrete surface.

33. Which party should apply a combination cure/sealer product?

34. Which party should infill tieback blockouts on a return trip to the project, after the tiebacks in a shotcrete wall have been de-tensioned?

35. Explain why it is important to identify the elevation of tiebacks for the formwork, rebar, waterproofing, and place and finish bidders.

36. Which subcontractor will be responsible for controlling concrete seepage through a wood formed deck, and which subcontractor will be responsible for seepage control through an unvented metal deck?

37. Which party is responsible for providing adequate lighting for concrete finishing activities extending beyond dusk?

38. What level of detail is provided in the construction drawings as to whether walls are to be shotcrete or cast-in-place with double-sided formwork?

39. Explain the merits of a shotcrete subcontractor's request for added compensation when shotcrete work is postponed the evening prior to mobilization to the site.

40. Describe the limits of the size of embeds in the face of a shotcrete wall.

41. Which subcontractor is responsible for providing scaffolding for the shotcrete work? Explain why.
42. Which four subcontractors will perform work associated with the shotcrete test panels?
43. How many shotcrete test panels are commonly required for a project?
44. What adverse effect might shotcrete splatter have on a waterproofing membrane that warrants protecting the membrane above and to the sides of each day's shotcreting application?
45. To ensure the layout work for an entire concrete structure is maintained under a single subcontractor's responsibility, which subcontractor should maintain responsibility for laying out the face of shotcrete walls?
46. Which party typically cleans out the keyways at the base of a shotcrete wall prior to shotcreting? When keyways are required to be formed at the top of a shotcrete wall, which party will be responsible for this work?
47. Which subcontractor will be responsible for hoisting all rebound concrete out of the excavation? What is suggested as the appropriate way to dispose of the rebound debris?

# Module Three

# 10 Structural Steel*

The success of a project can be influenced considerably by the structural steel erection. When properly planned and efficiently executed, the structural steel erection can be the most exciting and profitable aspect of a project. This is when a project begins to take shape. If not properly planned, inefficiently executed, or constantly halted by problems, the structural steel erection can compromise the successful performance of a project. An adverse financial impact can arise from substantial change order requests emanating from a poorly planned operation and an adverse impact on the schedule can result from a series of delays. The structural steel erection is characteristically the sole critical path schedule operation throughout its duration, so a day lost in erection is a day lost in the project schedule (Figure 10.1).

Since the structural steel operation has a relatively small crew and because the delay of a single day will extend the completion date by a day, it may be cost-effective to work the structural steel crew overtime in order to accumulate contingency time at a low cost early in the project. The cost of having the steel crew work overtime in order to gain a day on the schedule is significantly less expensive when the structural steel is the sole critical path activity than the cost of trying to make up a day when many subcontractors are on site late in the project and there are multiple critical paths in the schedule.

The structural steel procurement and fabrication often becomes a critical schedule issue from the onset of construction. For buildings with three or more levels below grade (assuming concrete below grade, but transitioning to structural steel at grade), there is a good chance that the duration of the excavation and concrete phases will allow sufficient time for the detailing, procurement, and fabrication of the structural steel. For buildings with two, one, or no levels below grade, the project may very well come to a halt awaiting the steel delivery. To avert steel delivery delays, begin by identifying the long lead-time steel members (usually special shapes have longer procurement durations) and place the order for those members immediately upon award of the structural steel subcontract. It is equally important to get the shop drawing production started immediately.

Steel work is extremely expensive, making it especially important to clearly define the scope of work to keep change order at a minimum. A clear division is always important, but usually difficult to establish, between the work of the structural steel subcontractor and the miscellaneous metals subcontractor. The basic differentiation is that the structural steel subcontractor is responsible for all primary steel and the miscellaneous metals subcontractor is responsible for all secondary steel. Secondary steel consists of that steel which does not directly function as a part of the building skeleton, e.g., steel that supports a building component, not the building itself.

---

* MasterFormat Specifications Division 5

**FIGURE 10.1**    Steel erection. (Photo by author courtesy of Hathaway Dinwiddie Construction Company and Tishman Speyer.)

**FIGURE 10.2**    Secondary steel.

A general rule of thumb is that any steel designed within the structural drawings is completed by the structural steel subcontractor and any steel that is not designed in the structural drawings is completed by the miscellaneous metals subcontractor. Establishing this distinction in reality is not as easy as it sounds, as this is only a general rule for which there are many exceptions. For example, some large secondary steel members can be much more efficiently erected in conjunction with the highly efficient structural steel operation rather than later during the miscellaneous metals operation (Figure 10.2). Such members are an exception to the rule and should be included with the work performed by the structural steel subcontractor. An especially arduous and tedious task consists of identifying and ensuring all miscellaneous steel components shown sporadically throughout the entire set of project drawings and specifications are identified, catalogued, allocated to the appropriate subcontractor, and tracked until completion.

Since the extent of miscellaneous (secondary) steel components throughout a project is so random there are no cut and dry rules to allocating this work, but there are some general approaches to this allocation that will be discussed in this chapter and in the chapter on miscellaneous metals. Another rule of thumb is that if the steel member in question is directly attached to the primary steel structure and the fabrication can be completed in the shop so that the member can be simply and quickly bolted and/or welded in the field, then the structural steel subcontractor should be responsible for the work. Otherwise, the work should be the responsibility of the miscellaneous metals subcontractor.

## SCOPE OF WORK ISSUES RELATED TO STRUCTURAL STEEL

1. Special permits will be required for lane closures, wide load trucking, and cranes tall enough to require a Federal Aviation Administration (FAA) permit. These permits may be pulled by the general contractor, but permits required specifically for the structural steel operation are more appropriately handled by the structural steel subcontractor. Be sure the steel subcontractor includes not only paying for the permits, but also all drafting, paperwork, and the timely process of obtaining the necessary approvals.

2. The Occupational Safety and Health Administration (OSHA) does not currently require 100% tie-off for structural steel erection crews. Despite this, many companies that are proactive about safety aggressively promote a 100% tie-off policy whenever workers are exposed to fall hazards above six feet in elevation. For the safety of the workers walking the high steel, a 100% tie-off policy should be established. This will create an additional cost and extended duration, so the owner needs to be supportive of the policy and be willing to accept the additional cost and duration. Even if the owner is opposed to the additional cost, the 100% tie-off policy should still be enforced. A human life is just too valuable to risk. Structural steel subcontractors who have been walking the high steel all their lives will often oppose this policy, but this should not be a negotiable topic.

3. Properly setting the anchor bolts is the responsibility of the formwork subcontractor, and the general contractor will have the project surveyor confirm the as-built locations of the bolts as part of their quality control program. In addition, the structural steel subcontractor will have their own surveyor as part of their own quality control program to verify the bolt locations and make any necessary adjustments to the baseplates to account for bolts slightly off the mark. It is a minimal cost to modify a baseplate in the shop, but if a baseplate error is discovered when a column is being boomed into position, the lost crew time and delay to the schedule to modify the baseplate will be exorbitant.

4. Quite often the structural steel subcontractors will bid a project based upon the general contractor (or others) setting the first leveling nut at each column, because this first nut must be set via a surveyor. The steel subcontractor will then set the others simply using a level, with the first leveling nut being the reference point. It does not really matter whether the steel subcontractor sets the first nut or whether it is done by the general contractor, as long as this layout cost is covered one way or another. Since the steel subcontractors have their own surveyors, as discussed above, this work is commonly allocated to them.

5. Carefully review the plumbing drawings with respect to sanitary sewer and storm drain lines running through the building. Check the slope of lines versus the ceiling clearances and determine whether the piping can be run below the steel beams, or if the space between the bottom of the beams and the top of ceiling system is so small that the drain lines will in some cases need to run through the beams (Figure 10.3). These beam penetrations can be quite expensive, especially if not planned for in the initial design, and this must then be done in the field rather than in the shop. At bid time it is very important to simply quantify the number of beam penetrations, but before steel fabrication is begun, the specific locations and sizes of the penetrations need to be identified. This MEP coordination information must be expedited and communicated to the structural steel subcontractor in time for the steel detailer to incorporate it into the shop drawings, so the fabrication crew can in turn cut and reinforce the beam penetrations in the shop.

6. Prime painting of the structural steel is a standard inclusion of structural steel subcontractors. Touch-up painting of the steel at welds or scrapes, on the other hand, is a standard exclusion. The structural steel subcontractors regularly argue that this work should be completed by the painting subcontractor, but this position cannot be logically defended. The most efficient

**FIGURE 10.3** Beam penetration. (Photo by author, courtesy of Hathaway Dinwiddie Construction Company and Tishman Speyer.)

means of completing the touch-up prime painting for the structural steel (or miscellaneous metals for that matter) is for the welders to carry a can of spray primer with them and quickly paint the welds after they are complete and have substantially cooled. There is not a union jurisdictional problem with this allocation of work as might be suspected.

7. For exposed structural steel that is finish painted a light color, as is common with modern construction, be sure a light-colored primer is used. Rust-red is a staple in the industry and will be used by the subcontractors unless otherwise specifically requested, though using an alternate color does not cost any more. Another benefit to using a light colored primer in lieu of rust-red is that any small rust spots on the iron will be clearly evident (not camouflaged by the rust-red paint) and will provide an opportunity to sand and touch-up those spots ahead of finish painting. These small rust spots will cause chips in the finish paint if not properly addressed and mitigated.

   Structural steel is not a clean operation. The steel may sit outside at the shop both before and after fabrication, often collecting dust and dirt for many months. Upon arriving at the site, the steel is often stepped on and continually covered in dirt. Prior to finish painting, this steel needs to be cleaned, commonly by the general contractor's own laborers to avoid the high cost of having this done by the structural steel subcontractor's iron workers. It is very important to clean the steel immediately or shortly after erection because the structure is still in a rough state in which the overspray and flooding from a pressure washer will not damage anything. If the steel is cleaned late in the project, immediately before painting is to take place, it will be far more expensive because hand cleaning will be necessary, as the mess from the pressure washer will damage the building finish components (gypsum board, paint, carpet, casework, etc.) already installed. Even steel that is exposed, but left prime painted, will need to be cleaned after it is erected.

8. Carefully examine and allocate the steel work within the elevator shafts (Figure 10.4). Due to the size of the members and relative ease of installation during the structural steel operation, it is suggested that the elevator hoist beams, separator beams, and guide rail supports be allocated to the structural steel subcontractor.

9. The structural steel subcontractors traditionally include hoisting of the metal decking in their proposals, which is very easy for the structural steel subcontractor to do. If for any reason the structural steel subcontractor has submitted a quotation on a project that has excluded the metal deck hoisting it should be strongly promoted as being part of this work to avoid the added cost for the metal decking subcontractor to provide their own crane for each decking delivery.

10. Furnishing and installing safety cabling at the perimeter of each deck, as well as constructing fall protection with steel posts and cable rails around openings in the metal decking, is most efficiently completed by the structural steel subcontractor. Railings are needed as soon as a fall hazard is known to exist. Since the properly equipped structural steel subcontractor will be on site when these fall hazards are created, it makes sense for them

**FIGURE 10.4** Elevator steel. (Photo by author, courtesy of Hathaway Dinwiddie Construction Company and Tishman Speyer.)

to complete the work. Note that this is a common exclusion by steel subcontractors, so be sure to specifically address it in the bid instructions.

Maintaining and eventually removing the rails after their initial installation is most appropriately included in the general contractor's scope of work, as the structural steel subcontractor will generally have demobilized for some time before it is time for the railings to be decommissioned.

11. The structural steel subcontractor needs to provide all the anchor bolts for their work FOB jobsite, for installation by the formwork subcontractor. The structural steel subcontractor should also furnish the steel templates for each set of bolts. Two concerns about quotations from the structural steel subcontractor are as follows:

   (a) There will be a temptation on the part of the structural steel subcontractor to provide paper templates for the formwork subcontractor to use in making wood templates. This creates additional work for the formwork subcontractor which they will not likely anticipate, resulting in the submittal of a change order request. It should also be recognized that wood templates constructed in the field are not as accurate as shop-made steel templates.

   (b) Structural steel subcontractors should be required to supply templates for each set of bolts. They might want to supply templates for multiple reuses, but after a single use (particularly after the stripping activity) the steel templates get damaged to the extent that reuse is not practical.

12. Structural steel subcontractors commonly exclude welding of rebar on site (as opposed to welding of rebar in the shop, which is rarely a problem), as do the rebar subcontractors. Since on-site welding of rebar is quite rare, rebar subcontractors do not generally keep welders on their payroll capable of this welding, therefore it is generally recommended that the structural steel

subcontractor be held responsible for this work. It is often difficult to locate a welder certified in welding rebar (a special and uncommon certification), so the responsibility might best be assigned to the subcontractor that has the ability to provide a certified welder and include it in their scope of work. If neither the rebar nor structural steel subcontractor has a welder with the necessary certification, the miscellaneous metals subcontractor may in fact be the most logical source for this work.

13. Steel supports for metal deck penetrations are most efficiently completed by the structural steel subcontractor, but they will commonly exclude this work. The biggest problem that structural steel subcontractors encounter when assigned this work is waiting for the actual field dimensions for the duct penetrations, pipe chases, etc. This information must be provided in cooperation with the MEP subcontractors via the MEP coordination effort. This information needs to be transmitted to the structural steel subcontractor during the shop drawing phase in order for this steel to be prepped in the shop and efficiently erected in the field. Though it is very important to handle this early in the project, in many cases the general contractors have not even begun the MEP coordination effort by the time the steel subcontractor needs this information. As a result, this can be a source of costly change order.

14. For large or odd steel configurations (such as high trusses spanning an arena) temporary shoring columns, guy cables, and other temporary erection aides may be necessary. The structural steel subcontractor will include these temporary structures in their bids, but items such as additional footings, setting anchor bolts, or setting embeds that are needed for these temporary structures will need to be picked up by other subcontractors, such as the formwork, rebar, and place and finish subcontractors.

15. Ledger angles for the metal decking are most efficiently constructed by the structural steel subcontractor, though this work is a common exclusion in the structural steel bids. Many general contractors choose to allocate the decking support angles to the metal decking subcontractor, but the fact is that metal decking subcontractors do not have the ability to self-perform this work due to union jurisdictions, so they would need to employ a lower-tier steel sub-subcontractor for this work. Regardless, the important thing is to be sure that these ledger angles are included in one of the bid quotations.

16. For metal decking that cantilevers at the perimeter (or interior) of a building beyond the edge of structural steel beam supports, a steel angle or bent plate is commonly installed. This angle or plate will support the cantilever and it also acts as the deck-edge form. Though the metal decking subcontractor will typically be asked to furnish and install all edge forms sized 10-gauge or less (the edge forms that lie directly on the top of beams and hold no significant vertical structural value), the structural steel subcontractor should be held responsible for the edge forms and deck supports that are greater than 10-gauge. Further, for decking with a substantial cantilever the steel bent plate regularly has angle braces below it kicked back to the bottom flange of the beam, and this should also be allocated to the structural steel subcontractor (Figure 10.5).

**FIGURE 10.5**   Deck-edge support.

17. The welding equipment for the structural steel operation requires a great deal of power, most often more power than is available on site. If the temporary power setup has insufficient power for the structural steel subcontractor's welding equipment, they will be required to bring their own generator.

18. It is always a good idea to get as much welding as possible done in the shop in lieu of the field. Whenever possible, have clips for curtain walls, stairs, façade panels or anything else welded in the shop by the structural steel subcontractor instead of being field welded by the respective subcontractor. This option will not always be feasible and could cause confusion when pressure is placed on the structural steel subcontractor to be on site, but nevertheless reviewing these options on a project-specific basis and implementing them when feasible is a good construction practice of the general contractor.

# 11 Metal Decking*

The metal decking operation is another item of work on a project that is relatively straightforward. As a general rule, few change orders are associated with metal decking. Due to the nature of this work, once the planning is properly completed this work generally progresses well with minimal oversight from the general contractor.

Setting down sheets of metal decking is a typical example of leading edge work. This means that the location of the unprotected edge keeps moving as additional sheets are laid down. Since the installation of guardrails is not realistic, some other type of fall protection is needed. In general, the best way to ensure the safety of the workers is to implement a 100% tie-off policy. Though there are several types of effective fall protection methods, in this case to increase mobility when laying down the metal deck sheets retractable lanyards are advised. Note that the OSHA regulations do not currently require metal decking (also structural steel or scaffolding) workers to tie-off when walking the high steel, but similar to the discussion in the chapter on structural steel, this policy of tying off is highly encouraged. A human life is entirely too important to put at risk. Decking subcontractors and their employees may be reluctant to comply with the requirement for 100% tie-off, but this is not a policy that can be justifiably relaxed or ignored. (Figure 11.1)

## SCOPE OF WORK ISSUES RELATED TO METAL DECKING

1. Decking that will either have the underside painted for an exposed ceiling or decking that needs to be fireproofed must be oil-free. The manufacturing process for metal decking leaves a slight oily film on the decking sheets which is detrimental to paint and fireproofing adhesion. In addition to cleaning the oily film, for exposed painted ceilings a treatment called paint-lock (also termed galvanneal) can be applied in the shop to the underside of decking to increase adhesion and subsequently yield a better finish.
2. Hoisting of the metal decking will generally be performed by the structural steel subcontractor in conjunction with the steel erection, but the decking subcontractor must properly coordinate the decking deliveries with the structural steel operation. With proper coordination this will be a seamless collaborative effort.
3. The metal decking subcontractor should provide all edge forms. Unless specifically indicated otherwise on the contract drawings, the deck-edge form should be galvanized sheet metal consisting of a maximum 10-gauge thickness.

---

* MasterFormat Specifications Division 5

**FIGURE 11.1** Metal decking. (Photo by author, courtesy of Hathaway Dinwiddie Construction Company and Tishman Speyer.)

(a) Any edge form greater than 10-gauge is considered structural steel and as such must be completed by the structural or miscellaneous steel subcontractors due to union jurisdictions. Note that 10-gauge is slightly thinner than 1/8″, so the division of work between gauge metals for trades such as metal decking or flashing and steel for trades such as miscellaneous metals and structural steel is that anything 10-gauge and thinner is considered a gauge metal, but anything 1/8″ or thicker is considered steel. Also, an important distinction with gauge numbers is that they decrease as the metal thickness increases, i.e., 10-gauge material is thicker than 12-gauge material. An interesting parallel to this non-intuitive measurement is with electrical wiring, where a 10-gauge wire is larger than a 12-gauge wire.

(b) These deck-edge forms should not be confused with structural deck-edge angles (which will be 1/8″ or thicker, most commonly either 1/4″ or 3/8″ thick), which are provided by the structural steel subcontractor as discussed in the chapter on structural steel. The first difference is that steel edge angles serve two purposes by supporting the deck vertically and containing the concrete placement horizontally, but the gauge metal deck-edge forms discussed above are simply used to contain the concrete placement.

4. Decking support angles are more efficiently installed by the structural steel subcontractor. Therefore, this work should not be included in the scope of work of the metal decking subcontractor. The decking subcontractor may be required by some general contractors to furnish and install the support angles as part of their system, but this is not advised because in many cases the angles can be much more efficiently welded to the structural steel members in the shop. Also, on union projects the union representing the metal decking work does not have jurisdiction over non-gauge metals (anything

**FIGURE 11.2**   Shooting studs. (Photo by author, courtesy of Hathaway Dinwiddie Construction Company and Tishman Speyer.)

thicker than 10-gauge). Thus, it makes more sense for the structural steel subcontractor to furnish and install these supports. If they are required to perform this work, the metal decking subcontractors often bring in a lower-tier structural steel subcontractor to complete the support angle work.

5. Concrete slurry seepage through the decking joints and/or loose edge metal can compromise the finish on the floor below during the concrete pour. It is very important to hose down and dilute this seepage material before it has a chance to harden. The responsibility for addressing this seepage issue may depend on the circumstances. For non-vented decking the metal decking subcontractor has complete control over the integrity of the metal deck forming system, and should therefore maintain full responsibility for hosing/cleaning any seepage through the deck. Since vented decking (decking with perforations which provide greater air circulation to expedite concrete curing) will invariably allow some seepage through the vents as an inherent part of the design that is out of the control of the decking subcontractor, the general contractor may decide to assume responsibility for protecting and cleaning any seepage associated with vented decks.

6. On union projects, the metal decking subcontractor traditionally has the jurisdiction for providing and installing the welded shear studs on top of the beams (Figure 11.2). For safety reasons (as required by OSHA), these studs cannot be welded to the tops of beams until the decking is laid in place and tacked down because the studs would otherwise form tripping hazards for the iron workers on the high steel.

7. As with the structural steel and miscellaneous metals operations, the welding equipment for the metal decking work requires significant power. Unless sufficient electrical power for this equipment is available on site, be sure the decking subcontractor furnishes adequate generator power to effectively perform their work.

# 12 Miscellaneous Metals*

The miscellaneous metals work is clearly the most tedious and difficult work to accurately scope in the bid instructions and the subcontract agreement. This scope of work basically consists of all the steel components that are not included in the bids of the structural steel subcontractor or various other subcontractors. A careful review of the contract drawings, specifications, and subcontractor qualifications is particularly important for this work, as a single item can be very expensive.

General rules of thumb:

1. If the steel in question is directly attached to the primary steel structure and can be fabricated in the shop to a point in which the member can be simply and quickly bolted and/or welded in the field, the structural steel subcontractor should have responsibility, otherwise the steel is more appropriately allocated to the miscellaneous metals subcontractor.
2. If the item in question is specifically identified in a specification section outside of Division 5, then it is a manufactured product more appropriately provided and installed by the trade responsible for the respective specification section. Tree grates, recessed floor mats, benches, stainless steel corner guards, and ornamental bollards commonly fall in this category.
3. Unless otherwise specifically addressed (and many items will be specifically addressed as exceptions to this rule)—any steel designed (not just shown, but actually structurally designed) on the structural documents is the responsibility of the structural steel subcontractor and any steel not designed on the structural drawings generally falls under the miscellaneous metals scope of work.

## SCOPE OF WORK ISSUES RELATED TO MISCELLANEOUS METALS

1. Miscellaneous metals subcontractors traditionally furnish and install the metal stairs as part of their scope of work. There are several items to be aware of when scoping the metal stairs.
   (a) The miscellaneous metals subcontractor must provide all embeds FOB jobsite for installation by the formwork subcontractor. This will include angles where the stairs attach to the concrete decks as well as potentially anchor bolts for steel columns supporting the landings.
   (b) The metal stairs are almost always a design-build scope of work requiring a deferred permit submittal. Design of the stairs by a registered engineer is not often a question, but be sure the miscellaneous

---

* MasterFormat Specifications Division 5

metals subcontractor maintains complete responsibility for obtaining
and paying for the related permit. Clearly specify the responsibility
for obtaining permits as the process can be involved, including paperwork, dealing with the city officials, and following up with responses
to permit queries.

(i)   Most supplemental permits need only to be stamped by a professional engineer who is registered in the state. This engineer is
commonly either a structural, civil, or mechanical engineer by
trade, but in reality, even an electrical engineer could stamp metal
stair drawings as long as they are licensed. Some jurisdictions are
more specific than this, so be sure to ask this question and get a
confirmation from the local building department before bidding
the project. This goes for all supplemental permits, not just metal
stairs. There will generally not be a problem with most supplemental permit designs, but more structurally significant scopes
of work (such as metal stairs and precast) could cause a problem. Many cities will specifically call for a structural engineer for
metal stairs, precast, and other items requiring significant structural considerations.

(c)   Traditionally the miscellaneous metals subcontractor will prime paint
the stairs and rails, then the painting subcontractor will complete the
finish painting. Priming is not often an issue, but be sure to clearly
identify the touch-up of scuffs, dings, and welds as work to be performed by the miscellaneous metals subcontractor.

(d)   Establish and closely monitor the schedule for the stair design and procurement. It is beneficial to everyone if the stairs can be installed floor
by floor immediately after each concrete deck pour. The general contractor benefits in that the metal stairs can be put to use for construction
access (Figure 12.1) and eliminate both the cost and inconvenience of
temporary stair towers. The miscellaneous metals subcontractor benefits in that hoisting the stairs into place with a crane as the building
is erected is considerably less expensive than manipulating the stairs
through the building to the point of installation on dollies and then trying to rig the stairs into place in a tight shaft. Further, unless deemed
unreasonable for a particular project, it is recommended that the miscellaneous metals subcontractor be held to a schedule that reflects the metal
stairs being erected floor by floor immediately after the concrete deck
pours. The subcontract should specifically note that any costs associated
with the rental of temporary stair towers due to delays in this effort will
be the responsibility of the miscellaneous metals subcontractor.

(e)   Be careful not to block construction progress with the metal stair
work. If the metal stairs are not on site in time for construction use
and a temporary stair tower is required, the best place for the temp
stairs always appears to be in the stairwells. Unfortunately, when the
metal stairs arrive, the temporary stairs will need to be removed prior
to erection of the permanent stairs, leaving the project without the

**FIGURE 12.1** 2x blocking in stair pan for construction use. (Photo by author, courtesy of Hathaway Dinwiddie Construction Company and California State University Northridge.)

respective stair access during the permanent stair installation. If temporary stairs are deemed necessary, be careful to properly locate them for maximum efficiency on the project.

(f) Since there are several different types of stair treads, there are an equal number of approaches to their construction.

  (i) For metal treads the work is easy to allocate as all work is to be performed by the miscellaneous metals subcontractor.

  (ii) For precast concrete treads the miscellaneous metals subcontractor should provide a flat plate support that will properly support a precast concrete tread with four bolts sticking out from the bottom of it. The precast concrete subcontractor should not have any welding or other steel work to perform, just the tread with four embedded bolts and associated nuts and washers.

  (iii) For cast-in-place concrete treads:

    (1) The stair pans and supports should constitute the complete formwork for the concrete placement.

    (2) The rebar subcontractor should provide the wire mesh in the pans. Note that this mesh may only be described in the miscellaneous metals specification and may not otherwise be indicated in the contract documents. There may be a temptation to have the wire mesh provided by the miscellaneous metals subcontractor and tacked into the stair pans in the shop, but do not forget that, for construction use, the stair pans should be filled with 2x blocking in lieu of the permanent concrete fill (to avoid scarring and staining the permanent concrete fill) and the wire mesh will be in the way of the blocking.

    (3) The placing and finishing subcontractor should complete the concrete placement.

      (4) As a general rule, the stair nosings for metal stairs should be set by the miscellaneous metals subcontractor. Some prefer that the formwork subcontractor include this in their work. The make and model of the nosings will need to be reviewed in order to make an educated decision. The important point, as always, is simply to be sure this work is covered.

      (5) Placing the concrete fill can be an extremely messy activity. Be sure the placing and finishing subcontractor protects the area and also cleans up. Concrete splatters on the risers, railings, and walls are difficult to clean and this can become a significant quality control issue.

2. All embeds must be delivered FOB jobsite for setting by the formwork subcontractor. Additionally, be sure the miscellaneous metals subcontractor provides a clear and concise embed-setting plan, clearly marks each embed delivered and ships them with clear and descriptive material tickets that allow for easy cataloging of the pieces on site by the formwork subcontractor. Further, the miscellaneous metals subcontractor needs to be held responsible for drilling nail (bolt holes for large embeds) holes in each embed for attachment to the formwork. These nail holes are easy to fabricate on a drill press in the shop, but this will be considerably more difficult if drilling is done by hand in the field.

3. Supplying and installing anchor bolts must follow the same protocol as for embeds. Further, the miscellaneous metals subcontractor must furnish steel templates for each set of anchor bolts that are provided. Since most templates get damaged after a single use, do not plan on reusing any templates. Most reputable miscellaneous metals subcontractors follow this protocol as a standard operating procedure, but others will simply provide paper templates or written dimensional instructions for use by the formwork subcontractor to construct the actual templates out of plywood. It is suggested that this practice be avoided as wood templates are not as accurate, sturdy, or durable as steel templates constructed in controlled shop conditions. The placement of anchor bolts must be precise and should not be jeopardized by a practice that saves a few dollars in exchange for lessened quality. Good quality control is essential to ensure the proper placement of bolts. The cost of correcting problems arising from poorly placed anchor bolts will be exponentially more expensive than the cost of the templates themselves.

4. It is generally recommended that the grouting of miscellaneous metals work be performed by the miscellaneous metals subcontractor. If they are not prepared to do it, the general contractor or the placing and finishing subcontractor are capable of completing this work. It is important that the grouting work is clearly identified. The miscellaneous metals work can be complicated, so the scope of work involving the grouting will also be complicated. Make sure this grouting is covered, but not double covered, by the general contractor and subcontractors.

5. Coiling doors regularly require varying types of steel framing for differing wall conditions. The framing must be provided by the miscellaneous metals subcontractor.

**FIGURE 12.2** Baseplate sticking out of a wall.

(a) For a stud-framed wall opening, tube steel jambs and header are standard. This steel is fairly straightforward, but keep in mind that the baseplates must be cut down in size so they reside fully within the stud wall. Baseplates that stick out the bottoms of walls (Figure 12.2) are common and constitute a costly error in construction. All steel beyond the tube steel jambs and header, including guide rail angles, should be provided by the coiling door subcontractor.

(b) For a CMU block wall it is generally preferred to allocate all steel components and their installation to the coiling door subcontractor, who will mount the coiling door housing and guide rails by use of expansion anchors.

(c) For a cast-in-place concrete wall there will be embedded angles at both jambs that will be furnished by the miscellaneous metals subcontractor FOB jobsite for installation by the formwork subcontractor. All steel work beyond these two embedded angles is most appropriately allocated to the coiling door subcontractor.

6. Make sure that the need for galvanizing and prime painting of steel is properly described in the contract and bid documents. Also, be sure that touch-up of primer and galvanizing in the field is covered by the miscellaneous metals subcontractor.

7. For the elevator shafts, the miscellaneous metals subcontractor should furnish the embedded sill angles and the embedded sump pit frame FOB jobsite for installation by the formwork subcontractor. Further, the miscellaneous metals subcontractor should both furnish and install the sump pit grating and elevator pit ladder.

(a) Confirm the necessity for the pit ladder, sump pit, and edge angles with the elevator design-builder.

(i) Many jurisdictions no longer require a sump pit. This must be ascertained.

    (ii)   Many elevator contractors do not need sill angles for passenger elevators, although the angles are a code requirement for freight elevators.

    (iii)   Some elevator contractors have begun supplying their own pit ladders. It does not really matter who provides the pit ladders, so long as they are covered, but not double covered.

    (iv)   The clearance between the elevator cab and the shaft wall is often too narrow for the ladder to be installed as per the code which stipulates a minimum dimension between the rung and the wall. A variance is regularly required to allow the ladder to be installed only a couple inches from the wall, but attaining this variance is not the recommended route. Be sure to account for this toe clearance when coordinating the elevator pit and shaft dimensions.

8. Steel corner or wall guards should be furnished by the miscellaneous metals subcontractor unless the corner guard is: (1) a manufactured product; and (2) not embedded in cast-in-place concrete. The more common plastic or stainless steel products simply attached to the walls with adhesive or screws are more appropriately furnished and installed by either the general contractor or the miscellaneous specialties subcontractor. Naturally, if the means of installation is embedment in cast-in-place concrete, the miscellaneous metals subcontractor will furnish the materials FOB jobsite for installation by the formwork subcontractor.

9. Caulking and sealants for the miscellaneous metals further complicate this scope of work. It is preferred that the miscellaneous metals subcontractor maintain complete responsibility for the caulking within and at the perimeter of their work. The miscellaneous metals subcontractor may not want this work, in which case a quantity survey needs to be completed by the general contractor and the work allocated to a caulking subcontractor.

10. Metal stairs are traditionally a design-build item and has already been discussed. There are often other miscellaneous metal items that require a design-build effort by the miscellaneous metals subcontractor. For example, large gates, catwalks, equipment platforms, railings, and pipe racks are common components of the subcontractor's design-build effort. In addition to the design cost, be sure the miscellaneous metals subcontractor is responsible for all time associated with the laborious tasks of submitting and following through with the city during their review, as well as for full payment of all associated permit fees.

11. For exterior railings it is important that end caps be installed on all open pipe or tube ends, including the ends that terminate immediately in front of walls or columns. Unless the railings are galvanized, water will eventually get into the tubes and rusty water will subsequently seep out and stain the adjacent wall or surfaces below.

12. The tops of non-bearing CMU walls will require steel angle clips for lateral support of the CMU wall (Figure 12.3). Ensure that the miscellaneous metals subcontractor has these included in their scope of work, as the masonry subcontractor will surely exclude them.

**FIGURE 12.3**   CMU wall clips. (Photo by author, courtesy of The University of Southern California.)

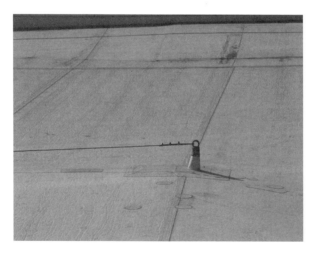

**FIGURE 12.4**   Tie-off point. (Photo by author, courtesy of The University of Southern California.)

13. Review the project conditions and determine the necessity for temporary tie-off points (Figure 12.4) for construction purposes. When they are only for construction purposes, these temporary tie-off points will not be detailed in the contract drawings. If required, they are generally not expensive. For example, additional tie-off points are commonly necessary for roofing work adjacent to low parapets, for skylight installations, and in elevator shafts. Although the tie-off points themselves are not expensive, the cost of delaying a crew who discovers there is no tie-off location as they are about to begin work in an area can be quite expensive and time-consuming. Since of

liability concerns, tie-off points will need to be designed and stamped by a licensed engineer prior to constructing the supports. Allow time for this engineering review.

14. The miscellaneous metals subcontractor should furnish and install all steel railings.

    (a) On union projects, the union jurisdiction of the miscellaneous metals subcontractor will not extend to many decorative materials, such as aluminum railings. Items such as aluminum railings must be furnished and installed by an ornamental metals subcontractor. Some general contractors choose to place the ornamental metals into the miscellaneous metals scope of work, which will require the miscellaneous metals subcontractor to allocate this work to an ornamental metals subcontractor via a lower-tier subcontract. This is a competent and widely accepted method of allocating and executing the ornamental metals scope of work.

    (b) Many architecturally significant railings will entail a flat sheet metal component, thinner than 1/8″, or an architectural mesh screen. Since these products are integral to the railings and customarily installed in the railing assembly in the shop, it is appropriate for the miscellaneous metals subcontractor to provide them. Since these items are installed in the shop (not the field), there should not be a union jurisdictional issue.

    (c) For exterior railings, it is preferred to layout and core drill for the railings rather than trying to set sleeves, as the sleeves are commonly moved and dislocated during a concrete pour. Core drilling provides for a much cleaner and more precisely located finished product. For interior railings, core drilling is not a good option for several reasons. For example, the coring will often cut through the vapor barrier, electrical conduits, or plumbing below the slab on grade and cut through rebar in the elevated decks.

    (d) To avoid unsightly baseplates, interior railings are best welded to embeds set in the concrete. If the design documents do not clearly indicate this method of construction, be sure to clarify this issue with the architect as the subcontractor may very well intend to employ the cheaper alternative of mounting the rails with baseplates and expansion anchors.

15. All gratings and their supporting structures (both interior and exterior) should be the responsibility of the miscellaneous metals subcontractor. The only exceptions would be manufactured products, such as tree grates or recessed floor mats, which are identified in specification sections outside of Division 5.

16. All steel ladders should be the responsibility of the miscellaneous metals subcontractor. Aluminum ladders are an exception to this rule, and are more appropriately allocated to the general contractor for self-performance.

17. Stair nosings (Figure 12.5) are frequently missed during the bidding phase. General contractors often require the miscellaneous metals subcontractor to provide all stair nosings FOB jobsite for installation by the formwork (or site concrete) subcontractor. Since stair nosings are a manufactured product,

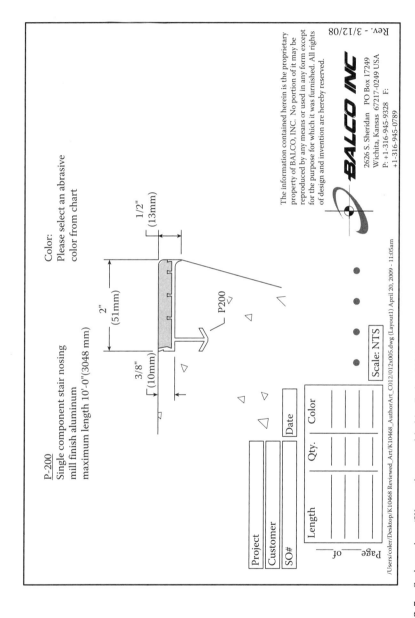

**FIGURE 12.5**  Stair nosing. (Illustration provided by Balco USA.)

they do not need to be provided by the miscellaneous metals subcontractor and can be furnished by either the general contractor or the miscellaneous specialties subcontractor, then installed by the appropriate concrete trade. The majority of interior nosings will be placed by the formwork subcontractor and exterior nosings will be set by the site concrete subcontractor. The exception to this rule is for nosings at metal stairs, which the miscellaneous metals subcontractor will in many cases furnish and install.

18. When gates are fabricated by the miscellaneous metals subcontractor, they must make the appropriate provisions for the finish hardware that will be furnished and installed by the door subcontractor (Figure 12.6). Some considerations for gates include the following:

   (a) The proper plate and accommodations for the lockset and deadbolt.
   (b) A screen around the lockset to prevent someone from reaching through the gate and opening it.
   (c) The proper plates and preparatory work for the panic bar, as well as the appropriately sized screen around the bar.
   (d) A mounting plate for the card reader.
   (e) A mounting plate, or post, for the handicap access button.
   (f) Mounting plate/accommodations for the automatic door closer.

19. For columns/posts at the roof, get the roofer to furnish roof jacks to the miscellaneous metals subcontractor before fabrication so the roof jacks can be slipped onto the columns/posts before the baseplate is welded to it. This will eliminate the undesirable use of split roof jacks.

20. Many general contractors have the miscellaneous metals subcontractor furnish low wall posts (Figure 12.7) and angle counter supports (Figure 12.8) FOB jobsite for installation by the framing subcontractor. It may be preferred to have the framing subcontractor furnish their own low wall posts and have the casework subcontractor provide the counter supports to the framing subcontractor. There is no set standard for these items, just be sure the items are

**FIGURE 12.6**   Door hardware at metal gate.

**FIGURE 12.7** Low wall posts. (Photo by author, courtesy of Hathaway Dinwiddie Construction Company and The California Institute of Technology.)

**FIGURE 12.8** Counter support angles. (Photo by author, courtesy of Hathaway Dinwiddie Construction Company and The California Institute of Technology.)

properly covered, but not double covered. Further, for the counter supports in metal stud framing be sure to set a standard as to whether the leg of the angles will face to the left or to the right because this will determine which way the framer faces the studs.

21. For dock levelers, the miscellaneous metals subcontractor will furnish the edge angles FOB jobsite for installation by the formwork subcontractor. They will also furnish and install the post (unless the post itself is actually a manufactured item identified outside of Division 5, in which case the dock leveler subcontractor will provide it) for the dock leveler controls and dock light, as this post is typically mounted with expansion anchors.

# 13 Expansion Joint Covers*

A building with an excessive width, or with an addition built onto an existing building, must be constructed with a sealed, continuous, uninterrupted gap between the independent structures. These gaps are termed expansion joints or seismic joints and are necessary to accommodate movement that occurs in each independent building structure caused by seismic activity. When expansion joints are designed to accommodate seismic activity, the width and number of such joints will vary depending on the historical record of seismic activity in the particular region and the height of the structure. California, with a history of considerable seismic activity, requires large and more frequent expansion joints than structures in regions with little seismic activity (Figure 13.1).

The construction of an expansion joint is quite elaborate. The expansion joints must be constructed so that each building structure acts fully independent of the other. Functionally, the expansion joint should permit the building components to move independently, while still maintaining the waterproof integrity of the building. At the same time, there must be continuity of building components and systems across the gap. For example, building elements such as conduit, ductwork, and piping will need to cross these joints, but can only cross these joints with appropriately sized flexible connections.

Expansion joint covers (EJCs), the focus of this chapter, function to cover the interior and exterior gaps between building structures. Conceptually, their installation is a simple task. If the proper planning and precautions are not taken in their installation, this simple task can become a tremendous hassle that can delay the schedule and generate numerous change order requests.

## SCOPE OF WORK ISSUES RELATED TO EXPANSION JOINT COVERS (EJC)

1. The scope of work included in the installation of EJCs must be thoroughly described. One common omission from EJC bid quotations is the fire barrier. This can be a costly omission. Fire barriers are not always required and, when they are specified, they may be bid as alternates or omitted altogether. Carefully evaluate the bid quotations to determine whether the fire barrier is included when required by the documents.

   The reason that EJC bids often omit the fire barriers is because they are very expensive. As a result, architects and general contractors will regularly abandon the prefabricated fire barrier in favor of a more cost-effective fire-rated assembly such as stuffing fire safing into the gap (Figure 13.2).

---

* MasterFormat Specifications Division 7

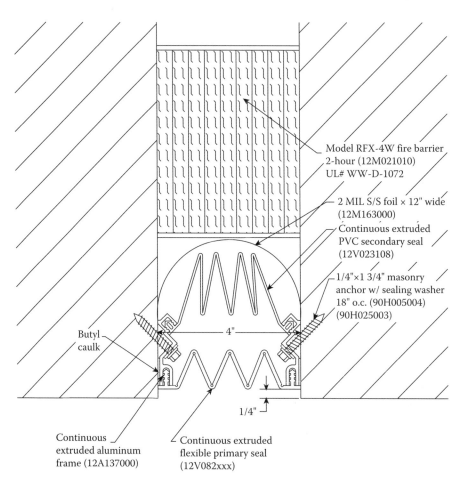

Model RFX-4W fire barrier
2-hour (12M021010)
UL# WW-D-1072

2 MIL S/S foil × 12" wide
(12M163000)

Continuous extruded
PVC secondary seal
(12V023108)

1/4"×1 3/4" masonry
anchor w/ sealing washer
18" o.c. (90H005004)
(90H025003)

Butyl
caulk

4"

1/4"

Continuous
extruded aluminum
frame (12A137000)

Continuous extruded
flexible primary seal
(12V082xxx)

**FIGURE 13.1**    Expansion joint. (Illustration provided by Construction Specialties, Inc.)

2. Review the method of installation of the specific expansion joint cover
(EJC) models for the project. Then determine if other trades or the project
schedule will be affected by the EJC installation for the manufacturer's
models that are specified.

(a)   Coordination of the various building systems is vital to the successful
installation of expansion joints. For example, interior floor joints
commonly need to tuck up behind the gypsum board walls (Figure
13.3). Schedulers commonly plan for expansion joints to be placed as
a finished product toward the end of the project, but this is often an
incorrect assumption. If not properly planned, the gypsum installation might be delayed until the expansion joint material is procured.
Otherwise, if the interior floor joint cover is to be installed behind an
existing gypsum board wall, the wall will need to be demolished and
replaced if the EJC is installed toward the end of the project.

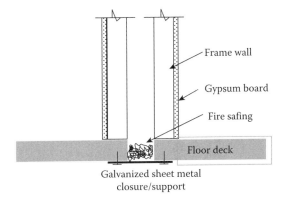

**FIGURE 13.2**   Fire safing in expansion joint.

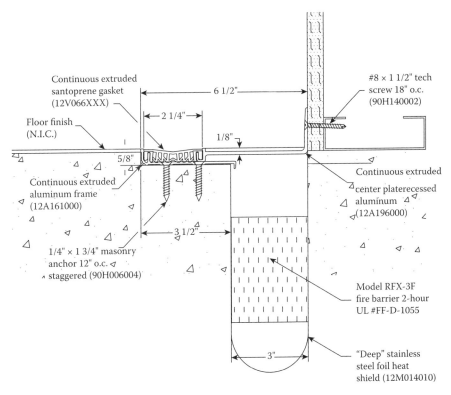

**FIGURE 13.3**   Floor to wall expansion joint. Note that architects will quite often require the up-leg be buried behind the gypsum board. (Illustration provided by Construction Specialties, Inc.)

3. EJC subcontractors do not typically provide sealant or flashing at the top of a roof-to-wall joint (Figure 13.4). This work would be the respective responsibility of the caulking and flashing subcontractors. The different expansion joint models vary as to whether a sealant joint or additional flashing is

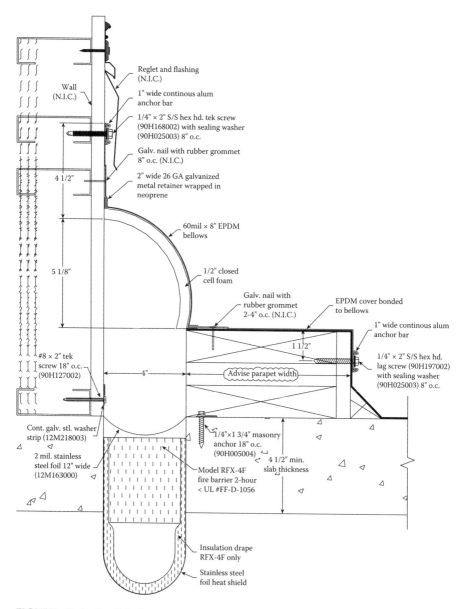

Reglet and flashing (N.I.C.)

Wall (N.I.C.)

1" wide continous alum anchor bar

1/4" × 2" S/S hex hd. tek screw (90H168002) with sealing washer (90H025003) 8" o.c.

Galv. nail with rubber grommet 8" o.c. (N.I.C.)

4 1/2"

2" wide 26 GA galvanized metal retainer wrapped in neoprene

60mil × 8" EPDM bellows

1/2" closed cell foam

5 1/8"

Galv. nail with rubber grommet 2-4" o.c. (N.I.C.)

EPDM cover bonded to bellows

1" wide continous alum anchor bar

1 1/2"

1/4" × 2" S/S hex hd. lag screw (90H197002) with sealing washer (90H025003) 8" o.c.

#8 × 2" tek screw 18" o.c. (90H127002)

4"

Advise parapet width

Cont. galv. stl. washer strip (12M218003)

2 mil. stainless steel foil 12" wide (12M163000)

1/4"×1 3/4" masonry anchor 18" o.c. (90H005004)

4 1/2" min. slab thickness

Model RFX-4F fire barrier 2-hour < UL #FF-D-1056

Insulation drape RFX-4F only

Stainless steel foil heat shield

**FIGURE 13.4** Roof bellows expansion joint. (Illustration provided by Construction Specialties, Inc.)

required. Thus, it is important to coordinate the specific EJC model proposed in the expansion joint bid with the caulking and flashing bidders, whenever applicable.

For example, one EJC bidder may submit a relatively high bid, but the EJC model being proposed may not require additional sealants or flashings. Another EJC bidder may submit a lower bid, but the proposed cover requires additional work to be performed by others. Review the entire scope of work involved as the higher initial bid may actually prove to be the best value.

4. The EJC subcontractor will typically include the fascia plates for the EJCs. Coordinating the color of these fascia plates with an adjacent curtain wall or metal panel system is a very important quality control issue and may not be considered by the bidders. Before entering into any formal agreement, be sure that the EJC subcontractor understands that there can be no significant color variations. The general contractor must follow through with diligent submittal review and quality control on site to ensure that color variations do not occur, as this can be a very expensive problem to correct.

    Note: Though this example is specific to EJCs, the issue of color variation between different subcontractors is common for all finish products. Color variations are particularly problematic for the exterior skin components. These differences are rarely discovered before the materials are on the building. Proper planning will help eliminate these problems before they occur.

    It is common to place responsibility for the fascia plates with the EJC subcontractor. It is best to make an educated and project-specific decision when allocating this work. If the expansion joints abut and are intended to seamlessly match an adjacent system, the fascia plates might best be furnished and installed by the metal panel or curtain wall subcontractor. This option might warrant a different product to be used and this will probably require a slight architectural revision that a simple request for information (RFI) should clear up. Another alternative would be to have the EJC subcontractor furnish the fascia plates to the adjacent subcontractor for painting, to ensure that the fascia plates receive paint from the same batch as the adjacent materials by being run jointly through the painting shop.

5. The roof parapet will be a coordination challenge as the roof EJC and exterior wall expansion joint cover will terminate on each side of a cap flashing. It is suggested that the architect provide an axonometric detail for this condition. The subcontractors involved (EJC, roofing, and flashing) may not provide adequate waterproofing barriers without specific and coordinated instructions.

6. When an addition is made to an existing building the new structure will always be separated from the existing structure by an expansion joint, even in regions with a negligible history of seismic activity. Since EJCs are designed for attachment to flat surfaces installing the EJC to an existing building will almost always require preparatory work, most of which should be the responsibility of the EJC subcontractor. This is a requirement that varies widely among individual projects, thus must be addressed on a project-specific basis.

    (a) An example of another subcontractor involved with the surface preparation may include the mason if brick protrusions need to be ground down in order to provide for a flat mounting surface for the expansion joint.

7. The general contractor must review the project schedule and scaffolding plan as it pertains to the EJC installation. The EJC subcontractor must then be informed if they will be able to install the joint covers from the project scaffolding or if a boom lift will need to be rented. The applicable costs must then be reflected in their base bid amount. Since EJCs are routinely installed very late in the project, the building scaffolding has often been dismantled before the EJC subcontractor arrives on site to perform their work.

8. Vertical wall EJC systems generally use either butyl tape as a sealant to the adjacent surface or a sealant joint. If butyl tape is used, the EJC subcontractor will generally assume full responsibility for the work. If a sealant joint is required, the general contractor must specify which subcontractor will be responsible for the work. This responsibility will vary from project to project and the general contractor must make an educated decision on this issue. Because the vertical exterior wall joint is an aesthetic feature (the roof joint is not in public view, therefore is only utilitarian or functional), the responsibility for this sealant joint can vary considerably. The following examples demonstrate this:

    (a)  If the EJC abuts a curtain wall on each side, it is best that the sealant joint be applied by the curtain wall subcontractor. This will help ensure that the sealant color will satisfactorily match all the way around the curtain wall. In this case the most important factor will be based upon aesthetics.

    (b)  If the EJC abuts a plaster wall on each side, the EJC subcontractor is probably the best party to perform the sealant work, primarily because there is not a sealant color-matching concern. This decision will be based primarily on keeping the responsibility for the waterproofing integrity of the joint with a single subcontractor.

    (c)  If there is plaster on one side of the EJC and a curtain wall on the other, it is beneficial for the curtain wall subcontractor to apply sealants on both sides of the expansion joint for uniformity. This is a very important distinction that the sealant be applied to both sides of the expansion joint. This procedure must be clearly described in the scope of work for the curtain wall subcontractor. Otherwise, they will assume that they are to apply sealants only where their work abuts an expansion joint. Thus, they will plan to apply sealant to only the one side where their system physically abuts the joint.

9. The length of the down leg of a roof expansion joint is a common point of debate between EJC manufacturers and waterproofing consultants. Waterproofing consultants will often detail an expansion joint cover with a 3″ down leg, but manufacturers commonly fabricate their product lines with only a 1-1/2″ down leg. Be sure this issue is fully resolved before awarding the EJC subcontract. Similar issues arise with prefabricated skylights, roof hatches, and many other roof mounted fabrications.

    A 3″ down leg of a roof EJC provides greater assurance that the joint will be waterproof. When construction tolerances (especially with a built-up asphalt roofing system) are taken into account, the reality is that the overlap from the down leg to the roofing may only be as short as 3/4″ to 1″. A remedy for this construction tolerance problem is to add an additional counterflashing under the down leg. This counter flashing will tuck all the way up behind the down leg and provide for a complete 1-1/2″ overlap.

This suggested counterflashing is costly, so plan for it at the onset of the project and prior to bid time. The reality is that during a roof inspection the inspector may notice the roofing does not extend a full 1-1/2″ up behind the down leg. This will be noted as a deficiency and the general contractor may be required to add this counterflashing as a remedial effort.

There is no clear industry standard protocol for the responsibility of this problem, so the responsibility for the associated costs has been known to vary from project to project based upon independent judgments made by each individual project team. The added cost might reduce the general contractor's contingency amount allowed for the project. An attempt might be made to have the owner cover it, as a claim might be made that the condition, as detailed, is not constructible. Consideration might also be given to issuing a backcharge to the roofing subcontractor.

10. An EJC on a tilt-up building and an EJC on a complicated structure (such as a Frank Gehry building) are very different issues. In terms of difficulty, most projects present challenges that lie somewhere between these extreme types of structures. The degree of difficulty is largely based upon the number of corners and turns that appear in an expansion joint as it wraps a building. EJC materials are generally straight and stiff. These materials do not lend themselves to creating aesthetically pleasing angles. Be sure to discuss the means, methods, and level of quality with the EJC subcontractor before awarding them the subcontract.

The EJC subcontractor will generally provide very attractive straight sections of expansion joints, but the appearance of the expansion joints will generally be less attractive when sharp turns or corners are encountered. At the sharp turns, extensive amounts of sealant are commonly used. If carelessly applied, these areas of joint bridging may not be aesthetically acceptable. It is important to plan for these sharp changes in direction of the EJCs and hold the EJC subcontractor fully responsible for a professional appearance.

# 14 Spray-Applied Fireproofing*

Structural steel, and sometimes metal decking, requires a fibrous spray-applied coating to prevent the metal from weakening and buckling in the event of a fire. This spray-applied fireproofing (referred to here as fireproofing) is ostensibly a simple activity, but the reality is quite the contrary. There are only a few issues that can negatively affect a fireproofing operation, but these issues seem to recur from project to project (Figure 14.1). Contrary to most trades, it is the scope of work of other subcontractors that typically impacts a fireproofing subcontractor, not the fireproofing subcontractor's scope of work itself.

Fireproofing is bid and awarded quite differently than other subcontracts. For example, the design documents will not provide a layout or fireproofing dimensions on the drawings. This scope of work will be bid and executed based primarily on the performance criteria in the specifications, the bidder's knowledge of the local building codes, and a review of the structural drawings. This is an industry-wide standard practice. Nevertheless, there are very critical trade coordination issues that must be addressed by the general contractor.

## SCOPE OF WORK ISSUES RELATED TO SPRAY-APPLIED FIREPROOFING

1. Spray-applied fireproofing will adhere to bare steel, but not to an oily, rusty, or painted steel surface. There are several confusing issues associated with fireproofing. If these are borne in mind, effective planning can resolve them.
    (a) Structural steel throughout a building is normally prime-painted. The steel that is scheduled to receive fireproofing must remain bare. The layout of the extent of the prime painting of the steel must be furnished by the fireproofing subcontractor. This must occur early in the project so the structural steel subcontractor has ample time to incorporate the criteria into their shop drawings.
        (i) If the structural steel subcontractor inadvertently paints a portion of the steel scheduled for fireproofing, the remedy is to attach lath to all surfaces of the painted steel with shot pins. The fireproofing will adhere to the lath. This is a costly remedial solution.

---

* MasterFormat Specifications Division 7

**FIGURE 14.1** Fireproofing.

        Be ever mindful that fireproofing will not adhere to rust. Bare structural steel that is left exposed to the weather must be fireproofed before it rusts. Also, note that water will ruin the fireproofing, so once the steel has been fireproofed, it cannot be left exposed to the weather. The dilemma is that leaving the steel exposed to the elements will result in the fireproofing not adhering to it. On the other hand, the application of fireproofing will be pointless if the steel is left out in the rain. There is obviously no perfect time to apply fireproofing. The general contractor must use good judgment and decide how to minimize the inherent risks. A common approach is to leave the structural steel exposed to the weather until the roof is watertight. After this, the fireproofing can be applied. Once the fireproofing has been applied, the general contractor should cover with visqueen any exposed fireproofing at the perimeter of the building and at any shafts open through the roof.

  (ii)  When the steel has been erected, but before the fireproofing is applied, all subcontractors with connection hardware (also termed clips) need to get those clips in place so the fireproofing can be applied around them. Otherwise, the subcontractors will need to scrape and patch the fireproofing at each clip. Scraping and patching fireproofing is tedious, time-consuming, costly, and messy (Figure 14.2).

 (b)  When metal decking is fabricated it is lubricated. It is common for metal decking to be shipped to the jobsite for installation covered with a slight oily film. For metal decking that is to be fireproofed, it is crucial for the metal decking subcontractor to obtain decking that is oil free. Otherwise, a tremendous effort will be required to clean the bottom of the decking to provide for proper fireproofing adhesion.

**FIGURE 14.2**   Scraped fireproofing.

2. Steel located in arenas, stadiums, excessively high atrium ceilings, and other high areas presents significant difficulty in accessing the fireproofing work. It is preferred that the fireproofing subcontractor arrange for their own means of access. Like shotcrete, fireproofing is an extremely messy operation that is known to thoroughly coat scaffolding members with overspray. This requires significant protection of the scaffolding and subsequent cleaning of any scaffolding members that were not properly protected during the spray application.

3. Spray-applying fireproofing is a very messy operation on a project. Without proper protection the fireproofing fibers will seemingly get everywhere. The fireproofing subcontractor will be responsible for all necessary protection measures to prevent this cleanliness problem. The general contractor must oversee the work to ensure the quality of the fireproofing application and ensure that the fireproofing subcontractor does not leave a mess of scattered fibrous materials stuck to surfaces not intended for fireproofing, especially overspray splattered on finish surfaces.

4. As already mentioned, it is important for all subcontractors that will make attachments to the steel through the fireproofing to get their connection hardware in place prior to the fireproofing operation. Otherwise, they will need to scrape and patch each attachment point. Even with a concerted effort by the general contractor and the subcontractors to get all of their clips in place, there will be some that will be missed. Scraping and patching fireproofing for those situations is inevitable. Be sure the subcontractors are held contractually responsible for scraping and patching the fireproofing that is damaged by their own work.

5. For exposed steel in high profile locations that require fireproofing, the architect will specify intumescent fireproofing paint. This is essentially a fire protective paint, or more appropriately termed a very expensive fire protective paint. When exposed to a fire, the intumescent paint will foam up

and expand exponentially beyond the originally applied thickness forming a layer of protective insulation. The protective layer reduces the surface burning characteristics of combustible materials and retards the penetration of heat, similar to common spray-applied fibrous fireproofing. The intumescent paint will fall within the painting subcontractor's scope of work. Be sure the painting subcontractor includes this work in their bid, as it will be listed in the fireproofing specification section, not the painting specification. Also, ensure that the fireproofing subcontractor knows where the intumescent locations are and that no spray fireproofing is needed on them.

Remember that the fireproofing subcontractor bases their proposal of work on: (1) a simple performance criterion in the specifications; (2) their knowledge of the building code; and (3) a review of the structural drawings. Thus, a note in the architectural drawings that calls for intumescent paint generally will not be found by the fireproofing bidders. The general contractor must specifically inform the fireproofing bidders of this clarification in the bid instructions.

# Questions—Module Three (Chapters 10–14)

1. What is the general division of work that distinguishes the work of the structural steel subcontractor from the miscellaneous metals subcontractor? If a steel member is directly attached to the primary steel structure and the fabrication can be completed in the shop so the member can be simply and quickly bolted and/or welded in the field, which subcontractor should be responsible for the work?
2. Which party should procure and pay for the FAA permits for cranes that will be solely used for the structural steel work?
3. Which party should be responsible for cleaning exposed structural steel immediately after erection? Which party should perform the finish painting?
4. Which party should be responsible for completing the elevator hoist and separator beams?
5. Which party should be responsible for hoisting metal decking?
6. What practice of the structural steel subcontractor will help ensure the quality and cost control related to the installation of structural steel anchor bolts?
7. Welding of rebar on site is generally an uncommon occurrence. When a small amount of field rebar welding is required, what can the general contractor do to expedite this work?
8. Which subcontractors often require more electrical power than is available from the temporary power service? Note that under these conditions, the subcontractor should furnish their own generator.
9. For what type of work would an exception apply for the requirement that workers comply with the 100% tie-off policy?
10. Which is thicker, 12-gauge metal or 10-gauge metal?
11. Which subcontractor should have the responsibility for the metal decking support angles?
12. Which subcontractor should provide the shear studs that will be attached on top of the structural steel beams? When should the shear studs be welded into place relative to the placement of the metal decking?
13. Should the miscellaneous metals subcontractor be responsible for furnishing and installing mass-manufactured steel benches? Explain why.
14. Which subcontractor should be responsible for prime painting the metal stairs and which subcontractor should perform the finish painting?
15. Which subcontractor should provide the edge containment for the stair pans? Which subcontractors are typically involved with the concrete pan filled metal stair construction?
16. Regarding the tube steel jambs for the support of a coiling door in a metal stud wall, which subcontractor should furnish and which subcontractor should install them?

17. For a coiling door at a concrete wall, which subcontractor should furnish the embedded steel angles for the coiling door guide rails and which subcontractor should install them?

18. Which subcontractor should furnish the stainless steel corner guards screwed to the corner of a gypsum board wall? Explain why.

19. Identify four items of work that are likely to require a design-build effort by the miscellaneous metals subcontractor?

20. Discuss the merits of assigning the aluminum railing work to the miscellaneous metals subcontractor on a union project.

21. Which party will furnish the angle frame for the dock levelers FOB jobsite for installation by the formwork subcontractor?

22. A building addition in San Francisco will need to be separated from the existing building structure with a continuous, uninterrupted element. What is this component called? How are these gaps sealed on both the interior and exterior of the building?

23. Which subcontractors will be involved in the roof to wall expansion joint covers commonly installed by the expansion joint cover subcontractor? Identify two additional subcontractors that sometimes have work associated with this type of expansion joint and why they may be involved.

24. Discuss the statement that an expansion joint cover between a new and existing structure will almost always require preparatory work to the existing structure.

25. What types of surfaces are ideal for attachment of expansion joint covers?

26. Which subcontractor should install the sealant joint at the sides of an expansion joint cover which is in the middle of a curtain wall system?

27. Under what conditions (on straight runs or at odd angle intersections) do expansion joint covers have a tendency to become aesthetic problems? Explain why.

28. What are the three basic criteria used in the scope of work for a fireproofing subcontractor for bidding and awarding the subcontract?

29. Name the types of surfaces to which spray-applied fireproofing will not adhere?

30. Which subcontractor should provide (and also eventually dispose of) the protection of finished surfaces from the fireproofing operation? Explain why.

31. In high profile areas, how can structural steel be treated to make it fireproof when it is also scheduled to be exposed steel?

# Module Four

# 15 Above-Grade Waterproofing*

There are various types of above-grade waterproofing. In this chapter, whenever the term waterproofing is used, reference is being made specifically to above-grade waterproofing, not below-grade waterproofing, though the same subcontractor is commonly responsible for both above and below-grade waterproofing work. The responsibility for this waterproofing may be assigned to different subcontractors, depending on a variety of circumstances. The primary types of waterproofing include building paper, roll-applied damp proofing, self-adhered sheet membranes (SASM), fluid applied deck waterproofing, and sheet product deck waterproofing. The duty of the waterproofing subcontractor is to complete all of the waterproofing elements which are not performed by any other subcontractors. The waterproofing subcontractor's scope of work can be described as being responsible for the project leftovers.

On many private projects and some public projects the below-grade waterproofing, above-grade waterproofing, and roofing are performed by the same subcontractor. Since this is not always the case, this chapter will discuss only the above-grade waterproofing. (Separate other chapters cover below-grade waterproofing and roofing.)

Assigning the responsibility for the above-grade waterproofing is not an intuitive allocation. The following examples will explain how allocations of work might be made:

A. Masonry facades are routinely constructed over the top of a secondary membrane, but the type of this secondary membrane varies. When a masonry façade is placed over building paper as a secondary waterproofing barrier, the building paper will usually be furnished and installed by the masonry subcontractor. If the secondary barrier consists of a self-adhered sheet membrane it may be the responsibility of the masonry subcontractor, but more commonly the waterproofing subcontractor. Because masonry subcontractors do not apply self-adhered sheet membrane very often, it is preferable that this waterproofing system be assigned to the waterproofing subcontractor. When the secondary barrier is a roll-applied damp proofing it is invariably the responsibility of the waterproofing subcontractor (Figure 15.1).

---

* MasterFormat Specifications Division 7

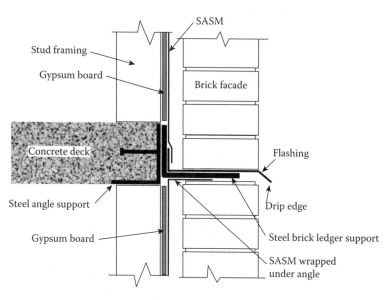

**FIGURE 15.1**   Brick façade over stud framing.

B.  For a plaster façade, the secondary barrier (whether it is building paper or self-adhered sheet membrane) will be the responsibility of the plaster subcontractor.

C.  For a panelized façade, such as metal panels or glass fiber reinforced concrete (GFRC), there may not be a secondary barrier. Though secondary barriers are always desired, there is a major reason for not using secondary barriers with panelized façades. When panels are installed, they are connected with numerous mechanical clips (or attachment devices) from the back side of the panels. A secondary barrier would be in the way of the installer, thus the installer would need to cut large access openings at each and every connection point, thus compromising the integrity of the membrane. Since these access openings and perforations caused by the connections themselves would compromise the effectiveness of the secondary barrier so significantly, this secondary barrier is often omitted. Instead of a secondary barrier, double sealant joints may be used between the panels. In the rare event that a secondary barrier is used, it will commonly be the responsibility of the panelized façade subcontractor.

D.  Deck membranes below paver systems, garden roofs, and topping slabs are similar to a roofing application, but these membranes will be the responsibility of the waterproofing subcontractor (Figure 15.2).

E.  Traffic toppings (frequently found on the upper level of a concrete parking structure or on a concrete balcony) will also be the responsibility of the above-grade waterproofing subcontractor.

F.  As a general rule of thumb, when a self-adhered sheet membrane is used for a total height/width of 18″ or less, this self-adhered sheet membrane is

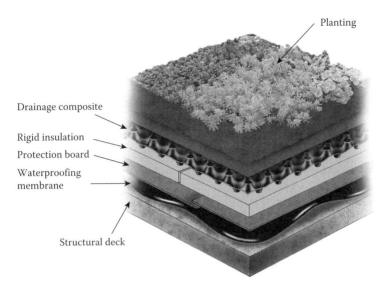

Planting

Drainage composite

Rigid insulation

Protection board

Waterproofing
membrane

Structural deck

**FIGURE 15.2**   Garden roof assembly. (Illustration provided by American Hydrotech, Inc.)

considered to be flashing. When it is considered to be a flashing, the application of the self-adhered sheet membrane will be by either the subcontractor
responsible for said system or the flashing subcontractor, as directed by the
general contractor in the bid instructions. The waterproofing subcontractor
performs work which protects full planes of wall or deck. That is, when
the waterproofing components are restricted to corners, edges, or trim they
are complimentary to another system. This complimentary waterproofing
is completed by the subcontractor responsible for said system or the flashing
subcontractor.

G. Water repellants, acrylic paints, and other roll-applied waterproofing systems are applied to vertical surfaces that are not covered by another system.
These waterproofing systems are invariably the responsibility of the painting subcontractor.

H. Interior floor applications of concrete sealer can be applied by the placing
and finishing subcontractor, the general contractor, or the painting subcontractor. While each of these parties is fully capable of performing this work,
it is most common for the general contractor to apply the concrete sealer.
An exception arises when the product specified is actually a combination
of concrete curing compound and concrete sealer. Such work should be
allocated to the placing and finishing subcontractor because they are always
responsible for concrete curing compounds.

I. Waterproofing beneath ceramic tile, even in showers, will be applied by the
ceramic tile subcontractor. The ceramic tile subcontractor is generally solely
responsible for the waterproofing integrity of the ceramic tile system.

J. Surface-applied membranes on the interior side (also termed the negative side) of below-grade walls are generally applied by the below-grade

waterproofing subcontractor. This situation most commonly occurs at elevator pits and at perimeter basement walls during building renovations. The surface-applied membranes come in various forms, including sprayed, rolled, or trowel applied.

## SCOPE OF WORK ISSUES RELATED TO ABOVE-GRADE WATERPROOFING

1. For horizontal membranes below pavers, topping slabs, or garden roofs, the waterproofing will be done by the waterproofing subcontractor. The area should be in a broom-swept condition before the waterproofing subcontractor begins the work. Any additional preparation that is needed, such as mopping or pressure washing, should be the responsibility of the waterproofing subcontractor. Some waterproofing bidders will exclude surface preparation. This is when the general contractor must make it clear to all bidders that they are responsible for any preparatory work beyond a broom-swept condition.

2. It is important to properly schedule the horizontal waterproofing work with the work of other trades. A waterproofing operation encompasses a significant area on a project. In fact, the waterproofing work will completely shut off other traffic through the area during the application. If the waterproofing is to be applied in a location that will hamper construction access, as is commonplace for plazas at main building entries, phasing the waterproofing application to occur in two operations will be necessary. The first operation will waterproof the left or right side of the main entry, permitting construction traffic to pass on the other side. These roles are reversed when the opposing side is waterproofed. Since this phasing will result in additional costs for the waterproofing subcontractor, this phasing plan must be clearly conveyed to the bidders. Note that the topping slab, pavers, or other system that will cover the waterproofing system will also require phased installations.

3. The horizontal waterproofing membrane is extremely vulnerable to trade damage. Once the waterproofing is in place, it must be protected from damage. For example, protection boards should be laid on top of the membrane before anyone other than the waterproofing subcontractor steps on the membrane. This membrane should not be touched by anyone except the waterproofing crew. Others walking across the membrane or moving materials or equipment across the membrane must do so via an approved protection board. Protection board is a common component of the waterproofing system, commonly consisting of a drainage composite membrane which serves the dual purpose of drainage and protection. If additional protection of the membrane beyond the components called for in the project drawings and specifications is required, the general contractor must specify this in the bid instructions for the waterproofing subcontractor. Otherwise, the general contractor must assume the responsibility for performing this protection work.

4. The general contractor must review the project schedule and scaffolding plan. After reviewing the waterproofing schedule activities, it can be determined whether or not the waterproofing subcontractor can complete their vertical wall work by working on the project scaffolding. If the scaffolding is not yet erected, or will have been dismantled, by the time the waterproofing subcontractor commences their work, the waterproofing subcontractor must make separate arrangements for a boom lift, scaffolding, or scissor lift.

5. The waterproofing subcontractor (not the insulation subcontractor) will be responsible for providing the rigid insulation located between the waterproofing membrane and the paver or garden roof system.

   Note the consistency in how this standard division of work has developed in the industry, whereas this allocation of work is typical to that of the insulation responsibility associated with below-grade waterproofing and roofing systems, as is discussed in these respective chapters as well.

6. Many waterproofing membranes scheduled to be covered with another system as a final product, are susceptible to damage from ultraviolet light. These membranes cannot be left uncovered for a significant amount of time, commonly 30 days or less. The waterproofing subcontractor must install their work shortly ahead of other trades, but they must avoid getting too far ahead. This might require additional site mobilizations and create significant inefficiencies for the waterproofing subcontractor, as they may need to periodically stop work while the follow-on trades catch up with them.

   At the sole discretion of the manufacturer, the amount of time a membrane may be exposed to ultraviolet light may vary by location and time of year. The manufacturer's literature should be reviewed regarding the allowable exposure time, as different products will have different allowable exposure durations. Any additional allowable exposure time granted by the manufacturer that is greater than indicated in their written product data must be clearly stated in writing. It is important that no allowable ultraviolet exposure is exceeded as this could be detrimental to the membrane and it could void the warranty.

   If the waterproofing subcontractor gets too far ahead of subsequent trades, temporary protection from ultraviolet light might be provided as a precautionary measure. For example, the membrane may be protected by hanging black visqueen over it. If this, or any other temporary protective measure is used, the manufacturer's written approval of the remedial action is necessary. It will inevitably be required that black visqueen is used, as the common translucent visqueen will not block the sunlight.

   When a protection board is used to cover a waterproofing membrane it will also protect it from ultraviolet light. Unfortunately, the edges of the membrane will still be exposed. For this reason, it is important to understand that the protection board does not extend the time before the membrane must be covered with a permanent material.

# 16 Lath and Plaster*

The plaster subcontractor relies heavily on the work performed by the framing subcontractor. Thus, if the framing work is performed properly, the plaster subcontractor should be able to proceed smoothly without interruption (Figure 16.1).

## SCOPE OF WORK ISSUES RELATED TO LATH AND PLASTER

1. The plaster subcontractor is responsible for those materials that pertain directly to the lath and plaster work. This will always include the commonly specified building paper used as a secondary waterproofing substrate. On more complex projects, a self-adhered sheet membrane (SASM) may be specified as a backing to the plaster. As a general rule, the plaster subcontractor should also be responsible for the SASM. Since personal preferences of different general contractors will vary, some prefer to allocate this SASM work to the above-grade waterproofing subcontractor. Each of these options is widely accepted in the industry. A decision simply needs to be made as to the SASM allocation. This decision must then be conveyed to the subcontractors bidding on the project to ensure that the SASM work is covered, but not double covered.

2. Several options exist for allocating the caulking and sealants scope of work for a plaster system. The common preference is for the plaster subcontractor to be held responsible for all caulking and sealants within their system. On the contrary, sealants at the perimeter of the plaster system are commonly the responsibility of the adjacent trade. This subject is discussed in further detail in the chapter on caulking and sealants.

3. Similar to concrete, plaster is a water-based product. As the water dissipates, voids are left and the density of the material changes. This often results in cracking of the plaster. There are many reasons for the formation of plaster cracks. There are also many reasons why the plaster cracks are unavoidable. The objective of the plaster subcontractor is to perform their work with a high level of skill in order to minimize the cracking to adhere to industry standard tolerances for allowable hairline fissures. The cracking may be reduced with the use of admixtures or additional reinforcing mesh. The general contractor must also allow ample time in the schedule for plaster application and curing. The general contractor should also be diligent in ensuring quality control during the plaster operation. One undesirable, but common, industry practice that compromises plaster quality is rushing the plaster subcontractor. This may result in the application of subsequent coats of plaster (for typical two or three-coat systems) before the earlier coat has

---

* MasterFormat Specifications Division 9

**FIGURE 16.1**   Plaster installation.

had sufficient time to cure. The general contractor must include ample time in the schedule for the plaster application and also for the plaster to cure. It is often difficult to allow time for plaster to cure when schedule pressures abound on a project, and job superintendents have been known to cut the curing period short in order to recover previously lost schedule time. This can become a costly mistake as the plaster quality may be seriously compromised. The plaster cure time must never be compromised as a means of recovering lost schedule time.

4. A fog coat may be specified as a final coat for plaster. A fog coat is a pigmented cement coating, used to either slightly alter the color of an integrally colored plaster that did not come out correctly or to provide a consistent finish around a plaster building that has been constructed with several different plaster color batches leaving slight color variations between each batch. The fog coat may be applied with a portable hand pump sprayer, an airless sprayer or a pressure tank-type sprayer. The fog coat is to be applied by the plaster subcontractor.

   If the plaster is specified to receive water repellants or a sealer, it is best to allocate these products to the painting subcontractor, since the painting subcontractor will also have the responsibility for applying all the other water repellants on the building.

5. When plaster bids are reviewed, make sure that the bids are not qualified with the use of inferior trim products. The bids should only reflect the use of materials as specified or an equivalent, approved product. A common bid error is when a plaster subcontractor will qualify their proposal with the use of an inexpensive plastic trim despite clear specifications requiring the use of an expensive aluminum trim. With clear specifications, the plaster subcontractor will seldom recover money via change order for this error. Judicious review of the proposals will avoid this conflict by addressing it prior to awarding the subcontract.

**FIGURE 16.2**   Thin brick on top of plaster base.

6. The plaster subcontractor will generally provide flashing reglets at the base of their system. These reglets are lapped several inches up behind the plaster. Flashings at the top of a plaster system are provided by the flashing subcontractor, not the plaster subcontractor. (Note that flashings are never used at the sides of a plaster system.) This is discussed in greater detail in the chapter on flashing and louvers.

7. Interior stone, exterior stone, or a thin brick system designed for installation with an adhesive, is typically applied over a one or two-coat plaster substrate. This is especially true of exterior systems (Figure 16.2). This plaster substrate is frequently excluded by plaster bidders, so be sure to look specifically for this in the bid proposals. This substrate will be virtually identical to the typical plaster system on the project. The only difference of this system is that there is no need for the finish coat.

8. Color consistency of an integrally colored plaster, as with concrete, is a complex quality control issue. If the color is not as it was expected, spot patching will not resolve the problem. If the color is rejected, the plaster subcontractor may be directed to fog coat the entire building. A good practice to help avoid this problem is to order all of the plaster components at one time and have it all delivered to the job site. These components include the cement, the sand, and the color additive. Multiple phased shipments may result in varied color batches and significantly increase the color inconsistencies.

   It must be recognized that plaster will invariably have some color inconsistencies. This may occur even when there has been stringent adherence to a well developed quality control program. At best, the objective is to minimize the color inconsistencies. The owner and architect may help mitigate this issue by specifying white cement or a fog coat in the design documents. These options may be costly, but will provide for a premium finish. White cement has a premium cost, but provides a much more consistent color base than the less expensive and more common grey cement.

# 17 Precast Concrete*

One of the main uses of precast concrete is as concrete structures in conditions or locations that are quite difficult to form in place. Coliseums and arenas are stereotypical structures that incorporate structural precast concrete elements. Some examples of where and why structural precast is used include:

A. High, odd-shaped beams that are more efficiently constructed on the ground (typically off-site) and then set in place with a crane, similar to setting a steel beam. Unlike steel beams, precast members offer considerably better fire protection, and thus do not require spray-applied fireproofing.
B. Repetitive elements with significant angles can be produced quickly and efficiently with a jig (Figure 17.1), in lieu of repeated formwork on site. Examples are stair steps or stepped seating sections common in coliseums and arenas.

Another common use of precast concrete is for decorative purposes (Figure 17.2). Such architectural concrete is constructed in the controlled conditions of a precast shop (Figure 17.3), typically using high grade wood, laminated plywood or fiberglass forms and highly skilled craft workers. This concrete has a much higher industry standard for quality than cast-in-place concrete work. The added expense of architectural precast concrete over that of typical cast-in-place concrete is driven solely by the expected higher aesthetic quality. Architectural precast concrete is commonly used in exterior locations where traditionally stone or glass fiber reinforced concrete may be similarly used, including uses such as wainscot, door and window trim, planter walls, and benches.

Quality control of architectural precast is very important. First and foremost, creating a clean monolithic finish (both in color and texture) on all panel faces and exposed edges requires a highly skilled and diligent workforce in the fabrication shop. Secondly, the essence of handling these large, heavy concrete panels with extreme care brings with it a high degree of difficulty. Some of the key quality control elements are

A. The design of supporting attachments (also termed clips) is an important consideration. The design must be such that in the finished product all steel components will be hidden from view. The most difficult hardware to hide includes the steel embeds used as pick points for the panels. Precast subcontractors will often plan to have pick points in the tops of the panels that

---

* MasterFormat Specifications Division 3

**FIGURE 17.1**   Precast concrete shop. (Photo provided by Walters and Wolf, Inc.)

**FIGURE 17.2**   Architectural precast. (Photo by author, courtesy of Hathaway Dinwiddie Construction Company and The California Institute of Technology.)

they later grout. This is acceptable, as long as the tops of the panels are not visible as a finished product. In other words, pick points in the tops of roof parapet panels will never be seen by the general public, but exposed pick points in the tops of wainscot panels protruding from the face of a wall will be visible and this must be addressed. Make it clear in the bid documents that pick points visible to the public in the finished product will not be acceptable.

B. Architectural precast will almost always entail an integral color. A large project could involve a hundred or more separate concrete batches (six to

FIGURE 17.3    Precast concrete plant. (Photo provided by Walters and Wolf, Inc.)

ten panels can be cast in a day and a new concrete batch is mixed every day). Keeping each and every batch consistent in color is very challenging. Precautions must be taken to ensure consistency. One step many precast subcontractors will suggest is using white cement in lieu of grey. Grey cement is the most common in the industry and significantly less expensive, but grey cement has inherent color variations that cause darkening and/or lightening of the integral color. Thus, even though identical measurements of the color additive in each concrete batch are made, the natural variations of the grey cement will increase the variations of the final panel colors. White cement, on the other hand, is quite consistent in color, helping to narrow the color range among the different concrete batches. Again, white cement is more costly than grey, so the value judgment as to whether or not to increase the quality standard for the precast in exchange for incurring this expense ultimately needs to be made by the owner. Spending the additional money on white cement in lieu of grey cement may be the final step for premium quality architectural precast concrete.

Though this is a design consideration for which the architect holds responsibility, it is very important for the general contractor to pose this value engineering suggestion to the owner for one very important reason. The owner of a project, as well as the architect, will expect a consistent color throughout the precast concrete. Neither of these parties are fully familiar with industry standards such as this, so at the end of a project if the precast panels have color variations the contractor will ultimately bare the blame and at that point have no way of determining if the color variations were caused by improper concrete batching for which they would be responsible, or if the problem was in fact caused by the inherent color variations of the grey cement. This indefensible situation can become a tremendous cost issue, as the remedial action is commonly to use a rub on colored slurry to correct variations. If that method fails to correct the problem, replacing many of the precast panels on the project will be necessary.

C. Care when handling the panels to avoid chipped edges begins with stripping the panels from the forms and does not end until the last punchlist item on the project is complete. The precast crews, in the shop and on site, must be properly trained in handling panels to avoid damage. The most common causes of damage to panels are rough trucking rides from the shop to the site and when manipulating them into final position.

D. Damage can be caused by other construction trades by running into the panels with forklifts, striking the edges with ladders, or with one of various tools/materials/pieces of equipment. Obviously, protecting the work in place is quite important. Mass production of panels achieves an economy of scale, but the fabrication and erection of a single replacement panel will be quite expensive in terms of both time and money because the formwork must be completely reconstructed and a small special concrete batch provided.

Architectural and structural precast will be provided by the same precast concrete subcontractor. Also, although precast concrete has two significantly different uses, the means and methods of fabricating and erecting the work are very similar. For these reasons both types of precast concrete are being discussed in this chapter.

## SCOPE OF WORK ISSUES RELATED TO PRECAST CONCRETE

1. In the case of a stadium or arena, some of the precast concrete may be too heavy for the tower crane to handle and a separate large crawler crane will be required. This crane and all associated permits, including the wide-load trucking permit and FAA permit, should be the responsibility of the precast concrete subcontractor.

   The large crawler crane and the associated trucking operation will dominate the project site. The impact of the use of this crane must be anticipated when developing the project schedule to avoid stacking trades or inadvertently creating any unsafe conditions in which precast elements are swung from the truck into position over the top of any trades workers.

2. Caulking and sealants are very costly and must be properly allocated. As a general rule, the precast concrete subcontractor should be responsible for the caulking and sealants within and at the perimeter of both the structural and architectural precast work. This will naturally vary from project to project based on the unique circumstances at the particular project. Regardless of the party ultimately being held responsible for this work, it is important that caulking and sealants are properly covered, but not double covered (Figure 17.4).

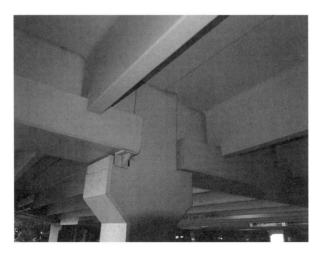

**FIGURE 17.4**   Structural precast connection.

3. The precast subcontractor should be held responsible for grouting all of their own connections. Structural precast regularly involves grouted connections and this will be detailed on the structural drawings. Since architectural precast concrete is predominantly a design-build effort, the necessity and extent of grouted connections is an unknown at bid time and only the precast concrete subcontractor can determine if and where grout-type connections will be necessary. The general contractor or the placing and finishing subcontractor may also perform the grouting work as long as the scope and extent of this work are clearly specified.

4. Grout tubes built into the cast-in-place concrete ledgers upon which the structural precast members are set will be required for connections that are inaccessible once the precast is set in place. These grout tubes and a clear, concise, setting plan should be supplied by the precast concrete subcontractor for installation by the formwork subcontractor.

5. The contract drawings will indicate the interpretation by the design team of the secondary steel needed for the architectural precast support. Since architectural precast is a design-build scope, there may be discrepancies between the extent of secondary steel indicated on the drawings versus the extent of secondary steel required by the specific subcontractor awarded the project. The precast subcontractor is the only party with the ability to make a final determination of the specific need for secondary steel and the extent of the secondary steel. This can vary due to the means and methods of different precast subcontractors. Generally, the precast subcontractor should be responsible for any secondary steel beyond that which is indicated in the contract drawings. The design team is responsible for diagramming attachment points for each of the four corners of each panel, which will show both the location and acceptable method of attachment of the precast panels to the building structure. This method of attachment to the primary building

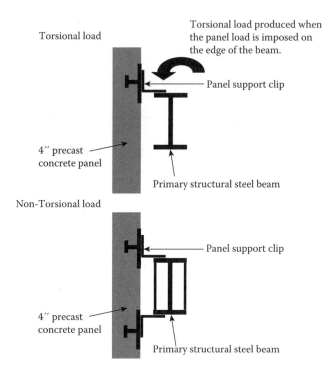

**FIGURE 17.5**   Torsional vs. non-torsional loads.

steel will be diagrammed by the architect with the assistance of the structural engineer to ensure that the weight of the panels bears at the centerlines on the structural beams and does not create eccentric (or torsional) loads on the beams, which will create twisting of the beams (Figure 17.5). The responsibility for any steel beyond this (such as intermediate supports) should be deemed a means and methods issue, which places the responsibility on the precast concrete subcontractor.

6. Many general contractors require the precast subcontractor to apply the sealer to all panels. This may not be the best decision as the painting subcontractor often has a higher skill level for this application and is possibly the better subcontractor to perform this work. Further, if there are additional building components requiring the same sealer, such as cast-in-place concrete or masonry facades, it is most sensible to place the sealer for all elements into a single scope of work, that being the painting subcontractor.

7. The design-build architectural precast work will be a deferred permit submittal. The arduous task of going to the city, waiting in lines, and following through with the permit process, and then finally pulling and paying for the permit is best handled by the precast concrete subcontractor. It is often best that subcontractors pull their own permits as the process seems to go much smoother when this is done. This puts the engineer who designed the system in personal contact with the city engineer who will review the design

so they can talk intelligently and collaboratively to work through the permit process. Nevertheless, this is a standard exclusion in many precast concrete bids, so be sure the associated costs of the time spent obtaining permits, as well as the permit fees, are accounted for in the final cost estimate.

8. The final dimensions for both architectural and structural precast concrete must be surveyed and documented as field measurements. These field measurements are necessary to ensure that the precast concrete work is altered as necessary to accommodate the final as-built conditions of the cast-in-place concrete and/or structural steel. High on a building, the actual positions of the structural members can vary by several inches, so the precast concrete work may need to be modified slightly to accommodate these standard industry construction tolerances. Taking field measurements is often excluded by precast subcontractors, but since they are responsible for the accuracy of their work, these field measurements must be included in their scope of work.

9. The general contractor must take responsibility for coordinating the work of the different trades. Included in this coordination is identifying and dimensioning of the miscellaneous penetrations and embeds for other trades that must be incorporated into the precast products. This can include embeds for railings, sleeves for pipe penetrations, electrical receptacles, etc.

   Note the discrepancy between similar trades, as discussed in the chapter on formwork, that MEP components such as pipe sleeves and electrical receptacles are installed by each respective MEP subcontractor in cast-in-place concrete. In the case of precast concrete these items are actually placed by the precast subcontractor. This is an industry standard that has developed due to the fact that precast concrete is fabricated offsite at remote locations and would otherwise require the MEP subcontractors to have personnel at the precast plant on a regular basis for this quite simple task.

10. The precast subcontractor must maintain responsibility for protecting their work until it is set in place and completed. The precast subcontractor should maintain responsibility to properly protect their work in place up to the time when they demobilize from the area. Once the precast subcontractor has left an area, the general contractor needs to assume responsibility for maintaining the protection until project completion, including the responsibility for removing and disposing of the protection.

# 18 Masonry*

The masonry subcontractor is responsible for a wide range of materials, each of which may very well require a different specialized crew from the same company. Full-size traditional red clay brick, thin-set brick tiles, concrete masonry units (CMU), glass block and stone are all quite different materials, but the installation of all these products falls under the expertise of the masonry subcontractor. The masonry scope of work is generally straightforward and there are not many areas that can cause confusion.

## SCOPE OF WORK ISSUES RELATED TO MASONRY

1. The industry standard practice is for the rebar subcontractor to furnish the reinforcing bars FOB jobsite for installation by the masonry subcontractor. It is very important that the responsibility for detailing the reinforcing bars be determined and properly reflected in the subcontractor bids. It is advisable to have the masonry subcontractor detail the rebar and provide a cut list to the rebar subcontractor. This way the masonry subcontractor will specify the bars exactly how they want them. Under this arrangement the masonry subcontractor dictates the work of the rebar subcontractor, but this generally works satisfactorily because this steel is extremely basic, usually a small quantity and these subcontractors are accustomed to working together.

2. There is usually only a little caulking associated with most masonry work, primarily at the expansion joints and at the perimeter of the system. The caulking work that falls within the body of the masonry work should be the responsibility of the masonry subcontractor, but the caulking at the perimeter of the system is best allocated to the adjacent trade. The caulking and sealants work is discussed more extensively in the chapter on caulking. Although most types of masonry work have little caulking, stone work quite often entails caulk joints in lieu of grout joints. Regardless, it is most appropriate for the masonry subcontractor to have responsibility for the caulking within their system, though this work is routinely excluded in their bid quotations.

3. Some general contractors expect the masonry subcontractors to install the hollow metal door frames at all CMU wall locations. The more common approach is to have the doors, frames, and hardware subcontractor install all door frames, including those at the CMU walls. Since the doors, frames, and hardware subcontractor has the highest level of skill in ensuring the

---

* MasterFormat Specifications Division 4

**151**

door frames are set plumb and level, they should be assigned to perform this work. Setting the hollow metal frames in CMU walls is a common exclusion by both the doors, frames, and hardware and masonry subcontractors, so it is most important that this work be allocated to one of the parties, i.e., it needs to be covered, but not double covered.

4. An often missed scope item is grouting of the hollow metal frames for both fire rating and acoustical purposes. Although the grouted frames may occur at concrete and stud-framed walls, as well as in CMU walls, it is most expedient for all the frames to be grouted in a swift and efficient operation by a single crew. Since the masonry subcontractor will already be set up with a grout pump on site, they will be well-positioned to perform all of this grouting work. The masonry subcontractors regularly include grouting of frames that occur in the CMU walls, but have a standard exclusion of grouting frames occurring elsewhere. It is important to be aware of the fact that grouting at one or more of the various wall type conditions may not be covered by any subcontractor, while one or more of the conditions may actually be double covered. This allocation of work must be addressed clearly in the general contractor's bid instructions, in time for the various trade bidders to properly reflect this work in their proposals.

5. Confirm the substrate and waterproofing method to be used for the masonry system and allocate each layer to the appropriate subcontractor. For masonry over metal stud construction, building paper, or plastic sheeting is most common and should be installed by the masonry subcontractor. Over a cast-in-place concrete substrate, damp proofing is most often specified, which is most appropriately allocated to the waterproofing subcontractor. In high-end buildings, a self-adhered sheet membrane (such as bituthene or ice and water shield) may be specified, which again is best allocated to the waterproofing subcontractor.

6. For a cast-in-place concrete substrate with a brick or stone façade, dovetail slots or other embeds will generally be required to be set by the formwork subcontractor. These embeds should be furnished by the masonry subcontractor and delivered FOB jobsite to the formwork subcontractor for installation with a clear and concise setting plan. There may be a large number of embeds for this work, so be sure the extent of this work is clearly conveyed to the formwork subcontractor for inclusion in their bid.

7. Masonry subcontractors generally have special scaffolding requirements. The basic requirement is that their scaffold be four feet wide, in lieu of scaffolding used by some subcontractors that is only three feet wide. Masonry subcontractors also may set the scaffolding back a short distance from the face of the wall to accommodate outriggers between the scaffolding and building face. Since other trades such as the framing and waterproofing subcontractors cannot efficiently work off of the masonry subcontractor's special setup (and vice versa), these subcontractors often prefer to set up and dismantle their own scaffolding. There would intuitively appear to be an obvious benefit for the various subcontractors to

share scaffolding, but this does not always happen due to the heightened efficiency the subcontractors see in their labor productivity by using their own scaffolding.

Many masonry subcontractors use platform lifts that can be raised and lowered by crank or (for the more expensive models) by motor. These platform lifts facilitate productivity increases, they are safer because workers are not required to bend excessively or work in awkward positions and the workers constantly work in a comfortable position. This improves the quality, safety, and speed of the work performance.

8. Waterproofing considerations commonly include neoprene pads, dry packing, weep holes, and other accessories integral to the masonry system. These are most efficiently completed by the masonry subcontractor. Since elements such as neoprene pads and dry packing most often occur at the perimeter of the masonry system, the masonry subcontractor will need to be informed about the extent of their responsibility for these items. For example, a dichotomy in the masonry scope of work is that they will be directed to exclude caulking and sealants at the perimeter of the masonry system, but they will include the neoprene pads and dry packing at the perimeter of their system.

9. CMU walls often occur in mechanical rooms and stairwells. It should be borne in mind that mechanical equipment, stairs, and other large items must pass through the building to their point of installation. When planning the paths of travel through the facility, it may be determined that several CMU walls are obstacles in the path. This is not a problem so long as the equipment is brought in and set before the CMU walls are constructed. A plan is also needed for leaving out some of the floor dowels so a forklift or other equipment has a clear route for transport. These omitted dowels will

**FIGURE 18.1** Glass block. (Photo by author, courtesy of The University of Southern California.)

need to be drilled and epoxyed after the equipment is set, which is most commonly done by the formwork subcontractor or general contractor's laborers due to union jurisdictions. Similar to the rebar subcontractor, the masonry subcontractor cannot perform concrete drilling work. Be sure to convey an estimated count of the dowels that will need to be set by drilling and epoxying. Also, include in this count a contingency for dowels that may be damaged and rendered useless due to other construction activities.

10. For both interior and exterior stone installations an important quality control issue occurs at the outside corners. The subcontractor must hone the exposed edges and ensure that there are no visible attachments.

11. The masonry subcontractor must perform all the glass block work. Though it would seem intuitive that glass block work is completed by the glass and glazing subcontractor, it is actually a stacked and grouted installation, thus always done by the masonry subcontractor (Figure 18.1).

# 19 Metal Panels*

The metal panels for a project could include countless products, from mass-manufactured corrugated metal panels for low-cost metal buildings to elaborate custom panels for I. M. Pei or Frank Gehry landmark buildings. The discussion in this chapter will include the full spectrum of these panels, as the installation of metal panels is quite similar regardless of how elaborate the panels themselves may be (Figure 19.1).

Metal panels are usually furnished and installed by an individual subcontractor specializing solely in metal panels. When the quantity of metal panels on a project is small, they might be furnished and installed by the glass and glazing subcontractor or by the flashing subcontractor. The following are rules of thumb to use when allocating this work:

A. If a project entails a significant number of metal panels, employ a subcontractor specifically for the metal panel scope of work.
B. If a project requires few metal panels and the panels are adjacent to a storefront or curtain wall system, the glass and glazing subcontractor should be held responsible for the metal panels. Glass and glazing subcontractors regularly perform metal panel work and are very capable of doing a quality job. On union projects, there are no jurisdictional disputes when glass and glazing subcontractors perform this work. The main reason the glass and glazing subcontractor should perform this work is that the waterproofing and secondary weeping systems of the metal panels and storefront framing systems must work together as a cohesive unit. This will also ensure that the metal panel and storefront framing systems are sent through the paint shop together to ensure color consistency.
C. If there are few metal panels, they are not adjacent to a storefront or curtain wall system, and there is not a metal panel subcontractor for the project, then the metal panels will not be included in the bid proposal of any subcontractor unless specifically directed in the general contractors bid instructions. In this case, it is most appropriate for the flashing subcontractor to provide them with the rest of the flashing subcontractor's eclectic mixture of work.

## SCOPE OF WORK ISSUES RELATED TO METAL PANELS

1. The metal panel subcontractor must be held responsible for all flashing work associated with the metal panels. This includes flashings within and at the

---

* MasterFormat Specifications Division 7

155

**FIGURE 19.1**   Metal panels.

perimeter of metal panels which will provide both primary and secondary waterproofing. This assigns responsibility for the waterproofing integrity to a single subcontractor and it eliminates difficult coordination between subcontractors. It also ensures that the flashings will be painted along with the panels yielding a consistent color among the components (Figure 19.2).

2. The metal panel subcontractor should apply all caulking and sealants within and at the perimeter of their panel systems. This will ensure that sealant matching the metal panels will be applied between adjacent panels, as well as between panels and the adjacent exterior skin systems or components. An example of a component would be a light fixture in the middle of a metal panel. It is important to inform subcontractors responsible for the adjacent systems or components that they do not need to provide caulking and sealants where their work abuts the metal panels. This will help to avoid both double coverage and the other subcontractors applying sealants to the metal panels that do not match the metal panels.

3. Metal panel work has emerged in the industry as a predominantly design-build trade. Most metal panel subcontractors have their own shops and their own way of doing things. The metal panel systems they provide are nearly identical in appearance, but the means and methods of how the panels are mounted, flashed, and weep will vary from subcontractor to subcontractor. Ideally, subcontractors should be given the opportunity to produce their work in the most economical way by using their own means and methods. This is to be encouraged as this approach results in a product that both meets the intent of the design documents and is the most cost-effective method. It is important for metal panel subcontractors performing design-build work to account for the following items:

    (a)   General building permits commonly list metal panels as a deferred submittal item because of the design-build nature of the work and the

**FIGURE 19.2**    Metal panels. (Photo by author, courtesy of Hathaway Dinwiddie Construction Company and California State University Northridge.)

fact that this design is not included in the architect's general building permit submittal documents. This means that before panels can be erected, a supplemental permit must be obtained. The subcontractor must include the procurement and payment of this permit. A common omission in the scope of work occurs when a general contractor holds subcontractors responsible for paying for a permit, but not for procuring the permit. Procurement of a permit is time consuming and this effort can be irritating. This is often a long process that begins by going to the building department and waiting in a long line to submit the drawings. For large projects a follow-up trip, or trips, to the building department are regularly required to meet with the plan reviewer and to answer questions. A final trip must then be made to hand the building department a check and to pull the permit. Each of these trips can consume a considerable amount of time. With many design-build subcontractors on a project, many different permits will be needed. If procuring the permits is not in the various subcontractor scopes of work, the general contractor would be making numerous trips to the building department. This is easily avoided by assigning permit procurement to each of the respective subcontractors.

(b)  Permit drawings must be stamped by a professional engineer registered (licensed) in the state where the project is located. Many large subcontractors perform work across the country, but for economical reasons will centralize their detailing and engineering work. It is quite common for a subcontractor's engineering staff to work on designs of projects in many different states. It is imperative that the engineer for a particular design be registered in the state for which the project design applies. For example, a project in California must have the design stamped by an engineer registered in California. Building departments will not accept a permit submittal that is not stamped by an engineer registered in the state where the project is located.

Note that most supplemental permits, particularly for exterior skin components, need only to be stamped by a professional engineer. This means that any type of engineer will be acceptable as long as they are registered with the state. Thus, a civil engineer stamping metal panel drawings would be deemed acceptable. Some jurisdictions are more restrictive than this, so it is important to understand the nuances of the different jurisdictions. It would be appropriate to obtain clarification on this from the local building department before bidding the project. This applies to all supplemental permits, not just the metal panel permit.

4. The metal panel subcontractor must assume all responsibilities associated with water testing when required by the specifications. They must also perform remedial work required by destructive testing, such as pull testing (when pulled until failure) of the sealant joints.

Quite often specifications will require an inspector to witness product testing. Since this testing is solely a design requirement and not mandated by the municipality, the subcontractor is frequently required by the specifications to obtain the services of an inspector. Despite this, it is more appropriate that the owner furnish all third-party testing and inspection services to effectively eliminate any conflict of interest whether or not the inspection is permit related. If a specification is written such that inspectors are to be employed by the subcontractor, it is recommended that the owner be asked by the general contractor to assume this responsibility. Experience has shown that owners will readily assume the responsibility for inspection fees because they greatly appreciate it when a general contractor foresees conflicts and alerts them about how to better ensure project quality control.

5. The metal panel subcontractor will be responsible for properly reinforcing panels, providing cut-outs and sealing all penetrations through the panels. Scuppers, hose bibs, electrical receptacles, and lighting fixtures are all common examples of penetrations for which the metal panel subcontractor will be responsible. This responsibility is typical to other exterior skin scopes of work as well.

It will be important for the general contractor to quantify the penetrations through the metal panels prior to bidding because not all metal panel subcontractors are skilled in reading MEP drawings and deciphering the expected quantity of penetrations. Similar to a duct penetration through a shear wall, as discussed in the chapter on reinforcing steel, identifying and quantifying the penetrations through metal panels will be a necessary task of the general contractor. This penetration count must be made prior to bid time, otherwise change order requests are virtually assured on the project.

6. The general contractor must review the project schedule and scaffolding plan in relation to the metal panel work. If the metal panel subcontractor will be able to use erected project scaffolding, the bid quotation for the metal panels will be reduced and an overall economy of scale will benefit

the project. On the other hand, metal panel bids will increase, and the economy of scale will decrease, if the subcontractors are required to include the considerable costs associated with providing their own access to their work (boom lifts, their own scaffolding, or scissor lifts). This is applicable to all exterior skin trades.

# 20 Roofing*

A watertight roof is one of the most important milestones to reach on a project. Once the ability to deflect water from the building is realized, the interior work can proceed. This includes the installation of wall insulation, gypsum board, duct insulation, and acoustical ceilings. These are all materials that can only be considered for installation after the building is dried in. Since so many critical path activities can begin after a building is dried in, completion of the roof is a major milestone on a project.

Though there are many types of roofing systems, they are all similar in terms of bidding and requirements for coordinating with other subcontractors.

## SCOPE OF WORK ISSUES RELATED TO ROOFING

1. Tapered roof insulation is used on most flat roof construction projects (despite the terminology, flat roofs do have a slight slope, commonly 1/4"–1/2" per foot). Roof insulation serves a dual purpose by creating the drainage slope for the roof and by providing thermal roof insulation. Thermal insulating properties of insulation are commonly expressed in terms of an R-value. Roofs are required by code to be constructed with a value of R-30. The R-value may be given as an average of 30 or a minimum of 30 (Figure 20.1). This distinction must be recognized and understood, as it carries a substantial cost impact.

    (a) There is a significant difference between a minimum and an average R-value in the thickness of roof insulation. To begin, note that all roof slopes begin at the roof drains. Roof insulation R-values vary among roof insulation products.

    For example, assume that a minimum R-value of 30 is specified. Also, assume that a thickness of five inches of a particular insulation yields an R-value of 30 (R-values are given per inch of thickness; therefore the roof insulation for this example has an R-value of 6 per one inch of insulation). To meet the specification, the entire roof would need to have a thickness of five inches or greater, meaning the thickness of the insulation will start at five inches at the roof drain and rise from there.

    On the contrary, if an average R-value of 30 is specified, this means the insulation will have no minimum thickness as long as the average thickness across the entire roof is five inches or greater. In this case, the insulation taper could begin at zero at the roof drain, as long as the

---

* MasterFormat Specifications Division 7

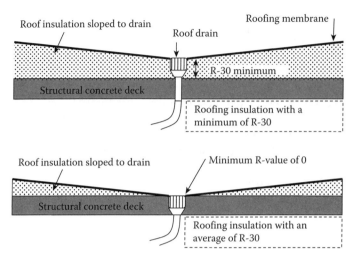

**FIGURE 20.1**  Average vs minimum of R-30.

overall average thickness is five inches or more. The latter provides a savings of up to five inches of insulation across the entire roof, which is a sizable monetary figure.

(b)  If the R-value of the roofing insulation is given without stating that this is a minimum or an average value, the roofing bidders will generally assume only an average R-value of 30 is required. If a minimum R-value of 30 is required, a sizable change order request can be expected to cover the cost of the additional insulation. This distinction should be provided in the contract documents.

2. An important distinction must be made regarding the responsibility for furnishing and installing the pipe boots, commonly called roof jacks (Figure 20.2). For asphaltic roofing, the pipe boots are typically lead or copper and are provided by either the MEP subcontractor who is penetrating the roof or by the roofing subcontractor for non-MEP penetrations. For PVC roofing, the pipe boots are a prefabricated PVC roofing product provided by the roofing subcontractor for all penetrations.

   Compatibility between the pipe boot and asphaltic roofing is important, as asphaltic roofing will not adhere to galvanized materials. It is important to clarify this in the bid instructions and to monitor the products put in place to ensure that this common error does not occur. Since galvanized material is commonly used for MEP sleeves, conduit, and ductwork, the MEP subcontractors get accustomed to using this material. They regularly send galvanized roof jacks to the project for installation not realizing that asphaltic roofing will not adhere to this material.

3. Cold-applied roofing is most commonly specified for re-roofing wood framed structures. This is preferred by many owners because of the potential fire hazard posed by hot-mopped asphaltic roof operations. Cold-applied roofing is also preferred by some owners for minor roof

**FIGURE 20.2**   Lead roof jack. (Photo by author, courtesy of Hathaway Dinwiddie Construction Company and California State University Northridge.)

patching on occupied buildings due to the lower cost and to avoid the pungent odors from a hot asphalt kettle. This type of roofing is rarely used for other conditions because the application of a cold-applied asphaltic roofing system creates a considerable mess which requires the use of significant protection and quality control measures. The roof membrane is laid in full beds of cold asphalt and goop runs out around the edges. This goop never dries completely and takes weeks before it is no longer sticky to the touch. When using a cold-applied asphaltic roofing system, vigilance will be required for several months after the application to keep workers off the roof. Otherwise, they will step into the asphaltic goop and then track it throughout the building because it is impossible to properly clean this gummy mastic off shoes and clothing. This is a common cause of ruined carpets.

4. When a hot-mopped asphalt roofing system is specified on an occupied building, or adjacent to an occupied building, the air handling units will need to be shut down to avoid drawing the pungent odors from the hot asphalt kettle into the building. It is most common for owners to require, via the contract documents, that roofing activities of this type be scheduled over a weekend to avoid disturbing the building occupants. This requirement may only be identified in Division 1 (the general conditions for the project, where most protocol and administrative requirements are addressed) of the specifications, which is not commonly read by roofing bidders because the roofing work is specified in Division 7. The overtime costs for the roofing crew must be taken into consideration, as well as the general contractor's weekend supervision costs. This premium time work should be clearly shown in the project schedule and reflected in the general contractor's bid instructions to ensure that it is properly accounted for by all roofing bidders.

5. An apparent disparity exists between the roofing subcontractor and the various subcontractors who perform vertical building envelope systems (such as metal panels, plaster, glass, glass fiber reinforced concrete, etc.) in regard to the roof flashings. This is because the roof flashings are not provided by the roofing subcontractor, but the flashings for vertical systems are provided by the subcontractors who install the respective vertical building envelope systems. Copings, reglets, and cap flashings at the roof will traditionally be furnished and installed by the flashing subcontractor.

   There are many exceptions to this rule. It is important to identify and properly allocate all of the flashings on a project prior to bidding. This is discussed in greater detail in the chapter on flashings.

   For example, a roof penthouse with plaster walls will have flashing from the plaster to the roofing provided by the plaster subcontractor. A parapet with plaster on the exterior side and roofing on the interior side will have a cap flashing spanning over the top of the parapet connecting these two systems and this cap flashing will actually be fabricated and installed by the flashing subcontractor (Figure 20.3).

6. The roof will be installed as early in the construction schedule as possible, because of the need to have the building dried in as early as possible. Protection of the completed roofing until the project is complete and turned over to the owner is extremely important. Since the roofing subcontractor will be off the project once the roof is completed, it is best

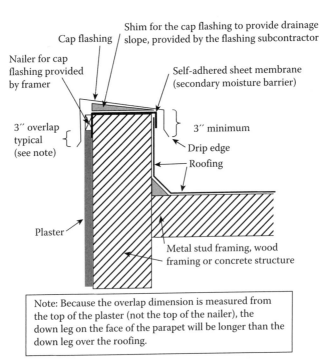

Shim for the cap flashing to provide drainage
Cap flashing / slope, provided by the flashing subcontractor

Nailer for cap
flashing provided
by framer

Self-adhered sheet membrane
(secondary moisture barrier)

3″ overlap
typical
(see note)

3″ minimum

Drip edge

Roofing

Plaster

Metal stud framing, wood
framing or concrete structure

Note: Because the overlap dimension is measured from
the top of the plaster (not the top of the nailer), the
down leg on the face of the parapet will be longer than the
down leg over the roofing.

**FIGURE 20.3**   Roof parapet.

for the general contractor to maintain responsibility for protecting the roof membrane.

In addition to a protection budget, an allowance for minor roof repairs should be included, preferably in the subcontractor's base bid. Even with good protection measures in place, minor damage is inevitable during the fast-paced construction process.

7. Due to the importance of roof protection and the fact that finding leaks is extremely difficult, it is recommended that a protocol of no backcharges be implemented for accidental damage to the roofing. Roof damage is extremely difficult to see with the naked eye and this will help ensure that roof damage will not be left unnoticed. The general contractor should provide several cans of orange upside-down marking paint on the roof for this purpose. Posted signage should notify workers that all they are asked to do when accidently damaging the roof is to circle the damage with orange paint. The workers will not need to tell their foreman, the general contractor's superintendent or anyone else. This "no name no blame" approach encourages workers to participate. The roofing subcontractor can then send a worker to patch the roof where orange circles mark the damage. This approach is cost-effective and helps ensure that multiple people will not be subsequently required to spend many hours trying to find small leaks.

   The term accidental damage is used for a very important reason, implying that this program is not applicable to intentional damage or damage occurring from negligent work practices. Damage that is caused intentionally or through negligence must remain the responsibility of the guilty party. This qualification should be clarified in the roof damage program.

8. Once the project is complete and the building is occupied, maintenance workers will be required to have access to air handling units, roof mounted fans, and other components on the roof. To prevent damaging the roof, walking pads adhered to the roofing membrane will be placed at strategic locations around the roof to provide an appropriate and durable walking surface (Figure 20.4). Walking pads necessary for maintenance workers are typically shown on the architectural drawings where it is necessary to gain access to items such as valves, filters, and other components which require routine maintenance. It is suggested that additional walking pads be added where necessary for construction traffic as well. These walking pads should be considered for placement around all sides of the equipment located on the roof, along large pipe racks, around roof screens, and at other areas where there will be considerable foot traffic during various construction activities. Walking pads are relatively inexpensive and they provide effective roof protection.

9. Condensate drains and small conduits are strewn across roofs. They are traditionally run across the roof on 4″ × 4″ pressure treated wood blocks that are adhered to a small piece of walking pad material, which is in turn adhered to the roof membrane. If this method is used, the roofing subcontractor should include delivery of additional walking pads, FOB jobsite, for use by the MEP subcontractors to create these supports.

**FIGURE 20.4**    Walking pads. (Photo by author, courtesy of Hathaway Dinwiddie Construction Company and The California Institute of Technology.)

10. For PVC roofing systems there is commonly a 1/4″ underlayment board, such as DensGlass, below the membrane and above the roof insulation. Confusion sometimes occurs because DensGlass is commonly the same product being used for the vertical sheathing at the parapet and penthouses. Clarification to the bidders that the roofer is responsible for all horizontal DensGlass (or similar material) and the framer is responsible for all vertical DensGlass is a good practice, though the majority of subcontractors will understand this distinction without clarification.

11. With high parapets, mechanical supports will usually be required to support the roofing run the full height of the parapet wall. This means backing will be required in the wall to receive the nail or screw fasteners. The framing subcontractor must be alerted of this work prior to bidding, because the backing is considered a contractor's means and methods issue and as such will not be shown on the contract drawings. This backing is considered atypical and will not be included in the framing subcontractor's proposal unless they are specifically directed to do so in the general contractor's bid instructions.

12. The roofing subcontractor should be responsible for all costs associated with water testing the roof; especially standing water tests which commonly last 24 hours. It must be recognized that roofs are not designed to hold a significant amount of water weight (except in harsh climates where roofs are actually designed to withstand heavy snow loads). In order to get standing water on a sloped roof water dams must be constructed intermittently up the slope. This is an elaborate and costly means of testing a roof. Make sure that the roofing subcontractor includes the costs for constructing and removing these water dams, as the cost is significant and they will likely exclude this work.

Make special note that this method of testing is expensive and the process can cause damage to the roof when the dams are removed. Through a value engineering study, owners may agree to a modified test that is significantly easier and less expensive, such as continuous sprinklers set on the roof for 24–48 hours. If a standing water test is specified, it is advised that this value engineering suggestion be presented to the owner, as experience has shown that most owners are willing to eliminate the standing water type testing because they agree the added cost and potential damage outweighs the benefit of the elaborate testing itself. It is imperative that this suggestion be made prior to bidding. Otherwise, this becomes a deductive change order issue that is generally not too favorable for the owner. Once the work has been bid and awarded, the subcontractors will be less inclined to make a full reduction for the cost of the test. Just as additive change order pricing has a reputation of being inflated, deductive change order pricing has a reputation of being deflated.

# 21 Flashing and Louvers*

The flashing and louvers subcontractor (referred to as the flashing subcontractor) is similar to the miscellaneous metals, caulking, and above-grade waterproofing subcontractors in that the flashing subcontractor assumes responsibility for the project leftovers—a variety of sheet metal work not otherwise performed by others. Allocating the flashing work on a project is an important and time consuming task. It is important that all of the flashing elements on a project are clearly included in the subcontracts and imperative that miscellaneous specific flashings found sporadically throughout the construction drawings are not overlooked or double covered.

Flashing work has many exceptions to the rules and can be extremely complicated. This chapter will explain this in greater detail. On some occasions, it may appear that there are more exceptions than there are rules, making it even more important that an educated and project-specific decision be made for each flashing element in the construction documents.

Assigning flashing work to the appropriate subcontractor is not an intuitive task. These assignments will be guided by standard industry practices that have developed and become widely accepted over the years. On union projects, the union jurisdictions have been developed within each of the different building trades; especially work pertaining to the building envelope systems (vertical and horizontal systems which waterproof the building, such as metal panels, plaster, brick, or roofing). Several examples will be given on which subcontractors are to be responsible for specific flashing work. Essentially, flashing work or louvers that have not otherwise been allocated to another subcontractor will comprise the flashing subcontractor's scope of work.

A. All subcontractors, except as specifically noted otherwise below, will provide their own flashings within the body of their work (for example, between two components of their respective system). However, discrepancies may develop at the perimeter of each system (for example, where their system abuts the system of another subcontractor).
B. Metal panel subcontractors will be responsible for all flashings touching their work. This helps to ensure the installation of a single composite waterproofing system from a sole source. There is also assurance of an accurate color match between the flashings and the panels.
C. Glass and glazing subcontractors (Figure 21.1) have responsibility for all exposed flashings surrounding their systems. These commonly include head flashings and exposed sill flashings. The flashing subcontractor

---

* MasterFormat Specifications Division 7

169

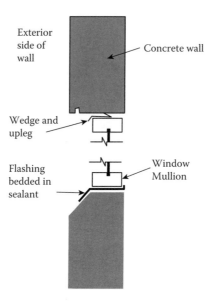

Exterior
side of
wall

Concrete wall

Wedge and
upleg

Flashing
bedded in
sealant

Window
Mullion

**FIGURE 21.1**   Window flashings.

is commonly responsible for all non-exposed flashings surrounding a
window system, including concealed window sill pan flashings and jamb
flashings.

D. A glass fiber reinforced concrete (GFRC) subcontractor will not provide
   any sheet metal work. Instead, the flashing subcontractor will provide all
   flashings for the GFRC system. While few flashings are typically involved
   with a GFRC system, they most commonly occur at the top and/or bottom
   of the system.

E. An architectural precast concrete system is similar to a GFRC system, with
   the precast subcontractor not providing any flashings. In the few instances
   where precast systems entail flashings, the flashing subcontractor will provide them.

F. A plaster subcontractor will provide a reglet at the base of the plaster system,
   but will not provide any flashings for the top or sides. The flashing subcontractor will be responsible for any flashings specific to the plaster system
   except the reglet at the base of a plaster wall.

   Note that the plaster subcontractor will provide plaster stops at all sides
   of the plaster system, which are commonly galvanized sheet metal or
   aluminum products. Though the material and location of these products is
   typical of flashing, they should not be confused with the flashings. Plaster
   stops are an integral part of the plaster system because they are used to
   provide a clean edge at the plaster stopping point. They are not flashings
   because they do not provide any waterproofing value.

G. A masonry subcontractor will not provide any flashings at the perimeter
   of their work and they rarely provide flashings within the body of their

work. One of the rare occurrences that a masonry subcontractor will provide flashings is for an inexpensive CMU or triple-wythe clay brick wall, when the masonry subcontractor provides thru-wall flashings. The flashings within an architecturally significant masonry system, such as a building façade, will have a high aesthetic value. The highest level of quality is best assured when this work is allocated to the flashing subcontractor because masonry subcontractors are not commonly skilled or experienced in decorative sheet metal work. The decision for this allocation of work must be made by the general contractor and specifically addressed in the bid instructions to ensure that it is neither missed nor double covered.

H. A roofing subcontractor will not furnish any sheet metal work. With few exceptions, the flashing subcontractor will furnish all flashings touching the roofing work. The exceptions to this rule include:

(a) The reglet at the base of plaster which flashes over the roofing will be furnished and installed by the plaster subcontractor.

(b) Roof jacks for an asphaltic roofing system are furnished and installed by either the applicable MEP subcontractor or by the roofing subcontractor for non-MEP penetrations.

(c) Counterflashing for mechanical, electrical, and plumbing penetrations will be provided by the MEP subcontractors. Counterflashing for all non-MEP penetrations, such as steel posts, will be the responsibility of the flashing subcontractor.

(d) Reglets and counterflashing for MEP equipment on the roof are typically furnished and installed by the respective MEP subcontractor. The cap flashings at MEP equipment pads on the roof (Figure 21.2), on the other hand, are most commonly furnished and installed by the flashing subcontractor, though the HVAC subcontractors are also capable of providing cap flashings at equipment pads and may also be allocated this work by some general contractors.

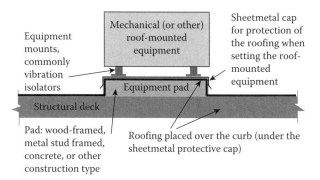

**FIGURE 21.2** Equipment pad cap flashing.

## SCOPE OF WORK ISSUES RELATED TO FLASHING AND LOUVERS

1. Prefabricated louvers and grilles are commonly furnished and installed by the flashing subcontractor, though some general contractors prefer to have these items provided by the HVAC subcontractor. Both of these subcontractors are capable of providing the louvers and grilles, therefore it is best to make an informed decision of which subcontractor is most appropriate for this task on a project-specific basis. This allocation of work will actually be driven primarily by the personal preference of the project manager writing the bid instructions.
2. Sill flashings for the louvers should be furnished by the louver manufacturer to ensure an optimal color match between these two elements.
3. The flashing subcontractor must coordinate with the painting subcontractor so the flashings are painted on the ground before installation. This is important for three reasons:
    (a) It is more efficient for the painter to paint all the flashings on the ground, and then simply touch up at the fasteners, along with any minor scratches that are incurred after they are in place.
    (b) Once the roof parapet cap flashings and other high flashings are installed, there is limited time for painting. For example, window sill flashings installed from inside the building (for example, there is no scaffolding) will be followed immediately by the window installation. The window installation will prevent access to paint the flashing. Another example is if the cap flashings are not painted on the ground the painter will need to lean over the parapet to paint them. This introduces a serious safety concern. A boom lift would provide adequate access, but this option is generally considered cost-prohibitive.
    (c) Drip edges flare out 1/8″ to 1/4″ at the bottom and the painter cannot effectively get up behind this drip edge once the flashing is permanently in position. As a result, the underside of each flashing drip edge will remain unpainted. Unfortunately, this galvanized silver surface will be visible.
4. The flashing subcontractor should be responsible for all sealants associated with their work. Other subcontractors providing flashings should similarly provide all sealants for their flashing work.
5. Roof scuppers are generally fabricated with galvanized sheet metal, copper, or other decorative material. These are commonly furnished and installed by the flashing subcontractor. An exception occurs with polyvinyl chloride (PVC) roofing systems where prefabricated PVC scuppers are specified and these are furnished and installed by the roofing subcontractor.

    Regardless of which subcontractor provides the scupper, the same subcontractor will also be responsible for sealant at the exterior side of the wall. This is a simple exterior sealant joint, but it can become expensive if no scaffolding is in place for access. The costly rental of a boom lift may become necessary to complete this simple exterior sealant joint.

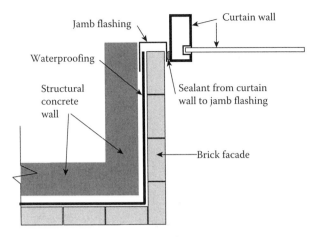

Jamb flashing

Curtain wall

Waterproofing

Structural
concrete
wall

Sealant from curtain
wall to jamb flashing

Brick facade

**FIGURE 21.3**   Jamb flashing.

6. The glass and glazing subcontractor will handle all exposed flashings at the head and sill of the window systems. On the other hand, the flashing subcontractor is commonly responsible for non-exposed jamb flashings at both sides of the window systems (Figure 21.3), as well as concealed sill pan flashings. Jamb flashings are usually only specified when the mullions abut a porous material, such as masonry. Sill pan flashings are used in a variety of conditions, normally acting only as a secondary barrier and tying into the secondary waterproofing membrane behind the exterior skin system. Note that the key differentiator between which flashings are provided by the glass subcontractor versus which are provided by the flashing subcontractor is whether or not the flashings are visible in the completed building, i.e., whether or not they need to be painted to match the window system.

7. Roof-mounted equipment set on pads will typically have a cap flashing shown on the contract drawings. This cap flashing is commonly furnished and installed by the flashing subcontractor. As mentioned earlier in this chapter, some general contractors prefer to allocate these caps to the mechanical subcontractor. Either allocation is acceptable. The important issue is that one (not both or neither) subcontractor is assigned this responsibility.

(a)   It is imperative that the subcontractor providing these caps is specifically instructed to use flat seams. Standing seams will conflict with the equipment mounts.

(b)   These sheet metal caps on the roof curbs and pads constitute a solid waterproofing barrier. An additional benefit is that they offer excellent protection from the equipment inadvertently nicking and tearing the roof membrane while being hoisted into position. For this reason it is a good construction practice for the general contractor to add sheet metal caps on the curbs and pads even when they are not specified on the contract drawings.

**FIGURE 21.4**   Secondary drain pan. (Photo by author, courtesy of Hathaway Dinwiddie Construction Company and The University of Southern California.)

8. As a general rule of thumb, any self-adhered sheet membranes that are 18 inches tall (or wide) or less will be the responsibility of the flashing subcontractor. Any self-adhered sheet membranes larger than this, commonly covering full planes of wall, are the responsibility of the above-grade waterproofing subcontractor.

9. When mechanical equipment, plumbing pipes, or fire sprinkler pipes are routed above electrical equipment, a secondary drip pan (Figure 21.4) is required by code. Fire sprinkler piping will always be routed to the electrical rooms to provide fire protection, but other piping should be routed around the electrical rooms unless absolutely unavoidable. It is suggested that the flashing subcontractor furnish and install these pans. Again, the mechanical subcontractor could also perform this work. The project manager of the general contractor who is writing the bid instructions should make this decision by considering the project circumstances. A few guidelines concerning drip pans include the following:

   (a) Due to MEP coordination is the general contractor's responsibility, drip pans are generally not shown on the contract drawings. Since these pans can be costly, they must not be overlooked. It is important to quantify these drip pans with a cursory review of the MEP documents prior to bid time.

   (b) The drip pans will typically be fabricated and installed by the flashing subcontractor (or the mechanical subcontractor if the general contractor so decides), but these drip pans will require a drain line. The flashing subcontractor must construct each pan with a prefabricated drain fitting to receive a standard pipe fitting. The plumber will route the drain line to its termination point.

(c)   Many city building departments have differing requirements for drip pans. These variations may include the following:

   (i)   There may be varying stipulated minimum widths of the drip pans, measured from each side of the mechanical equipment or piping and each piece of electrical equipment. This will be qualified such that either the piping itself or the edge of a drip pan must be a specified distance from the electrical equipment.

   (ii)  There may be varying stipulated minimum heights of the sides of the drip pans.

   (iii) There may be requirements that the drain be piped to the sanitary system. Others may permit the drain line to be simply daylighted (terminated about two inches beyond face of wall or ceiling) into a corridor. A drain line daylighted into a corridor will make a mess in the event of a leak, but this is the recommended routing for one very important reason, namely that the electrical room is not regularly visited. When the electrical room is visited, the maintenance workers normally do not check overhead piping for small leaks. Consequently, a leak in the piping may not be noticed for a significant amount of time. If the drain is daylighted into a heavily trafficked corridor, as soon as any water starts dripping out of that pipe, it will be noticed by someone. This will invariably receive prompt attention and the leak will be quickly repaired.

# 22  Glass and Glazing*

The glass and glazing (often referred to simply as glass) installation is routinely the last step in completing the building enclosure and the last step necessary before the air conditioning and heating systems can begin operation. This is an important milestone as subcontractors that require a controlled work environment can begin interior work once the building is completely enclosed. With a dry and temperature-controlled environment humidity-sensitive work can begin, including gypsum board taping, wood flooring installation, and painting. Depending on the region, time of the year, and respective local climate, these finish activities are commonly postponed until the building enclosure is completed and the building temperature is stabilized. This makes any window, storefront, and/or curtain wall installation a major critical path activity (Figures 22.1 and 22.2).

There are many varieties of windows. The most common windows in commercial construction are set in rectangular aluminum frames. These aluminum framed systems are typically known as storefront windows when they are less than about 15 feet high and as curtain walls when they are over about 15 feet high. The distinction between these terminologies is that because curtain walls are tall they will require significant structural steel within the aluminum frame, whereas storefronts are self supporting by their aluminum frame alone.

## SCOPE OF WORK ISSUES RELATED TO GLASS AND GLAZING

1. Curtain wall work is typically a design-build effort requiring a supplemental permit, which is procured and paid for by the glass subcontractor. As with all permits, it is important to verify that all the engineering work is stamped and signed by an engineer registered in the state where the project is located. The supplemental permit procurement process for the glass and glazing work is identical to the process described in the chapter on metal panels.
2. The glass subcontractor should furnish and install all exposed flashings at the head and sill of their system. This will ensure that the flashings are painted in the same run through the paint shop as the frames, providing a perfect color match.

   Jamb flashings are commonly used when aluminum mullions occur at porous surfaces such as brick walls, but rarely for other conditions. The jamb flashings are primarily concealed and are designed to wrap behind the brick wall before the windows are installed. In contrast to the head and sill flashings, these flashings are traditionally furnished and installed by

---

* MasterFormat Specifications Division 8

**FIGURE 22.1**   Glass shop. (Photo provided by Walters and Wolf, Inc.)

**FIGURE 22.2**   Curtain wall installation. (Photo provided by Walters and Wolf, Inc.)

the flashing subcontractor. Jamb flashings are not normally painted. If the flashings are scheduled to be painted, the glass subcontractor should provide them to ensure proper color consistency.

Window sill pan flashings are not usually exposed and they are installed prior to the window system. As a result, they too are commonly provided by the flashing subcontractor.

3. Self-flashing windows require extensive self-adhered sheet membrane to be applied behind and on top of the window fins. This is a step-by-step process for which the glass subcontractor should be fully responsible.

4. The general contractor is responsible for providing a project schedule for the subcontractors during the bidding phase. By analyzing the schedule, it is possible to evaluate when the scaffolding will be erected and dismantled. The general contractor must inform the glass subcontractor if they can perform their work with the use of the project scaffolding. If not, the glass subcontractor will need to include the cost of providing their own scaffolding, boom lifts, or scissor lifts to access their work. This information could be provided in the bid instructions or it could be incorporated into a detailed construction schedule. This information should be provided in the bid instructions when the project schedule has not yet been developed to schedule work in such detail.

5. There are countless variations of window construction, as each individual project will have a different architectural design and its own unique exterior skin configurations surrounding the windows. This is particularly evident at the top and bottom corners of windows where various exterior skin systems from multiple subcontractors will intersect from different angles. The work of many other subcontractors comes in contact with installed windows. Ensuring that each subcontractor has the correct scope of work covered in their base bid is an arduous, but important, task. Architectural drawings will always provide head, jamb, and sill details for each different window, storefront, or curtain wall condition. The design documents do not commonly provide details of how the top and bottom corners are constructed. These are important details that are often left to the general contractor's exterior skin coordination effort. It is important for the general contractor to coordinate these conditions with each of the respective subcontractors, as these conditions can become the source of controversy with regard to responsibility for the various components. Full construction coordination of these conditions is not feasible during the bidding phase, but coordinating the work to the point that a clear division of responsibility is made is feasible and necessary (Figure 22.3).

6. The glass subcontractor will be responsible for all glass and glazing work occurring within the hollow metal doors and frames. This also includes sidelights and vision lights. Glass for prefabricated wood doors is often set in the doors at the factory, but not always. The general contractor must thoroughly review the door schedule and manufacturer's product data to properly allocate the glass work within the wood doors.

7. Glass within casework will be furnished and installed by the casework subcontractor. This includes glass in casework doors, mirrors, shelves, and countertops. The casework mirrors stand out from the other components as they are often excluded by both the casework and glass subcontractors. The casework subcontractor can generally be persuaded to include mirrors and are the commonly preferred source for this work. This can be arbitrary as the mirrors could just as well be included in the glass subcontractor's scope of work. The main point is to be sure that any casework mirrors are covered, but not double covered.

**FIGURE 22.3**   Curtain wall.

8. Mirrors not occurring within the casework and not identified in the toilet accessories specification are the responsibility of the glass subcontractor. The glass bidders often exclude mirrors, but they should be held responsible for all mirrors not otherwise covered by the casework or toilet accessories subcontractors.

9. The glass subcontractor will be responsible for all caulking and sealants at the interior and exterior sides of their work. This includes any caulking and sealant work within and at the perimeter of the windows, storefront, and curtain wall systems.

10. Furnishing hardware for storefront doors is a common exclusion of the door and glass subcontractors. This is despite the fact that the glass subcontractor will always install the hardware. With the exception of the hinges, also termed pivots, it is advised that the door subcontractor furnish the remainder of the door hardware. This will ensure that the hardware for the storefronts matches the rest of the hardware in the building. While it may not be intuitive, the glass subcontractor should furnish only the pivots, while all other door hardware is furnished by the door subcontractor. This common industry practice does not cause a matching problem with other door hinges because the pivots for a storefront door will never be the same product as a wood or metal door hinge. The finish (such as brushed aluminum, polished aluminum, stainless steel, etc.) of the hinges needs to be coordinated with the finish of the storefront components provided by the door subcontractor.

11. Assume that a project requires handicap door operators at the storefront entry (and not at any other entries). The responsibility for these operators should be placed with the glass subcontractor. On the other hand, if the project also has handicap door operators at steel doors or other openings, it is suggested that all of the handicap door operators be assigned to the door subcontractor. It is best for all of the door operators to be furnished by a

**FIGURE 22.4**   Curtain wall head support.

**FIGURE 22.5**   Curtain wall intermediate support.

**FIGURE 22.6**   Curtain wall base support.

single party, even if they are installed in the systems of several different subcontractors. (Handicap door operators are discussed in greater detail in the chapter on doors, frames, and hardware.)

Regardless of who furnishes and installs the handicap door operators, the glass subcontractor will be responsible for preparing the aluminum framing system to properly accommodate the handicap door operator installation, including any mullion-mounted push buttons.

12. Each manufacturer of aluminum storefront and curtain wall systems has slightly different sizes of aluminum framing members. The architect will choose one of the systems to show in the drawings as an example, or basis of design. The dimensioning of all work shown in the drawings will be based upon this specific model. At the same time, to promote competitive bidding, the specifications will deem acceptable the products of multiple manufacturers. Thus, when a bidder chooses to use a different product than the one shown in the design drawings, it is important to check the dimensions of the product that is to be used. The dimensions of the aluminum mullions must be comparable to the dimensions of the product depicted in the drawings or design changes may be required. Although most alternate products will be very close in size, this is an important aspect to verify. Any necessary design changes to accommodate size differentials can be extensive and costly with the ultimate financial responsibility lying with the subcontractor.

13. The glass and glazing subcontractor will be responsible for all costs associated with water testing of their systems and any patching required after destructive testing of their sealant joints. This requirement is similar to that discussed in the chapter on metal panels.

14. Where glass is scheduled to receive a window film of any sort, this application work should be allocated to the glass subcontractor. It is a good practice for the window film to be applied in the shop before the glass is shipped to the site for installation. This will ensure that the edges of the window film will tuck under the mullion gaskets. If the window film is applied after glass installation, the film will need to be neatly trimmed around all of the edges. This field trimming is problematic. First, the workmanship of the trimming is likely to leave small, but noticeable, gaps between the edge of film and the mullion gasket. Secondly, these raw film edges are prone to curling or other damage. Deterioration of the film edge is also more likely to occur after a period of time (Figures 22.4, 22.5, and 22.6).

# 23 Caulking and Sealants*

The caulking and sealants work (referred to here as caulking) on a project will be completed by various subcontractors. Although allocating this scope of work is a difficult process, it is manageable when certain guidelines are followed (Figure 23.1).

This chapter is unlike most others in that the caulking work is not being allocated to the caulking and sealants subcontractor nearly to the degree in which it is being allocated to other subcontractors. This is typical of caulking and sealants work. As a result, even on large projects, there is often no separate subcontract issued directly by the general contractor for caulking work. Only when there is difficulty in allocating certain elements of the caulking and sealants work is the assignment of work to a separate caulking subcontractor the most viable option to take.

To be thorough in the allocation of this work, it is advisable for the general contractor to prepare a matrix showing all of the caulking and sealant applications (Figure 23.2). The applicable subcontractors with their assigned responsibilities are then indicated on the matrix. This matrix is made available to all bidders so they all understand their scope of work, as well as what is covered by other subcontractors.

The caulking and sealants within a subcontractor's work is defined as any caulking or sealants applied between one element of the subcontractor's work and another element of the same subcontractor's work. Caulking and sealants at the perimeter of a subcontractor's work is caulking and sealants applied between an element of the subcontractor's work and an element of a different subcontractor's work. Some standard allocations for the scope of work of caulking and sealants will be described. Some of these allocations are quite intuitive, while others are not.

A. The glass and glazing subcontractor should be responsible for all caulking and sealants within and at the perimeter of their work. This applies to both the interior and exterior.
B. The metal panel subcontractor will also be responsible for all caulking and sealants within and at the perimeter of their work.

These examples involving glass and glazing subcontractors and metal panel subcontractors may appear simple, but there is a potential conflict in these scope allocations. Since the metal panel and glass subcontractors are each responsible for the sealants at the perimeter of their work, the conflict arises when metal panels abut a curtain

---

* MasterFormat Specifications Division 7

**FIGURE 23.1**  Sealant application. (Photo by author, courtesy of Hathaway Dinwiddie Construction Company and The University of Southern California.)

wall system. This conflict must be resolved in some way, or both subcontractors will be responsible for the same work. There are two possible ways that this can be resolved:

    (a)    If the sealants are the same color for both the glass and metal panel systems, the general contractor can make the determination by virtually any criteria. It will not make a significant difference in terms of the final work product.

    (b)    The issue is more complex when the sealants are a different color on each of these systems, and the drawings do not clearly indicate which color sealant to use at the juncture of these systems. As soon as this dilemma becomes known, clarification should be sought from the architect. The architect should dictate which color sealant should separate these systems. The architect's decision will resolve the allocation of the work.

  C.  The caulking and sealant applications within the masonry work should be performed by the masonry subcontractor. If unsuccessful in getting the masonry subcontractor to agree to this work, the caulking subcontractor is another viable source for this work. A masonry subcontractor will typically exclude all caulking and sealants. This is not generally a major concern as the masonry work rarely has many joints requiring sealants within their system. The caulking and sealants at the perimeter of the masonry system are most appropriately completed by the adjacent trade.

    (a)    A unique situation arises when the masonry work consists of a panelized stone system (not a common brick and mortar system) with a sealant joint between each panel. For such a system, it is best for the masonry subcontractor to be responsible for all sealants within and at the perimeter of the stone panel system. The masonry subcontractor

Caulking and sealants responsibility matrix

| Area | Caulking within system | Caulking @ perimeter of the system |
|---|---|---|
| Plaster | Plaster | Adjacent trade[a] |
| Brick | Masonry | Adjacent trade |
| Glass exterior | Glass | Glass |
| Architectural precast | Precast | Precast |
| Structural precast | Precast | Precast |
| Pavement | Site conc. | Site conc. |
| HM Frames—exterior | Painter | Painter |
| Glass interior | Glass | Glass |
| HM Frames—interior | Painter | Painter |
| Louvers | Flashing | Flashing |
| CMU | Masonry | Adjacent trade |
| Plumbing fixtures | N/A | Plumbing |
| Flashing | Flashing | Flashing |
| Firestopping (incl @ CMU & other walls) | Framing | Framing |
| Acoustical caulking (incl @ CMU & other walls) | Framing | Framing |
| Miscellaneous metals items | Misc metals | Misc metals |
| Expansion joints | Exp. joint | Exp. joint |
| Metal panels | Mtl panels | Mtl panels[b] |

[a]Plaster sub also has caulking & sealants at plaster/masonry intersection
[b]Metal panel sub will not have caulking/sealants at glass or HM frame intersections

**FIGURE 23.2**  Caulking and sealants matrix.

will inevitably employ a lower-tier caulking subcontractor to complete this work. Obviously, the general contractor can just as well allocate this work to the caulking subcontractor.

(b)  Unique conditions also arise with fire caulking and fire safing at the heads of CMU walls. Masonry subcontractors regularly exclude this work and it is difficult to get other subcontractors to include it in their work. Most often the framing subcontractor will perform this work, as they are already providing fire safing and fire caulking at the framed walls. The caulking and insulation subcontractors could also combine forces to perform this work, but this is generally not very efficient.

D. The plaster subcontractor will be responsible for caulking and sealants within their system. The caulking and sealants at the perimeter of their system, on the other hand, are best performed by the adjacent trade.

The last two examples might result in another problem. This concerns a situation that may result in a gap in the scope of work. When a common brick and mortar type masonry system meets a plaster system, it is typical that neither subcontractor will assume responsibility for the sealant joint between them. It is most common to allocate this joint to the plaster subcontractor, but this is not a strong trend. This joint could be allocated to either the masonry, plaster, or caulking subcontractors. The important point is making sure that one, and not more than one, of the subcontractors has allowed for this work in their bid.

E. Expansion joint cover (EJC) subcontractors traditionally provide their own sealants. To ensure there are no inconsistencies in the sealant colors, a determination must be made about the merits of having the EJC subcontractor provide the sealants. If the expansion joint cover abuts a curtain wall or metal panel system the sealant should match the curtain wall or metal panel system, not the expansion joint. Clear communications must alert the EJC subcontractor to exclude this sealant joint. The subcontractor performing the adjacent work must be informed about the need to include this work in the interest of achieving proper sealant color matching.

F. The glass fiber reinforced concrete (GFRC) subcontractor will be responsible for all caulking and sealants within and at the perimeter of their work.

G. Precast concrete subcontractors generally prefer to exclude caulking and sealants. As with the GFRC scope of work, it is best to have the precast subcontractor responsible for all caulking and sealants within and at the perimeter of their work.

H. Both above and below-grade waterproofing subcontractors will perform the sealant applications within and at the perimeter of their work.

I. Flashing subcontractors will perform the sealant applications within and at the perimeter of their work.

J. Site concrete subcontractors often exclude caulking and sealants. Despite this, they are quite capable of filling their joints, so this exclusion should not be accepted.

K. It is best to hold the miscellaneous metals subcontractor responsible for any caulking within or at the perimeter of their work on the interior of a building. This work could also be effectively allocated to the caulking subcontractor or the painting subcontractor. The adjacent trade should maintain responsibility for the respective sealants at the exterior of the building. For example, a tube steel canopy support might penetrate a metal panel. In this case, the joint sealant would be applied by the metal panel subcontractor to ensure that the waterproofing integrity of the system is under the full responsibility of a single subcontractor.

L. The casework and millwork subcontractor will be responsible for all caulking and sealants within and at the perimeter of their work.

M. Ceramic tile subcontractors, and other interior flooring trades such as Terrazzo, will complete the caulking and sealants within and at the perimeter of their work. This is routinely included in the flooring subcontractor bids.

N. Mechanical, electrical, plumbing, and fire sprinkler subcontractors always perform their own caulking, sealants, and fire stopping. This is a common inclusion in their bids.

O. The food service subcontractor must be held responsible for all caulking and sealants in any way touching their work. This work will be subjected to greater scrutiny since it must fulfill the requirements of the health department. This work consists of an extensive list of miscellaneous joint filling required to prevent bacterial growth in tight crevices. Food service subcontractors are quite accustomed to performing this work and will expect this work to be noted in the bid instructions and subsequent subcontract agreement. This work is discussed in greater detail in the chapter on food service equipment.

P. The framing subcontractor should be responsible for all acoustical caulking and fire stopping necessary to complete the sound and/or fire rating of their walls. Framing bidders will usually include this work in their base bid, but this should still be verified. This work is sometimes either excluded or listed as an additive alternate in framing bid proposals. This can result in a costly change order if this work is not allocated during the bidding phase.

Q. The painting and caulking subcontractors are both good sources for minor interior caulking not otherwise provided by other subcontractors. This includes door frames and other caulk joints that are shown on the contract drawings.

When a general contractor reviews the contract drawings for their own bidding and planning purposes, it is important that they make a conscientious effort to identify and clearly allocate responsibility for all obscure caulking, sealants, and fire stopping elements that do not fit into one of the scenarios noted above. Any obscure items of work not clearly allocated at bid time will be potential sources of conflict during construction.

## SCOPE OF WORK ISSUES RELATED TO CAULKING AND SEALANTS

Once the complex allocation of caulking and sealants work throughout the project has been successfully accomplished, there are not many common problems associated with the execution of the caulking and sealants scope of work. A few of these issues will be described:

1. One of the most common change order issues on a project relates to the premium cost of a custom color. Though this is a design issue, and this book is directed towards contractor issues, it is an extremely common design issue of which general contractors need to be aware, therefore it will be

discussed here. This problem commonly originates because the color of building finishes are not usually confirmed prior to bidding a project and on such projects the specifications often state that caulking and sealant colors are to be either a standard or custom color as directed by the architect. Such specifications are often contested as they require subcontractors put a lump sum competitive bid on a variable cost. Without clarification, it is not possible to accurately price this work item. It is clearly not a fair specification. The architect may contend that the subcontractor should include the most expensive product in their base bid. There are several problems with this approach:

(a)  Owners will feel cheated if subcontractors include money in their bid for a custom color, but ultimately use a standard color. There is also no opportunity to receive a deductive change order when a standard color is used in this situation. The owners will feel they have paid for something that they did not receive.

(b)  Subcontractors are awarded contracts on the basis of submitting the lowest bid. With an ambiguous specification, a subcontractor may decide to take a chance and include only enough money for a standard color to keep costs down. Therefore, subcontractors who included enough money in their bids for the custom color may lose the job.

   The low bidder in this case may include a qualification in their proposal stating that only standard colors are to be used. This qualification may not be caught by the general contractor. As discussed above, the caulking and sealants work is complicated because there are a great number of different subcontractors; most are lower-tier subcontractors that will be providing caulking and sealants for the project. Making sure each and every one of them has the correct scope is at times a futile task. Clearly, a specification written in this manner will result in bids that are not comparable.

2. It is important to verify during both bidding and quality control that subcontractors are not using lower caliber products than are described by the specifications. This is a problem that occurs most frequently with building envelope systems and is rarely caught.

   For example, the specifications may stipulate that sealants are to be premixed at the factory. Despite this, a subcontractor may have submitted a bid with a qualification that ignores the requirement. Such a subcontractor might subsequently be observed mixing a powdered sealant product with water on site, even if they did not qualify their bid as such.

   Another problem may arise with a specification that requires the use of a two-part sealant. Again, a subcontractor may effectively rewrite the specification with a qualification in the bid, or simply ignore the specification altogether, by using a one-part sealant product on site.

3. When the general contractor employs a caulking subcontractor for exterior skin systems, access to the work is crucial. For example, a building with a stone panel façade that has a sealant joint between the panels will have the stone set by the masonry subcontractor from a scaffold.

To avoid prolonged rental costs, the masonry subcontractor will plan to dismantle their scaffolding immediately after setting the stone. Unless another arrangement is made, this will necessitate newly erected scaffolding, boom lifts, or a swing stage for the caulking subcontractor to access their work. Having access to the work is commonly overlooked during the bidding process by many of the smaller subcontractors. This can be an expensive oversight.

# Questions—Module Four (Chapters 15–23)

1. Which subcontractor commonly performs the above and below-grade waterproofing scopes of work in addition to their traditional trade?
2. Which subcontractor commonly provides the building paper that is installed behind a masonry façade as a secondary membrane? Which subcontractor is responsible for a membrane constructed of roll-applied damp-proofing? Which subcontractor will be responsible for a secondary membrane constructed of self-adhered sheet membrane?
3. What is the common practice regarding the use of secondary membranes behind a panelized façade?
4. Which subcontractor typically provides a 12" wide strip of self-adhered sheet membrane?
5. Which subcontractor will provide the waterproofing membrane below the ceramic tile in a shower?
6. Which party typically provides the trowel-applied waterproofing membranes applied to the negative side of a below-grade wall?
7. What should be the condition of the substrate for the waterproofing subcontractor?
8. Which party will provide the insulation between a waterproofing membrane and paver system?
9. Discuss the statement that waterproofing membranes scheduled to be covered with another system as a final product are commonly susceptible to damage from ultraviolet light if left exposed to the sun for a significant amount of time.
10. Which party should provide the self-adhered sheet membrane behind a plaster system?
11. Discuss the wisdom of trying to make up for lost schedule time by slightly shortening the curing duration for each coat of plaster.
12. Which subcontractor is responsible for the fog coat over a plaster system? Which subcontractor is responsible for water repellants over a plaster system?
13. Give two different reasons for using a fog coat over a plaster system.
14. Describe two premium items the owner and/or architect might add to the plaster scope of a project in order to help ensure a consistent plaster color is attained.
15. Give two reasons for using structural precast concrete in lieu of traditional cast-in-place concrete.
16. Describe two methods of correcting excessive color variations in precast panels.
17. Explain why the recasting of a single precast panel is, or is not, completed at the same unit cost as the precast panels cast in mass production.

18. Do different subcontractors provide architectural and structural precast items?

19. Which parties should provide the caulking and sealants within and at the perimeter of the precast concrete work?

20. Which party should furnish the grout tubes for structural precast connections? Which party should install them?

21. Which one of the following will never be responsible for pulling a permit in the performance of their work? (Precast subcontractor, miscellaneous metals subcontractor, site concrete subcontractor, site utilities subcontractor, place and finish subcontractor, or elevator subcontractor.)

22. Which subcontractor should be held responsible for as-built field measurements prior to fabrication of the architectural or structural precast?

23. Which subcontractor should be responsible for furnishing a domestic water pipe sleeve in a cast-in-place concrete shear wall? Which subcontractor should install it? Which subcontractor should provide a domestic water pipe sleeve for a precast concrete panel, and which should install it?

24. Identify four different material types for which the masonry subcontractor will be held responsible.

25. Which subcontractor is ideally responsible for the grouting of hollow metal door frames at a metal stud wall? (Assume that this subcontractor has other work to perform on the project site.)

26. Discuss the statement: It is rarely economical to erect joint-use scaffolding from which both the masonry and framing subcontractors can work.

27. Which subcontractor will be responsible for the caulking and sealants at the perimeter of masonry work? Which subcontractor will be responsible for the neoprene pads at the head of the masonry walls?

28. On union projects, which of the following parties cannot drill for dowels due to union jurisdictions? Identify two from: masonry subcontractor, general contractor, formwork subcontractor, and rebar subcontractor.

29. Which subcontractor is responsible for all glass block work?

30. Which three subcontractors may furnish and install metal panels on a project?

31. For a project in which there are very few metal panels, which party should provide the metal panels which abut the curtain wall system?

32. Which party should provide the flashing at the head of the metal panel system?

33. Which party should provide and install the sealant around a steel support penetrating a metal panel?

34. Discuss the statement that while water testing of metal panel, roofing, and curtain wall systems is required only as a quality control measure by the contract documents and is not required by the permit, it is still most appropriate for the owner to have this testing observed under direct contract with a third-party inspector.

35. Describe in detail the difference between a minimum value of R-30 and an average value of R-30 with regards to roof insulation.

36. Which party should provide roof jacks for MEP penetrations through an asphalt roof? Which party should provide the roof jacks for MEP penetrations through a PVC roof?

37. What materials are commonly used to make roof jacks for an asphaltic roof?—Give two.

38. Give three advantages of using cold-applied asphalt roofing for minor roof patching.

39. What is the primary disadvantage of cold-applied asphalt roofing?

40. Explain why subcontractors should not be backcharged for accidental damage to the roofing membrane.

41. Discuss the statement that it is not advisable to add roof walking pads to the project for construction purposes due to their high cost.

42. How willing are owners to eliminate standing water type testing of roof membranes? Explain why.

43. Identify three subcontractors commonly held responsible for a variety of work that is simply not included by any other subcontractor, for example, the project leftovers.

44. How intuitive is the allocation of flashing work?

45. Which party should provide all exposed flashings around a window system? Who should provide all non-exposed flashings around a window system?

46. Which two of the following systems will require all flashings to be provided by the flashing subcontractor?—GFRC, precast, metal panels, and plaster.

47. Discuss the statement that the masonry subcontractor will never provide through-wall flashing for a CMU wall.

48. Which subcontractor should provide the counterflashing for MEP items that penetrate the roof? Which subcontractor should provide counterflashing for non-MEP items that penetrate the roof?

49. Which two subcontractors might provide cap flashings at MEP equipment pads on the roof?

50. Which subcontractor provides louvers?—Name two possible subcontractors.

51. Give three reasons flashings should be painted prior to installation.

52. Which subcontractor should provide sealant at a flashing?

53. Aside from the cost benefit, explain why daylighting the drain line from a drip pan below sprinkler piping in an electrical room into a corridor is preferred over piping it to the building sewer system?

54. Will a curtain wall and/or a storefront have integral steel framing built into the mullions?

55. Which subcontractor will provide the glass in the transom of a hollow metal door? Which subcontractor will provide the glass in a cabinet door?

56. Which subcontractor will provide the mirrors identified in the toilet accessories specification section?

57. Which subcontractor will provide the caulking at the interior perimeter of a storefront that resides within a gypsum board wall?

58. Discuss the statement that pivots (hinges) for a storefront door will always be furnished by the glass subcontractor, even when the other door hardware

is furnished by the door subcontractor. If the door subcontractor furnishes the hardware for a storefront door, who will install it?

59. When should window films be installed, relative to setting the glass in the frames?

60. Discuss the statement that on a large project the general contractor will always directly employ a caulking subcontractor.

61. For which of the following examples is caulking and sealants not <u>within</u> a subcontractor's scope of work:
    (a)    Sealant between metal panels
    (b)    Sealant from a metal panel to a plaster system
    (c)    Sealant from a metal panel to a metal panel flashing
    (d)    Sealant at a masonry expansion joint

62. Which subcontractor should perform the work associated with the sealant joint between a storefront and a metal panel, for which both systems have the same color sealant? Which subcontractor should perform the sealant joint work between the masonry and plaster systems?

63. Which subcontractor should perform the sealant work around a canopy support penetrating a GFRC panel?

64. Which subcontractor should perform the work of fire caulking at the head of a rated gypsum board wall? Which subcontractor should perform the fire caulking around an electrical conduit penetrating a rated gypsum board wall?

65. Discuss the statement that when subcontractors are required to bid the caulking and sealants work based upon a qualification that either a standard or custom color may be used there is never a problem because standard and custom colors are the same cost.

# Module Five

# 24 Framing and Drywall*

Framing and drywall (referred to here as framing) are two of the most important trades on a project. The work of these schedule-driven trades is commonly completed by the same subcontractor and has a tremendous influence on the success of a project. They will be on the critical path from the moment the order is placed for the first batch of studs to the moment the last bit of taping mud has dried sufficiently for painting (Figure 24.1).

Framing subcontractors are capable of self-performing a number of different tasks. A framing subcontractor may also be requested by the general contractor to employ lower-tier subcontractors to perform specific tasks because they are not capable of completing them with their own labor force. The trades included in the framers scope of work are dependent on several factors, including the size of the project, capabilities of the framing subcontractors in the region, and the degree of difficulty. Both private and public projects regularly require the framing subcontractor to perform the drywall and taping work in addition to the stud framing itself. At times, particularly on hard-bid public works projects, it will make sense for the general contractor to employ a separate subcontractor to complete the drywall and taping work. Conversely, at other times it will be logical for the framing subcontractor to perform spray fireproofing, building insulation, or even plaster work. All of these various combinations are common in the industry. The frequency of the combinations that are the most common will vary greatly in different regions of the country. The local industry and local subcontractors will dictate the local standard practices. For instance, one large and reputable framing subcontractor in California is even capable of performing metal decking and roofing work on a large scale with their own labor force. These various, optional responsibilities of the framing subcontractor may seem cumbersome and confusing at first glance, but are in fact quite easy to devise when allocating the work of a project. For the purposes of this discussion we will assume the most frequent arrangement, for the framing subcontractor to only provide the framing, gypsum board, and taping work.

## SCOPE OF WORK ISSUES RELATED TO FRAMING AND DRYWALL

1. Exterior metal stud framing is typically a design-build scope of work on concrete and structural steel buildings. As such, the framing subcontractor must procure and pay for the supplemental permit. (The supplemental permit process is discussed in greater detail in the chapter on metal panels.)

---

* MasterFormat Specifications Divisions 5 and 9

197

**FIGURE 24.1**   Stud framing.

(a)   Exterior stud framing commonly requires a supplemental permit due to the heavy loads it will carry from exterior skin systems. Interior stud framing does not commonly require a supplemental permit because the loads an interior wall will carry are traditionally very low. In the rare occasions that interior stud framing must bte designed to carry heavy loads, such as walls clad with a 2″ thick interior stone system, the interior framing may also require a supplemental permit. The necessity for permitting each of these framing systems will be noted in a list of deferred permit submittals, found within the first few sheets of the architectural drawings.

2. PVC roofing systems commonly have a gypsum board product, such as 1/4″ DensDeck, between the roofing membrane and the top of rigid insulation. Although it is a sheet of gypsum board, this product is part of the roofing system that will be furnished and installed by the roofing subcontractor. On the other hand, a similar roofing substrate at the vertical parapet and penthouse walls will be furnished and installed by the framing subcontractor. The framing and roofing subcontractors are accustomed to this division of work, but it should still be reiterated in the bid instructions and subsequent subcontract language. Contract language should stipulate that the roofing subcontractor is to provide the gypsum substrate within the roofing system at all horizontal conditions. Similarly, the framing subcontractor is to provide the gypsum substrate within the roofing system at all vertical conditions.

3. The framing subcontractor is responsible for in-wall backing (Figure 24.2) for the myriad of wall-hung items. The general contractor often maintains responsibility for shopping lists of specialty items throughout the project that will be identified, quantified, and priced. This is not the case with required backing as identifying, quantifying, and pricing in-wall backing is the sole responsibility of the framing subcontractor. This includes backing for toilet

**FIGURE 24.2**    In-wall backing. (Photo by author, courtesy of Hathaway Dinwiddie Construction Company and The California Institute of Technology.)

partitions, casework, handrails, MEP control cabinets, marker boards, and any other items mounted on walls or a gypsum board ceiling. Creating further confusion, backing itself is not often shown on the architectural drawings because it is deemed a contractor's means and methods issue. Framing estimators have become versed in reviewing construction drawings, understanding the basics of components provided by other trades, and identifying the components which will require in-wall backing. Intuitively this would seem to be a responsibility of the general contractor, but the industry standard practice has in fact evolved such that the framing subcontractor has this responsibility.

Subcontractors representing one trade commonly claim they are not responsible for reviewing the subcontracted work of other trades prior to submitting a bid. For example, formwork and rebar subcontractors routinely exclude deck or wall penetrations for ductwork unless they are specifically shown on the structural drawings, even though these penetrations are clearly located on the mechanical drawings. Similarly, carpeting subcontractors will not include elevator carpeting if it is only shown on the elevator drawings and a painting subcontractor will not include the painting of ductwork if it is only shown on mechanical drawings. In contrast, the framing subcontractor will be responsible for reviewing the work of other trades and including the necessary backing. This review by the framing subcontractor relieves the general contractor of a tedious burden.

As a professional courtesy the framing bidders should be informed whenever obscure backing requirements are noted during the general contractor's review of the bid documents. Such good faith efforts are not only a moral obligation, but are also important for developing quality working relationships in the construction industry.

An example of obscure backing is roofing that runs up a high parapet wall. When roofing is run more than about three or four feet (this varies by specific roofing product line) vertically, it will often need to be held up with mechanical fasteners. These mechanical fasteners will require backing. Because this is not a common occurrence on projects, framing subcontractors often will not know to anticipate it. Consequently, the framing subcontractor may easily overlook this backing. If such backing was overlooked during bidding, the framing subcontractor will almost certainly submit a change order for this work. They will contend that this work is not intuitive and should have been specifically shown on the architectural drawings.

This discussion on backing and the above examples are somewhat counter-intuitive. Nonetheless, that is the nature of the in-wall backing scope of work. Experience is the best teacher for developing the ability to effectively identify, bid, and execute the in-wall backing scope of work.

4. There are two steel components integral to the framing systems that are commonly completed, at least in part, by the miscellaneous metals subcontractor due to union jurisdictions. First, low height walls isolated away from other full height walls often require tube steel posts (Figure 24.3) in line with the wall framing for lateral support. Secondly, countertops are often supported by in-wall steel supports (Figure 24.4). These elements are similar in nature and are both installed by the miscellaneous metals subcontractor. The primary question pertains to which subcontractor will furnish these items. Both interior and exterior metal stud framing is predominantly a design-build trade. As such, because the posts are for the wall assembly, it is commonly the framing subcontractor's responsibility to design, fabricate, and furnish the low wall posts. Conversely, because the counter supports are for the counter assemblies, they are commonly designed, fabricated

**FIGURE 24.3**   Low wall posts. (Photo by author, courtesy of Hathaway Dinwiddie Construction Company and The California Institute of Technology.)

**FIGURE 24.4**   Counter support angles. (Photo by author, courtesy of Hathaway Dinwiddie Construction Company and The California Institute of Technology.)

and furnished by the casework subcontractor FOB jobsite, but will still be installed by the miscellaneous metals subcontractor.

(a)   These rules are not always followed. Other arrangements can also be successful. On a non-union project the framing subcontractor could furnish either of these items if so instructed. The framing subcontractor could also install the low wall posts and the counter supports. Another variation would be to have the miscellaneous metals subcontractor furnish these items themselves. As always, it is important that this scope is covered by a competent party, but not double covered.

(b)   Regardless of who furnishes or installs the counter supports, it is important to coordinate fabrication of the support angles with the metal stud framing work. For example, the framer must be informed whether the horizontal leg of the angle will point to the left or to the right. This impacts how they will position the studs. Without this information, the framer may need to reconstruct the framing. This will consist of reversing the studs so the flat portion of the stud matches the flat portion of the angle bracket.

5. The framing subcontractor must maintain responsibility for all caulking, sealants, and fire safing at the top, bottom, and ends of walls. This will help ensure that the fire and acoustical ratings specified by the contract drawings are met. Most framing subcontractors will automatically include this work, but this must be verified through a careful bid review. Some framing bidders will actually list this work as an additive alternate to their base bid. Not catching this exclusion is costly for the general contractor.

6. Typically, the framing subcontractor will be responsible for scaffolding, scissor lifts, and any other means employed to access their work. Often there are multiple subcontractors with significant work that needs to be completed in an area by using the same scaffolding. In this case the general contractor

may decide to furnish scaffolding for use by all trades. This decision needs to be made on a project by project basis. When scaffolding is provided, this must be clearly conveyed to the bidders to ensure that a substantial reduction in the respective base bids is fully realized.

(a)   There is another common, but convoluted, approach whereby the general contractor provides scaffolding for use by various subcontractors, but this is provided with a catch. To compensate for the use of the scaffolding, each subcontractor who will be performing work with the use of the scaffold is required to include money in their base bid to compensate for their share of the scaffolding cost. Allocating, and subsequently negotiating, the scaffolding cost among a myriad of subcontractors in this manner requires a significant effort in management to ensure that the cost of the scaffolding is not only fairly distributed, but is also fully recovered. To avoid this management nightmare, this approach is not advised.

7. As with the in-wall backing, the framing subcontractor is responsible for quantifying and installing all access doors (Figure 24.5). The subcontractors furnishing the access doors will vary. Typically, the subcontractor whose work is being served by the access doors will provide the access doors to the framing subcontractor. The framing subcontractor will box out the framing and install the access doors. Most access doors are for MEP work and, consequently, they will be supplied by the MEP subcontractors.

8. Construction is a massive and fast-paced operation. Understandably, it must be realized that minor trade damage is part of construction and needs to be accounted for on every project. The framing subcontractor should include an allowance in their bid for repairing the inevitable cases of minor trade damage. An allowance in the form of a specific number of worker hours and the cost of materials should be established by the general contractor.

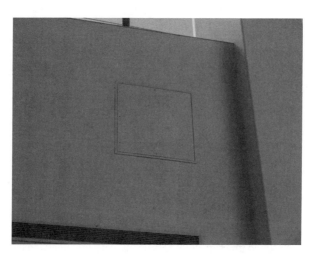

**FIGURE 24.5**   Access door.

The framing subcontractors can then compute the appropriate value of the allowance to include in their bids.

(a)   For best cost control the framing subcontractor's bid should stipulate the number of hours allocated for the repair work, not a fixed dollar value. It should be recognized that subcontractors tend to use different labor rates when bidding a project than they do when submitting change orders. When bidders are asked to include a stipulated number of hours in their base bid as an allowance for repairs, they will include these hours with their base bid hourly rates. On the other hand, if they provided only a dollar amount as the allowance, they will invariably deduct hours from the allowance by using their higher change order labor rate.

(b)   Additionally, by including the labor and material value of this allowance in the framing subcontractor's base bid the subcontractor will already have the taxes, insurance, overhead, profit, and other markups on this work included in the bottom line of their base bid estimate. Therefore, the time and materials charged to this allowance will be done so without any additional markups. Though repairing trade damage is an indirect cost of a project it is still an actual and necessary cost on a project and as such should be viewed as a standard base bid responsibility just as with the direct work.

9. Trash chutes are commonly bid as materials only by the chute manufacturers, then delivered FOB jobsite. The framing subcontractor will accommodate access for the HVAC subcontractor to install the chutes by constructing the back and side walls of the chute shaft prior to installation of the chutes, and then return to construct the front face of the shaft after the HVAC subcontractor has completed the chute installation. The framing subcontractor will also have the responsibility for installing the trash chute doors in conjunction with framing the front wall. The trash chute installation is surprisingly complex and will be discussed in the chapter on miscellaneous specialties.

10. The substrate for ceramic tile is a cement board furnished and installed by the framing subcontractor. The cement board is not always specified or shown on the drawings as having mesh tape at the joints and corners. This taping will be necessary for a quality ceramic tile installation and should be considered part of the system as a standard construction practice. This component of the substrate is frequently overlooked by the framing subcontractors; therefore, it is a good practice to inform the framing subcontractor about their responsibility regarding the mesh taping for ceramic tile prior to bidding. This will avoid any subsequent confusion because even though this is commonly considered a standard construction practice, framing subcontractors do not always view it this way. This work is a common point of contention on many projects.

# 25 Building Insulation and Fire Safing*

The building insulation and fire safing (referred to here as insulation) scope of work is managed in different ways. The insulation work is commonly performed by a second-tier subcontractor under an agreement with the framing subcontractor. This is not always the case, as the general contractor may also decide to contract directly with an insulation subcontractor. For residential or office buildings with nearly all of the insulation within framed walls, the framing subcontractor traditionally assumes responsibility for the insulation work. Some structures will require a considerable amount of insulation in locations other than within the stud-framed walls and ceilings, such as rigid insulation in the ceiling of a high bay warehouse. A structure might also require black-faced sound-absorptive rigid panels surrounding a theatrical stage or a substantial amount of fire safing at expansion joints. Another project might require a significant amount of spray-applied insulation. Under these various conditions that require a considerable amount of insulation other than within the framed walls and ceilings, it is logical for the general contractor to directly employ an insulation subcontractor. This division of work should be made by the general contractor by considering the various unique circumstances of the project.

Most work of the insulation subcontractor will be contained within the framed walls, in the ceiling framing, or in very large sections identified on the room finish schedule. Due to this, defining the scope of work of the insulation subcontractor is generally quite simple. Occasional problems are presented by obscure locations requiring insulation. In general, the insulation scope of work entails a relatively low cost. If a few locations requiring batt insulation are not included in the bid proposals, it is a comparatively inexpensive problem to resolve.

## SCOPE OF WORK ISSUES RELATED TO BUILDING INSULATION AND FIRE SAFING

1. On public works projects, insulation bidders may provide separate bids for acoustical insulation, thermal insulation, and fire safing. To avoid subsequent problems, the bids should be carefully reviewed to ensure that all aspects are covered.

   For example, it is generally agreed that when a bid proposal for any trade appears to be quite low in comparison to other bids for the same scope of

---

* MasterFormat Specifications Division 7

**205**

work, the general contractor is morally obligated to question it. In the case of an insulation subcontractor's low bid, it may be for only thermal insulation while the other subcontractors' bids properly included the acoustical insulation and fire safing work. In this case, if the acoustical insulation and fire safing scopes of work are minimal on the project, the error may not be clearly evident based upon the value of the bid proposal and will be difficult to catch with a cursory review of the proposal. This example exemplifies why it is important to thoroughly read each subcontractor's qualifications prior to awarding a subcontract.

2. A commonly missed project insulation element is fire safing at the floor-to-floor expansion joints. This fire barrier is sometimes completed as a manufacturer's supplied component of the expansion joint. At other times it is a fire safing product. In addition to verifying the construction type of the fire barrier, always verify how this fire safing is held in position. Sometimes it will simply be wedged into place, but other times it will be supported by a piece of bent sheet metal. The flashing subcontractor will be responsible for providing the sheet metal support when required (Figure 25.1).

3. It is generally preferred that the framing subcontractor perform the fire safing work at the heads of CMU walls, but sometimes this work is allocated to the insulation subcontractor. Either allocation is acceptable.

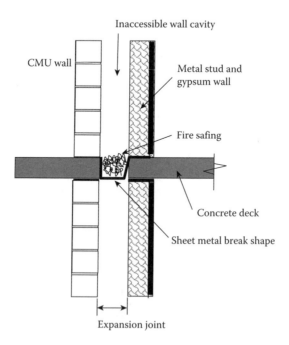

**FIGURE 25.1**   Fire safing at expansion joint.

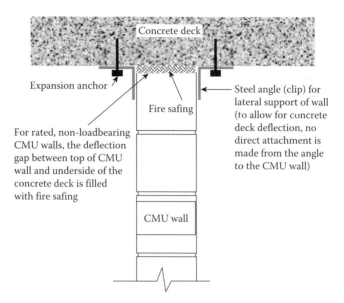

**FIGURE 25.2**   Fire safing at head of CMU wall.

The fire safing at the heads of CMU walls is never allocated to the masonry subcontractor, who is always quite reluctant to perform this work (Figure 25.2).

4. Some roofs are constructed with rigid insulation above the structural roof deck. The insulation subcontractor will not furnish or install this roofing insulation, as this insulation work is always part of the roofing system. On the other hand, the insulation subcontractor will always furnish and install the horizontal roof insulation below the structural roof deck. This applies to either batt or rigid insulation.

A similar division of work occurs at below-grade perimeter walls. When insulation is specified on the exterior side of a below-grade perimeter wall, it will be part of the waterproofing system. This will be provided by the below-grade waterproofing subcontractor. When the insulation occurs on the interior side of the walls, it will be provided by the insulation subcontractor.

The term horizontal roofing insulation differentiates it from the vertically installed insulation in conditions such as parapet and roof penthouse walls. The vertically installed insulation will be provided and installed by the insulation subcontractor.

5. Consider rigid insulation occurring on top of a waterproofing membrane and below plaza deck pavers (Figure 25.3). This insulation will be furnished and installed by the waterproofing subcontractor. This insulation serves a dual purpose of providing thermal insulation and by acting as a protective board for the waterproofing membrane.

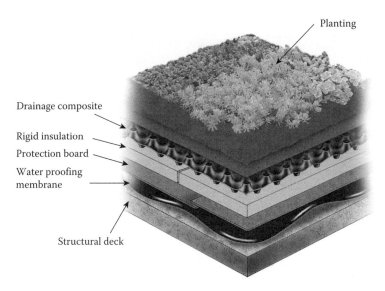

**FIGURE 25.3**   Garden roof assembly. (Illustration provided by American Hydrotech, Inc.)

6. Spray-applied insulation is commonly used in locations such as parking garage and mechanical room ceilings. This application will be performed by the insulation subcontractor. Since not all insulation subcontractors have the ability to install a spray-applied system, be sure the bidders list consists of companies that are in fact capable of performing this work.

# 26 Doors, Frames, and Hardware*

The door scope of work includes doors, frames and hardware. This work is traditionally bid by and awarded to a single subcontractor on private projects, but will likely include multiple subcontractors on public works projects. On a public works project, various bidders will submit bids for the various doors, frames and hardware components and due to public bidding laws the lowest responsive and responsible bidder for each component must be awarded the contract. This means different subcontracts may be awarded for steel doors, hollow metal frames, wood doors, door hardware, and specialty doors.

The confusion and coordination complications of managing numerous subcontractors are reduced on private projects where public bidding regulations do not prevail, i.e., this work is commonly bid and awarded as a single package. Though the private projects may not receive the lowest combination of bids, it is generally considered that the award of a single subcontract provides the best overall financial value. That is, it is much simpler to work with a single door subcontractor rather than an assortment of several subcontractors. In addition, a significant savings in reduced management is realized.

## SCOPE OF WORK ISSUES RELATED TO DOORS, FRAMES, AND HARDWARE

1. There are varied industry practices with regards to the installation of hollow metal frames in stud walls. This work is sometimes performed by the door subcontractor, but other times by the framing subcontractor. Generally, the most prudent approach is for the door subcontractor to install their own frames as they are ultimately responsible for making sure that the installed door is plumb and square.

    (a) In order for the door subcontractor to install the frames, the framing subcontractor must construct the stud wall and leave loose studs at one side of the frames. The door frame is first installed by the door subcontractor, then the framing subcontractor will finish up the framing, and finally the door subcontractor will plumb the frame and secure it to the completed stud framing. This coordinated effort is necessary because the stud framing tucks into the head and jambs of the hollow metal door frames (Figure 26.1). This coordination effort is

---

* MasterFormat Specifications Division 8

eliminated on some projects where the framing subcontractor will be asked to install the frames in lieu of the door subcontractor. Both approaches of installing the frames are widely used and successful. On union projects, there are no jurisdictional problems as the work of the framing and door subcontractors are both in the carpenters' union jurisdiction.

2. Hollow metal frames in CMU walls are commonly installed by the masonry subcontractor. As with hollow metal frames set in stud walls, there is a general preference to have the door subcontractor install these frames. Some even contend that masons are not accustomed to the high level of precision with which door frames must be installed. Nonetheless, hollow metal door frames have been successfully installed by both door subcontractors and masonry subcontractors.

   (a) If the general contractor wants the door subcontractor to install the door frames, they will need to coordinate this work with the masonry subcontractor. This is not a difficult coordination issue, but it requires that all parties remain fully informed about when their services will be needed.

   (b) Hollow metal door frames in CMU walls are often sound-rated or fire-rated openings, which means that the frames must be grouted to the CMU. Regardless of who installs the frames, this grouting is best performed by the masonry subcontractor simply because the grout pump used to fill the CMU cavities can also be used to grout the frames, so they are already set up for this work. Be sure to pay special attention to this work because grouting of the hollow metal frames is a common exclusion by both the masonry and door

**FIGURE 26.1**  Door frame. (Photo by author, courtesy of Hathaway Dinwiddie Construction Company and California State University Northridge.)

subcontractors. This is a common gap in the door scope of work that is to be avoided.

    (i)   It must be clear in the subcontract that the masonry subcontractor is responsible for pour stops around the hollow metal frames to ensure that there is a clean and professional grout joint on all sides of the frames.

3. The door subcontractor should exclude all glass and glazing within the doors and frames. Glass items, such as sidelights and door vision panels, should be included in the scope of work of the glass and glazing subcontractor. This is a common allocation in the industry, but care must still be taken to ensure that it is not excluded in the glass bids.

    (a)   There is a difference in the treatment of glass between the scope of work of casework subcontractors and door subcontractors. Whereas all glass doors, mirrors, and shelves in cabinets are routinely provided and installed by the casework subcontractor, glass in doors is the responsibility of the glass and glazing subcontractor. The rationale for this distinction is that the casework glass is installed in the shop, whereas the door and frame glass is commonly installed in the field. Even on unionized projects, the casework shops are not commonly unionized so jurisdictional issues do not arise between the millwrights and glaziers. In the case of the door and frame glass, a jurisdictional dispute would arise on a union project if the door carpenters attempted to install the glass claimed by the glaziers.

4. Storefront hardware should be included in the scope of work of the door subcontractor, with the exception of the pivots (hinges), which will be furnished by the glass subcontractor. The storefront doors and frames will always be furnished and installed by the glass and glazing subcontractor. The storefront hardware is also installed by the glass and glazing subcontractor, but this hardware is actually furnished by the door subcontractor. Since the door subcontractor is not furnishing or installing the storefront doors and frames, they may not recognize the need to provide the hardware for those doors. Since the storefront hardware typically matches other hardware on the project, it is preferable for the door subcontractor to furnish these items to the glass subcontractor FOB.

    (a)   When one subcontractor furnishes items to another, be sure to clearly stipulate whether the items are to be delivered to the receiving subcontractor's shop or if they are to be delivered to the jobsite. If this is not clearly communicated, the receiving subcontractor may expect the hardware to be delivered to their shop, but later find that it has been delivered at the jobsite. This will add a cost for the recipient subcontractor to retrieve the materials from the project site. The distance to the shop can be considerable, for which the hassle and added financial burden can be avoided by clearly communicating the routing destination of materials.

5. Similar to the storefront hardware, hardware for exterior steel or chain link gates is commonly not included in the scope of work of any subcontractors,

but this is most appropriately allocated to the door subcontractor who will both furnish and install it. (Hardware installation at gates is discussed in greater detail in the chapter on miscellaneous metals and the chapter on chain link fencing.)

6. Coiling doors and grilles often require a keyed cylinder for the manual locking mechanism. The furnishing of these keyed cylinders is commonly excluded by both the coiling door and door subcontractors. This work should be allocated such that the door subcontractor furnishes these items for installation by the coiling door subcontractor. By having the door sub-contractor furnish the keyed cylinders, this arrangement will ensure that the cylinders at the coiling doors and/or grilles are keyed similarly to the remainder of the building. For maintenance purposes, it is inefficient for an additional key to be required for any specialty item, and the coiling doors and grilles are no exception.

7. There should not be a noticeable gap between the painted hollow metal frames and the painted gypsum board walls. Visible gaps commonly occur at about half of the hollow metal frames and the other half of the hollow metal frames fit tight enough to the wall that caulking is not necessary. It is best that the painting subcontractor be responsible for caulking these gaps. Though the extent of this caulking work is a product of construction toler-ances and as such not shown on the construction drawings, it is not clearly quantifiable at bid time. Since it is minor in nature and the 50/50 ratio of frames that do/do not require caulking is fairly constant from project to project, the painting subcontractors are generally agreeable to including minor caulking of these gaps in their lump sum base bid. While these gaps do not often become significant issues on a project, it is worth addressing them in advance with the door, framing, and painting subcontractors. The door and framing subcontractors will be diligent to construct their work with few gaps, and the painting subcontractor will generally agree to per-form the minor caulking work necessary to make up for minor construction tolerances.

8. The door subcontractor is routinely responsible for staining the transpar-ent finish wood doors, but will only prime the hollow metal frames, wood doors, and steel doors which are scheduled for a painted finish.

There are other important distinctions to recognize that pertain to stain-ing and painting. Casework subcontractors will provide a complete finish on cabinets whether they are painted or stained. Door subcontractors will provide a complete stained finish on doors, but will only prime doors that are to be painted. Millwork subcontractors will not provide any finish work for a stained finish and will only prime paint millwork that is scheduled to be painted. These are similar scopes of work but each has evolved with a distinctly different industry standard. The primary factor in the assignment of work pertaining to staining and painting relates to whether the work is performed on site or off site.

9. Handicap door operators may occur at storefront doors, steel doors, wood doors, or other types of doors on a single project. This work will invariably

be performed by a lower-tier subcontractor who specializes in low-voltage items such as door operators. The lower-tier subcontractor may be employed directly by the door subcontractor, the glass and glazing subcontractor, the electrical subcontractor, or the general contractor. Regardless of the sub-contracting arrangement, it is important to assure this work is covered, but not double covered. It is suggested to allocate this scope of work in the following order of precedence:

(a)  If the handicap door operators occur only at storefront doors, it is preferable for the glass subcontractor to include these items. If operators occur only at steel or wood doors, there is a preference to have the door subcontractor perform this work. If they occur at the storefront and on steel or wood doors, the door subcontractor should probably be assigned this work. As illustrated here, allocation of this work is highly subjective and there is no single correct method. So, the general contractor should make an educated, project specific, decision in this regard.

(b)  In some project-specific cases it might be difficult to assign the operators to one selected subcontractor. When this occurs, the general contractor will directly employ a low-voltage electrical subcontractor for this work. An example of this would be a single handicap door operator at a steel door on a public works project where there are several subcontractors responsible for the doors, frames, and hardware scope of work.

10. Handicap door operators are a relatively simple scope of work, but they do require coordination with other trades. This is especially true when these door openers occur at storefront doors. This coordination is a comparatively minor issue, but to avoid potential change order issues later in the project it is important that this is addressed before the associated subcontracts are awarded.

(a)  Verify the size of the push button and the location where it is to be mounted. Also, determine the size of the electrical back box necessary to house the push button. The electrical subcontractor will provide conduit raceways for the controls wiring and a junction box to mount the push buttons.

(b)  In addition to the raceways, the electrical subcontractor must provide a single point of connection for 120V power to a power booster for the handicap door operator. Furnishing and installing the power booster and all wiring beyond the power booster should be the responsibility of the subcontractor installing the handicap door operator. A step-down transformer to convert the supplied 120V power to 24V low-voltage power should be integral to the power booster.

(c)  The push button for the door operator may be shown as being placed on a storefront mullion, which could be a complicated coordination effort if the mullions are too narrow to readily accommodate the push-button back box. In this case a short post is commonly added in front of the storefront for mounting the push button. When discovered late in a project, this post will become both an aesthetic blemish and a change order issue. Consequential costs will surely be encountered for demolishing and replacing the hardscape, cutting into walls, and possibly cutting into interior ceilings.

(d) The handicap door operator subcontractor should be held responsible for providing the code-required signage associated with the door operator. The signage is typically simple, commonly just a sticker or two, but if it is not in place for the final building inspection it can result in a failure of the final building inspection.

(e) When the door operators are wired to the same door as a card reader, coordination of these two devices is important. The door operator must be wired such that it will not open the door unless a valid card is first swiped.

 (i) For example, assume that the security subcontractor installed, programmed, tested, and commissioned the card readers. Then, much later, the handicap door operators were installed and wired directly to the door, thus inadvertently bypassing the card readers. This would make the card reader ineffective for security purposes, as simply pressing the handicap door operator button will unlock and open the door. This type of problem can be avoided with a little coordination effort.

11. Lack of coordination between the door and security subcontractors commonly produces gaps in the scope of work with regards to the request to exit devices (Figure 26.2). Without a request to exit device (commonly referred to as a Rex), which informs the door security sensor when it is clear for the door to open and that no alarm is necessary, the door security sensor will sound the alarm every time the door is opened. The security subcontractor will always provide both the card readers and door security sensors, but responsibility for the request to exit device can vary. There are three different scenarios to be aware of when scoping this work, one scenario when entering a secure room and two scenarios when exiting a secure room. Each scenario will be described:

(a) When a person approaches a security door from the public side (from outside the building entry or in a corridor), they will first swipe their card key. This action signals the card reader to do two things. First, it will open the door. Secondly, it will send a signal to the door security sensor that this is an authorized opening of the door to avoid sounding an alarm. The wiring, terminations, and commissioning of this signal work must be performed entirely by the security subcontractor.

(b) When a person exits a secure room by pushing a panic bar, there is no need for a card reader. They simply open the door and walk out. Somehow, because the door security sensor only recognizes whether the door is open or closed and it cannot recognize from which side the door has been opened, the door security sensor must be sent a signal that this is an authorized opening of the door. This is when a Rex

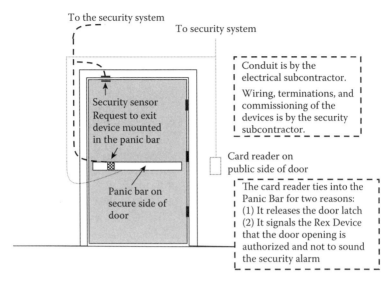

To the security system

To security system

r-------------------------1
¦ Conduit is by the        ¦
¦ electrical subcontractor.¦
¦                          ¦
¦ Wiring, terminations, and¦
¦ commissioning of the     ¦
¦ devices is by the security¦
¦ subcontractor.           ¦
L_____J

Security sensor
Request to exit
device mounted
in the panic bar

Card reader on
public side of door

Panic bar on
secure side of
door

r-------------------------1
¦ The card reader ties into the ¦
¦ Panic Bar for two reasons:    ¦
¦ (1) It releases the door latch ¦
¦ (2) It signals the Rex Device ¦
¦ that the door opening is      ¦
¦ authorized and not to sound   ¦
¦ the security alarm            ¦
L_____J

**FIGURE 26.2**   Card reader raceways.

needs to be installed in the panic bar. When the panic bar is depressed, the door will open and the Rex will send a signal to the door security sensor. This signal informs the door security sensor that the door opening is authorized and no alarm is necessary.

(i)   The door subcontractor should furnish and install the Rex devices within the panic hardware to avoid compatibility problems fitting this device into the panic hardware. It is worth noting that the make and model of a Rex device is not a significant issue, as Rex devices are compatible with any card reader, so coordinating these products is not a scope issue.

(ii)  If the security subcontractor furnishes and installs the Rex devices in the panic bars, the general contractor must ensure that the panic bar devices provided by the door subcontractor are prepped for the Rex. The panic bar devices must be prepped electronically and they must also have a mounting location inside them large enough for which the specific Rex model provided by the security subcontractor can be mounted. Additionally, in this case the door subcontractor will need to work closely with the security subcontractor in the costly coordination task of disassembling and reassembling the panic bar devices on site.

(c)   When a person exits a room which is equipped with a simple lockset (not a panic bar), there is not sufficient room inside the small lockset device for a Rex. This is when it is a common practice to have a ceiling-mounted Rex motion sensor (Figure 26.3). This motion sensor will send a signal to the door security sensor if it detects movement (a person) on the secure side of the door. This signal communicates to

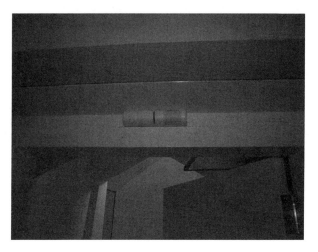

**FIGURE 26.3**   Ceiling mounted request to exit device.

the door security sensor that the door is being opened from the secure side so there is no need for an alarm. In this scenario the security subcontractor will furnish and install a Rex motion sensor, thus eliminating the need for coordinating with the door subcontractor. While it is not common practice, a ceiling-mounted Rex could also be used with a panic bar to avoid the coordination problems discussed in the previous scenario.

12. Card readers require electrified hardware such as panic bars, door hinges, and raceways within the doors and frames for the control wiring. Each installation is slightly different, so prior to bidding it is important to obtain a copy of the product data and single line diagrams for the specific products identified in the project specifications. Educated efforts can then be devoted to ensuring the work is properly allocated.

(a)   In addition to the divisions of work described, make sure the electrified hinges have been allocated to one subcontractor. Electrified hinges are most often furnished by the door subcontractor, but sometimes by the security subcontractor. Regardless of who furnishes them, the electrified hinges will be installed by the door subcontractor. It is advisable that the door subcontractor furnish the electrified hinges. This will eliminate any coordination problems between the subcontractors, and the door subcontractor will also maintain the best quality control in ensuring that the electrified hinges aesthetically match the other door hinges.

13. It is suggested that the door subcontractor provide all "Not an Exit—Alarm Will Sound" signage. These are typically simple stickers and are regularly forgotten until the end of the project. Specifically including these signs in the door subcontract will probably save a punchlist item and needless effort in trying to find and obtain these stickers at the end of the project.

14. The door subcontractor should be held responsible for preparing the hollow metal doors and frames for the door security sensors. These simple sensors often require nothing more than drilling one hole for the wiring and four predrilled screw holes for mounting the sensor at the head of the frame and top of the door. If this minor coordination item is not addressed prior to fabricating the frames, the field drilling will likely become the source of a change order request.

15. Verification of the responsibility for final keying is very important.

    (a) Sometimes the general contractor will be fully responsible for final keying of a building. At other times the owner will opt to self-perform this work. Often the owners prefer to do the keying to mitigate the security risk of additional keys being made and sold to criminals, keys being stolen, or any other potential security breaches. The door hardware specification section will identify the responsibility for this work.

    (b) When the owner performs the final keying, they will often require the door subcontractor, via the general contractor, to furnish blank cylinders FOB jobsite. Other times, they will simply procure the blank cylinders themselves. Clearly conveying this responsibility to the bidders will ensure that the subcontractors include the cylinders when required, but do not have additional money in their bids for cylinders that will not be required.

    (c) Some projects require the door subcontractor to develop the keying schedule, also referred to as the keying tree. Nonetheless, it is preferred for this to be prepared by the owner. The owner is the ideal party for this task as no other party truly understands how the building is intended to function. If the door subcontractor is required to develop the keying schedule, the owner must provide clear criteria for the development of this schedule. Regardless of who prepares the keying schedule, the important thing is to be sure it is clearly allocated prior to bid time.

16. Dents and dings to installed doors and frames are inevitable on a project. A plan should be in place for protecting doors and frames, as well as for making minor repairs. It is preferred that the protection be installed by the door subcontractor, with the maintenance and removal of the protection performed by the general contractor. Since the door subcontractor will not have a presence on the project after completing the door installation, it is not considered appropriate to hold them responsible for maintaining the protection or for damage caused by others.

    (a) Though this example is specific to the door trade, similar concerns exist with other installed items that might be damaged by different trades. If subcontractors are asked to protect their own work, it is important to also address the, often costly, maintenance and removal of this protection. Be sure to also address the inevitable minor repairs necessary due to trade damage throughout the trades, particularly the trades providing the building finishes.

# 27 Coiling Doors and Grilles*

Coiling doors and grilles are commonly employed where a standard hinged door will not be of sufficient size for access. Typical locations for coiling doors and grilles include parking entries, loading docks, concession counters, MEP equipment rooms, and trash rooms. With proper pre-planning, the coiling doors and grilles subcontractor (often referred to as the coiling door subcontractor) will quickly and efficiently perform the work. This work is performed so quickly that little involvement is required from on-site management of the general contractor.

## SCOPE OF WORK ISSUES RELATED TO COILING DOORS AND GRILLES

1. The most confusing coordination issue for coiling door installations concerns low-voltage wiring raceways, because these raceways are not often depicted on the construction drawings. On union projects, these raceways fall under the jurisdiction of the electrical subcontractor. The work is routinely allocated between the coiling door and electrical subcontractors via bid instructions distributed by the general contractor. The instruction should describe the following components (Figure 27.1):
   (a) The electrical subcontractor will provide a single point of connection for medium-voltage power to the door motor. This will be clearly shown on the electrical drawings.
   (b) The electrical subcontractor will provide a 3/4" conduit and wire from the door motor to the switch location(s).
   (c) The electrical subcontractor will provide a 3/4" conduit and wire from the door motor to the door interlock. The door interlock is a motor protection device at the coiling door mechanical lock that signals the motor when the door is manually locked. This prevents the motor from starting, which prevents the motor from burning out.
   (d) The coiling door subcontractor will provide a step-down transformer at the motor.
2. The support steel for the coiling door housings and door guides will be omitted from the coiling door scope of work. There are various ways

---

* MasterFormat Specifications Division 8

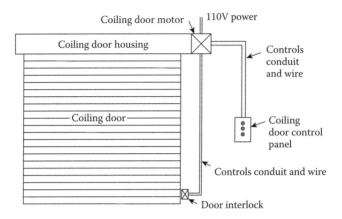

**FIGURE 27.1**   Coiling door raceways.

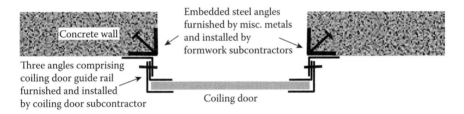

**FIGURE 27.2**   Coiling door guide rails.

that this will be resolved as there are differing coiling door installation conditions:

(a)   For coiling doors at a concrete wall, the miscellaneous metals sub-contractor will provide the embedded edge angles (Figure 27.2) for the jambs, which will be set in place by the formwork subcontractor prior to concrete placement. All work beyond installing these embedded steel angles will be the responsibility of the coiling door subcontractor. This type of jamb generally has three additional angles attached to the embedded angle to support the door guide rail, all of which will be provided by the coiling door subcontractor, not the miscellaneous metals subcontractor as is commonly misunderstood. Establishing responsibility for providing the non-embedded rail components is a common dispute between the miscellaneous metals and coiling door subcontractors. This division of work must be clearly defined in the bid documents.

(b)   For a coiling door to be set in a CMU wall, all work beyond the CMU wall itself is most appropriately performed by the coiling door subcontractor. This condition entails no embedded angles or other supporting steel provided by other subcontractors. The coiling door in this condition will be mounted solely with expansion anchors. The coiling door subcontractor can easily perform this work in its entirety.

(c) For a coiling door set in a stud-framed wall, there are two similar mounting conditions. The differences relate to the door size. In either case, the coiling door subcontractor will be responsible for all work beyond the basic frame support.

   (i) For a large coiling door, a tube steel frame will be required. This frame will be installed in line with the stud framing. This is commonly a 4"×4" tube steel frame spanning from the floor deck to the deck above and constructed in line with a four-inch or a six-inch metal stud-framed wall. This tube steel frame will be installed by the miscellaneous metals subcontractor.

   (ii) Small coiling doors are light enough to be supported by the stud framing itself, though the gauge of the framing may need to be increased and the stud spacing decreased. The supporting structure in this case will be the full responsibility of the framing subcontractor.

3. For stud-framed walls that are rated for fire and/or sound, the coiling door subcontractor is responsible for maintaining this rating at their penetrations. This is particularly important where the supports for the coiling door penetrate the gypsum board between the guide rails and framing. It is imperative that the coiling door subcontractor includes fire and/or acoustical caulking to patch any penetrations. This is commonly missed in both the bidding phase and during construction. This oversight is frequently discovered when the building inspector logs a deficiency.

4. When a factory-applied color finish is identified for the coiling doors, ensure that the coiling door subcontractor includes this requirement. Also, notify the painting subcontractor that there is no requirement to paint the doors. Since this requirement may only be noted in the coiling door specification section, which the painting bidders typically will not review, a diligent general contractor will need to specifically bring this issue to the painter's attention.

(a) When the specifications stipulate that the coiling doors are to receive a factory finish it might be interpreted in different ways. It could mean that only the door itself will receive a factory finish. It might also mean that this requirement includes the door housing, motor shroud, and guide rails. Review the documents carefully for the specific requirements to be sure this work is properly incorporated by both the coiling door and painting subcontractors.

5. The miscellaneous metals specification will commonly require all steel exposed to the exterior to be galvanized, but galvanizing will not likely be specifically addressed in the coiling door specifications. Since coiling door bidders typically will not read the miscellaneous metals specification, the coiling door subcontractor may miss this requirement and will plan on installing guide rails that have not been galvanized. Communicate this

information during bidding and verify compliance during construction to avert any problems.

6. Coiling doors and grilles are regularly specified with a keyed lock and the coiling door subcontractor will be responsible for furnishing the locking hardware. It must be compatible with the hardware described in the door hardware specification (or the hardware schedule), because the lock cylinder and final keying will be provided by the door subcontractor. While the cost of the lock cylinder is negligible, the remedial work and time lost when such items are overlooked can be costly. The allocation of this work must be clearly communicated to the coiling door subcontractor and the door subcontractor.

   (a) Coiling door subcontractors would not bid or design a coiling door with a mechanical lock without also including an interlock with the motor. This interlock is a failsafe device that will keep the motor from running when the mechanical lock is engaged, thus, preventing the motor from burning out. Inclusion of the interlock for doors with mechanical locks is a standard practice. It is still wise to verify that the bidders have accounted for this component and, as discussed above, coordinate the low-voltage signal wiring work with the electrical subcontractor.

7. Coiling grilles have relatively loose grille construction. Grilles wider than about thirty feet can often be pulled up in the middle by one person while a second person slides underneath, which negates the purpose of a security grille. The construction of the grille is a design-build task, and the coiling door subcontractor should be held responsible for providing an effective design. Wide grilles should be designed with a bottom bar that is sufficiently rigid to withstand this type of security breach (Figure 27.3).

**FIGURE 27.3**   Coiling grille with heavy bottom bar. (Photo by author, courtesy of The University of Southern California.)

Note that some coiling door subcontractors may attempt to use a light-weight bottom bar, but put a mechanical lock down, or hook, in the center of the opening. Unless specifically allowed by the contract documents, which is very rare, this method will not be an approved method of construction.

8. For the installation of very long interior coiling doors, the traditional construction sequence may need to be altered. For example, a clear path of travel from outside the building to the point of installation must be maintained. Walls and other elements may need to be left out to maintain a clear path for the coiling doors to be moved through the building. The general contractor must make sure the framing, CMU, painting, flooring, ceiling, MEP, and other affected subcontractors are informed prior to bidding the project if they are to allow for a return trip to the project to accommodate this phased work. This change in traditional work sequence should be reflected in the project schedule.

    For simple interior renovations, also termed tenant improvement projects, sufficient time may not be available to procure the coiling doors prior to walls being erected. This is unlike new construction where the general contractor and coiling door subcontractor will have ample time during the structural phase to make arrangements to acquire these doors and store them in the area of installation before any walls are constructed. On tenant improvement projects in particular, the general contractor must clearly address this time issue in the coiling door bid package. The coiling door subcontractor must be held financially accountable for meeting the scheduled deadline.

9. An additional benefit of the early installation of the coiling doors at exterior walls is to secure the building once finish work has begun. To avoid installing the permanent exterior hinged doors early in construction and leaving them susceptible to construction damage, the general contractor will commonly construct lockable plywood doors at the main building entrances. Since large coiling door openings are too large to secure with temporary measures, these permanent doors must be installed as early as possible and properly protected.

# 28 Casework and Millwork*

The scope of work for a casework subcontractor is clear-cut. Casework is custom fabricated for each installation on a project. These subcontractors generally have very efficient operations which they apply both in their fabrication shop and on site. This efficiency is achieved through diligent planning and meticulous execution.

Through extensive experience perfecting their trade and by routinely providing quality products, the operations of casework subcontractors have emerged as a predominantly design-build effort. Architects will typically specify a performance criterion for the casework subcontractors to bid. The casework subcontractors will then develop the completed design and execute the work. Their means and methods are rarely questioned, and these subcontractors are frequently sought for advice by architects and general contractors during the design phase of a project. Casework subcontractors have established a reputation of intelligence and competence which has resulted in architects and general contractors placing a high level of confidence in them.

As with casework, the millwork trade is easy for the general contractor to manage and coordinate. Although walls, floors, and ceilings frequently have imperfections related to alignment, these minor tolerance variations are no major challenge for the millwork subcontractor. In fact, the millwork often conceals the mistakes of the other trades. This is because the millwork trim is field measured and cut to fit on site.

There are few coordination issues the casework subcontractor will have with the other trades, and the coordination issues they do have are not only quite simple, but are also quite consistent from project to project. Some of the coordination issues the casework subcontractor will have with other trades include:

a. Cabinets must be installed precisely in preset locations near the end of a project. Although cabinets often abut adjacent finishes, there is generally little need to coordinate this aspect of the work with other trades. Since field measurements will be taken by the casework subcontractor immediately prior to fabrication, any minor dimensional deviations will be resolved without conflict.
b. Often sinks, as for lavatories, will be integrated into the casework. While the installation of sinks is the plumbing subcontractor's responsibility when installed on site, these are at times installed in the shop by the casework subcontractor. This coordination effort is typically easy and routine for both parties, as will be discussed.

---

* MasterFormat Specifications Division 6

   c. Electrical wiring will frequently be run through the cabinets for lighting, receptacles, or other electrical devices. Due to the frequency of this occurrence, this coordination effort is also quite simple and requires little effort.

   d. The locations where cabinets attach to the wall must be coordinated with the framing subcontractor, who will provide in-wall backing for structural support of the casework. Since this is also a regular occurrence, the coordination effort consists of little more than handing the casework shop drawings to the framing subcontractor.

Since the casework and millwork scopes of work are commonly undertaken by the same subcontractor and have many similarities, they are discussed in a single chapter. Thus, the terms casework subcontractor and millwork subcontractor are considered synonymous in this book.

## SCOPE OF WORK ISSUES RELATED TO CASEWORK AND MILLWORK

1. All glass within the casework is commonly included by the casework subcontractor. This includes glass shelving, glass in the cabinet doors, mirrors, and any other glass products. Glass and glazing subcontractors are accustomed to this allocation of work and while glass and glazing unions claim glass work, they do not claim glass within casework. This is not a contentious area because the glass in casework is installed in the casework subcontractor's shop, not on the project site.

2. The casework subcontractor will include paint and stain finishing of all cabinets. On the other hand, they include prime painting of all millwork scheduled for paint, but will exclude the finish painting and stain finishing. This is common in the industry, but the scope of work must still be properly coordinated between the casework/millwork subcontractor and the painting subcontractor.

    A common problem for color matching may arise when millwork trim abuts a cabinet. A considerable challenge is encountered when stained upper cabinets are scribed to the ceiling with an identically stained crown molding that wraps around the entire room. As described above, the upper cabinets would have a shop applied finish, whereas the crown molding would be left bare in anticipation of a field finish by the painting subcontractor. To minimize color matching problems, the suggested method of construction would be to have all crown molding shop finished in conjunction with the casework. Further, the nail holes resulting from installing crown molding should be filled and finished by the casework/millwork subcontractor with the same batch of stain or varnish originally used. This will help avoid any spotting of the crown molding that might otherwise be caused by a new and different batch of finishing products supplied by a different party. These imperfections and color matching would certainly be issues that would arise if the molding was finished by the painting subcontractor. The division of work needs to be properly allocated at the onset of the project and clearly relayed to the bidders.

3. Wood base is commonly completed by the wood flooring subcontractor where it occurs in conjunction with wood flooring. At all other locations, the millwork subcontractor would be held responsible for wood base trim. In a room or area with only a single type of flooring it is generally a simple matter to resolve the scope of work regarding wood base trim. It is slightly more complicated when wood base is to be provided in an area that has both wood flooring and other floor finishes. Since precise color and dimensional matching might be a problem, it is not advisable to change responsibility for wood base halfway around a room. The most important factor is to make sure the wood base matches the wood flooring. Thus, it is most logical for the wood flooring subcontractor to furnish and install all wood base trim in those areas.

   Additionally, it is important to coordinate the height of the wood base in these areas and clearly convey this to the bidders. For example, wood flooring is commonly 1″ to 1-1/2″ thick, but carpeting with a pad is about 3/4″ thick. To accommodate this differential, the flooring subcontractor will transition between these surfaces with a reducer strip, leaving the mean elevation of the carpeting about 1/2″ lower than the wood flooring. Since the base sits on top of the carpeting and wood flooring, the same 1/2″ transition drop will occur in the wood base unless proper planning is instituted. The base should be sized so that the top of the wood base remains at the same elevation throughout the room, while the bottom of the wood base trim will step up or down as necessary to remain flush with the top of each type of flooring.

4. Due to the use of quality jigs and skilled craft workers employed in the production of cabinets, the units will arrive at the project in near-square condition, but not perfectly square. Likewise, the gypsum walls that the cabinets are scheduled to abut will be within industry tolerances, but they may also not be perfectly square. To accommodate small irregular gaps between the cabinet box and the wall, either a scribe piece or caulk joint will commonly be required. Regardless of whether the tolerances are off in the gypsum walls, the cabinets, or both, the work to resolve the imperfections is best performed by the casework subcontractor.

   (a) Scribes and caulk joints are commonly used for base cabinets and the sides of upper cabinets, but the tops of upper cabinets are slightly different. When slight variations exist between the ceiling and the top of the upper cabinets a quality control decision will likely need to be made, especially if the ceiling is slightly sloped. For example, if the ceiling height varies 1/4″ and the upper cabinets are installed level there will be a noticeable gap sloping from 1/4″ thickness to zero at the head of the cabinets. There are different acceptable approaches to remedy this problem.

(i) If the gap is truly 1/4″ on one side and zero on the other it may be possible to split the difference by cheating the cabinets to slope 1/8″, leaving only a 1/8″ blemish at the head of the cabinets. Though a 1/4″ gap would be very noticeable, neither the 1/8″ gap nor 1/8″ cabinet slope would be evident to the naked eye.

(ii) A scribe or crown molding could be used to cover the gap. If this is not in the construction drawings, a change order debate may arise and the responsibility will be difficult to establish. The general contractor will generally need to determine at each location whether or not the gypsum board and casework are each installed within standard industry tolerances. This determination may very well be different at each location, thus complicating the issue further.

5. It is advisable for the casework subcontractor to furnish and install all countertops, even the solid surface and stone tops. Though stone countertops reside under the masonry subcontractor's description of work, it is best for the casework subcontractor to complete this work. First of all, the casework subcontractors do this work regularly and it keeps full responsibility for the cabinetry under a single management entity which helps eliminate coordination problems. Secondly, even though the stone countertops are claimed by the masons union, the masonry subcontractor employed for the brick façade, CMU, or stone paving, will not likely be experienced in stone countertops. Since this has become such a specialty trade, a great number of subcontractors exist who work exclusively with stone countertops. Due to this, the masonry firms in the industry who do brick work have stayed away from this specialty. Even if the general contractor contracts with a masonry subcontractor for the countertops, they will probably perform this work through a lower-tier subcontract with a second masonry subcontractor.

6. Cut-outs for sinks in the countertops must be made with precision. This is well understood by casework subcontractors who devote special attention to this work. Sinks are generally provided with paper templates to be used for making the sink cut-outs. Some casework subcontractors try to ensure the accuracy of their cut-outs by requesting that the actual sinks be delivered to their shop. They mount the sinks in the countertops under the controlled shop conditions prior to shipping the countertops to the jobsite. This method has grown in popularity and has proven to provide an excellent finished product. Many architects are so pleased with this approach that they have begun writing specifications that specifically require all sinks to be mounted in the shop. Determine prior to bidding if the casework subcontractor will only need the templates or if the sinks themselves must be delivered to the casework shop, as the plumbing subcontractor will need to know whether or not to include the cost of mounting the sinks in their estimate.

(a) A casework subcontractor's shop is rarely unionized. On union projects, because the casework subcontractor mounts the sinks in their shop, not on site, this should not create a jurisdictional dispute between the millwrights and plumbers unions. This allocation of work is similar to the division of glass within the casework described earlier.

**FIGURE 28.1**    Cabinet base. (Photo by author, courtesy of Hathaway Dinwiddie Construction Company and The California Institute of Technology.)

(b)    Whether the casework subcontractor requires the templates or the sinks themselves, it is important to procure the sinks early in the project. The templates typically are delivered with the sinks, not separately. Since the templates are larger than the standard 11″ × 17″ pieces of paper that will fit in a printer, they cannot be downloaded on the Internet. A common cause for project delays is late delivery of the countertops resulting from late procurement of the sinks.

7. Thin flooring systems, such as resilient sheet vinyl and carpeting, will often run below the base cabinets. For wood flooring and ceramic tile it is a most efficient and standard practice for the casework subcontractor to set the cabinet bases first (the bottom 4″ platform, commonly forming the toe kick) (Figure 28.1). Then the flooring subcontractors will complete their work up to the bases. Lastly, the casework subcontractor will return to set the cabinets. This takes the guesswork out of layout for the flooring subcontractors by allowing the ceramic tile to be grouted right to the bases and allows the wood flooring to be installed uniformly along the bases for the best finished product.

(a)    The casework subcontractors generally prefer to set the upper cabinets during the same trip as the cabinet bases. This is a good construction practice, because with the upper cabinets installed ahead of the finish floor the additional risk of construction damage to the flooring is minimized.

# 29 Painting and Wall Coverings*

The painting and wall coverings (referred to here as painting) can transform a project almost overnight. The buildings quickly change from the drab colors of rough construction to the vibrant colors of an exciting completed project.

Most of a painting scope of work is clearly allocated in the contract documents, but it is still important to verify that the detail work is properly covered. There are several industry standard divisions of work related to painting that are counter-intuitive and must be specifically addressed by an experienced general contractor in their bid instructions.

## SCOPE OF WORK ISSUES RELATED TO PAINTING AND WALL COVERINGS

1. Intuitively it might be assumed that sheet wall coverings would be the responsibility of a specialty subcontractor, not the painting subcontractor. Most painting subcontractors are quite capable of self-performing wall covering work, though some painting subcontractors will partner with a lower-tier specialty subcontractor for this work. When a lower-tier subcontractor is employed, the relationship tends to be a close partnership and the wall covering subcontractor is a well-managed extension of the painter. A general contractor or owner might think that it would be beneficial to employ the wall covering subcontractor under a direct contract, but because many of the painting bidders will in fact perform this work in-house, it is advised to package the wall coverings with the painting. Further, painting subcontractors are generally judicious in trying to keep their base bids as low as possible. The management of this particular lower-tier subcontractor is sufficiently easy that they typically include a minimal markup on the wall covering bids. As a result, it is in the overall best interest of the project to keep the sheet wall coverings in the painting subcontractor's scope of work.

2. Similar to the miscellaneous metals and caulking trades, the painting work will include various building and site elements requiring a field-painted finish. Painting bidders will thoroughly review the entire set of construction documents and assume financial responsibility for completing all painting work. In good faith, the general contractor is ethically responsible to

---

* MasterFormat Specifications Division 9

flag odd items requiring painting that they discover while reviewing the drawings.

3. The painting subcontractor, not the above-grade waterproofing subcontractor, will be responsible for the application of the exterior water repellants.

4. The painting subcontractor will also be responsible for graffiti protective coatings. If both water repellants and graffiti protection are specified for a project, it is extremely important to verify that the two products are compatible. Since both of these products repel by nature, it is possible that these products will repel each other. This could result in peeling of the graffiti protective coating.

It is important to understand the disparity in the treatment of doors, casework, and millwork. The allocation of work for these topic areas is dramatically different, despite the similarity of these work tasks. There are reasons for these disparities, primarily dealing with union jurisdictions and whether the finish is applied in the shop or in the field. The industry standards have developed individually for each of the different elements of work. These will be described in items five to seven, as follows.

5. Doors, including hollow metal doors, wood doors, coiling doors, rolling doors, and large stage doors, that are scheduled to have a painted finish, will be primed at the factory by the respective door subcontractor. The finish painting is done in the field by the painting subcontractor. Conversely, the doors scheduled for a stained or transparent finish will have the finish applied in the factory by the door subcontractor themselves. This standard practice has developed in the industry because factory-applied stained and transparent finishes on doors are of higher quality than field finishing. The factory provides controlled shop conditions where the finish work can be done efficiently and to a higher quality standard. Since wood doors are shipped in a finished state, protection of the doors is particularly important once the doors arrive on site. Patching a stained door finish will invariably leave a blemish on the final product, so proper protection will avoid the need for any remedial surface repairs.

    (a) Particularly on high-end projects, there are exceptions to this rule that will be identified in the project specifications and/or door schedule. Not every door that is to receive paint is painted on site. Doors scheduled to receive baked enamel, Kynar (a hard, polyvinylidene fluoride resin that is commonly applied on metal roofing, siding, and window/door frames), or other specific shop-applied finish will arrive on the project pre-finished. High-end finishes of this sort will be performed in the factory by the door subcontractor because it is not possible to apply these types of finishes on site.

6. Casework will invariably be finished in the casework subcontractor's shop. This applies to painted or stained finishes. This standard has developed because casework is completely assembled in the shop, is not modified in the field and it can be quickly and efficiently finished in the shop. The controlled shop conditions also contribute to ensuring an optimal finished product.

7. The millwork finishing work differs from casework. If millwork will be stained, it is installed as unfinished material. If millwork will be painted, it is shipped to the project with a coat of primer on it. After the millwork subcontractor has installed their work, the painting subcontractor will complete the staining, varnishing, or painting as required. Since millwork consists of components that are cut, sanded, and fabricated on site, it does not offer the same potential for enhanced quality as if the materials had been previously finished in the shop.

   (a) Unlike casework, millwork is cut, trimmed, and sanded in the field. Due to of the mode of installation, it will contain exposed nail holes that need to be plugged and sanded. If this material was previously finished in the shop, it would require extensive touch-up work. Normally, the painting subcontractor will fill and sand any nail holes after the material is in place, and then apply the final finish.

8. Hollow metal door frames are designed with "throats" that fit tightly around the gypsum board walls (Figure 29.1). With a combination of manufacturing and construction tolerances, the frame may extend from the wall 1/8″ or so. This is a noticeable gap and needs to be caulked. Since this work is a construction issue it is not found on the construction documents, so the quality of such conditions will be a variable at the time of bidding. The painting subcontractors regularly include the cost of performing minor caulking work at the door frames in their lump sum bids with the general expectation that about 50% of the frames will require caulking and about 50% will not, as this ratio has been found to be fairly constant from project to project. This particular caulking work is typically easy and it can be performed quickly. Even considerable deviation from the estimated quantity will not amount to a significant sum. As a result, in practice, such variances will not often result in change order requests. Nevertheless, it is important

**FIGURE 29.1**  Door frame. (Photo by author, courtesy of Hathaway Dinwiddie Construction Company and California State University Northridge.)

to address this caulking work when awarding the painting subcontract to avoid confusion about the responsibility for it.

9. Simple roll-applied epoxy flooring may be completed by the painting subcontractor. If the quantity of epoxy flooring is small and a roll-applied method is specified, the painting subcontractor should perform the work. This will avoid having to bring in a specialty epoxy flooring subcontractor for a minimal scope of work. This suggestion pertains only to roll-applied epoxy flooring that does not involve any mesh, troweled base, or other specialty work. Epoxy flooring systems vary greatly from a simple roll-applied product to a multi-part trowel-applied system. Thus, the general contractor will need to review the specific product scheduled for use, make project-specific decisions, and inform the bidders.

    (a) Complicated epoxy floors, such as a trowel-applied system, will always be completed by a specialty epoxy flooring subcontractor. Traditionally, and most appropriately, this subcontract will be awarded directly by the general contractor.

10. Painting of the walls typically occurs when many other subcontractors are also finishing their respective work obligations. It is not surprising that the large number of subcontractors on the project will result in some damage to different portions of the project. For example, scrapes, scratches, and dings in completed walls are a common reality on construction projects. This is essentially unavoidable, as the cost of fully policing the project to prevent such damage is not a realistic objective. It is a far simpler, and more realistic, solution to touch-up the damage that occurs. To account for the cost of touching up this minor damage, the painting subcontractor should include an allowance in their bid proposal for this work. The value of the allowance should be established by the general contractor by stating the number of hours and the cost of materials to be allocated for touch-up painting.

    (a) As reiterated in other chapters, this type of allowance is preferred to a fixed monetary amount. By including the hours in their base bid, the subcontractor will price these hours with their bid rate. If only a monetary amount is held for this allowance, the time will be charged against the allowance at the subcontractor's change order rates, which are invariably higher than the labor rates used for their base bids. The same principal applies for the inclusion of a monetary amount in their base contract for materials. As the allowance is consumed on the project, the materials will be deducted at cost with no additional markups because the subcontractor will have already calculated and included the markups for this amount in the bottom line of their initial estimate. These terms should be clearly outlined in the subcontract agreement to avoid any misunderstandings.

    (b) Many general contractors choose to use a very different technique in allocating the cost for repairing trade damage. They begin by totaling the cost of repairing all trade damage at the end of a project. This total sum is then divided by the number of subcontractors on the project. The result is an amount that represents a unilateral deductive change order for each of the project subcontractors. This technique is not recommended as its basis relies on many unfounded assumptions. This effectively punishes subcontractors who have been diligent and conscientious about not damaging any in-place construction work. On the other hand, this approach financially commends or rewards the subcontractors who have acted negligently and inflicted the majority of the damage.

11. Not all painting will be allocated to the painting subcontractor. This is especially true of painting on concrete surfaces. For example, the painting subcontractor will not be expected to paint parking stripes, arrows, handicap symbols, parking curbs, and many other items found on the ground. This applies to both interior parking garages and exterior parking lots. This work will typically be performed by a specialty striping subcontractor, working as a second-tier subcontractor under the asphaltic-concrete paving subcontractor. On projects with little or no asphaltic-concrete paving work, the general contractor will likely elect to subcontract directly with a striping subcontractor.

12. Projects in city centers or other dense urban environments will have pedestrian barricades that must be painted at the onset of the project. These barricades must be maintained with touch-up paint to cover graffiti and other damage that might be incurred. Some general contractors prefer to self-perform all of this painting work with their own laborers, but it is more common for the painting subcontractor to initially paint the barricades. Regardless of which party paints the barricades, the general contractors often elect to maintain the barricades and perform any necessary touch-up painting with their own labor forces. The general contractors prefer this method in many instances because they will be on site every day, whereas the painting subcontractor will not. This will avoid requiring the painting subcontractor to make unscheduled, and costly, trips every time someone applies graffiti on the barricades. Graffiti can be a very frequent occurrence when working in a city center.

13. CMU blocks have a rough and porous surface requiring a block filler to provide a suitable substrate for the finish painting coats. This block filler should be the responsibility of the painting subcontractor. The painting subcontractor proposals should be carefully reviewed to ensure that this work has been included, as it is occasionally omitted.

14. Some wall surfaces receive wallpaper or fabric. These wall coverings need to be applied prior to casework, wainscot, base trim, crown molding, fire extinguisher cabinets, or any other wall-mounted items. The reason for this is to ensure good quality work. The wall-mounted items will cover and protect the paper or fabric edge, and prevent peeling (Figure 29.2). This is an important quality control issue for the general contractor to manage. If the wall-mounted items are installed first and the fabric or wallpaper is applied later, it will result in the edges of the fabric or wallpaper abutting up against the wall-mounted items with a raw edge. This will often result in many imperfections along the edges that will be quite visible.

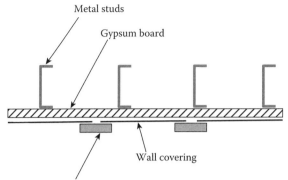

Metal studs

Gypsum board

Wall covering

Millwork trim should be used to protect the edges of
wall coverings whenever possible. Applying the
wall covering after the millwork trim has been installed
can leave rough and raw edges in the wall covering.

**FIGURE 29.2**   Wall covering protection.

15. As a good construction practice, and commonly as required by the project specifications, the gypsum board behind wall coverings should be primed. This is done so that when the wall coverings are removed in the future they will pull cleanly off the top of the primed surface. If the surface is not primed the wall covering adhesive will adhere directly to the paper face of the gypsum board and pull the gypsum board apart with it.

16. The design documents may not specifically state that the pattern of the wall coverings must be continuous and uninterrupted. Despite this, good construction practice dictates that these patterns should be matched to give a continuous appearance. The painting subcontractor must complete the wall covering work in a professional manner such that the walls (or ceilings) appear seamless.

17. As each coat of paint is applied, the work must be coordinated with other subcontractors. Painting is typically a three-coat system consisting of one prime coat and two finish coats.

   (a)   Since it is quite difficult to identify imperfections on a bare wall that has only been taped, the painting subcontractor will provide a sandable primer for all gypsum board walls. Once the walls have been primed, the monolithic color helps the naked eye in spotting wall imperfections. In essence, the primer coat is used as a tool for the taping crew to efficiently locate wall imperfections. The taping crew can then correct the flaws before the painter continues with the first and second coats of finish paint. With this procedure the painter will need to apply additional primer to touch-up the various spots where the tapers have made modifications. This touch-up work must be considered when the painting subcontractors formulate their bids. Though the amount of touch-up is a variable, painters regularly include this in their base bid. This is similar to caulking as it applied to the door frames, where the quantity and potential variation in the amount of work also is typically very small.

(i)  This fine tuning of the substrate is especially important when applying thin wall coverings, as there is no second chance to repair the substrate once the covering has been applied. Be sure walls have their imperfections corrected and the repair work primed and touched up.

(b)  Similar to wallpaper and fabrics, the first coat of finish paint should be applied before the installation of casework, acoustical ceiling grids, fire extinguisher cabinets, corner guards, and anything else that mounts on the wall. This will help ensure a consistent and complete finish. It is also much easier to apply a coat of paint to a surface before these various items are attached to it. When wall-mounted items are installed ahead of the finish paint, there is invariably a slight line of unfinished wall that will be visible around each item regardless of how good of a job the painter does in taping off these items. In addition, it is common for specifications to require all concealed surfaces to receive a minimum of one finished coat. This can only be met if the first coat is applied prior to attaching anything to the walls.

(c)  Slight dings and scratches in the wall can be expected to occur during the installation of casework and other wall-mounted items. These can be readily addressed if the second, and usually final, coat of finish paint is applied after these items are installed. This means the painting subcontractor will need to tape off all the casework and wall-mounted items before applying the second finish coat. The painting subcontractor must be fully informed of this procedure prior to bidding. Otherwise, the painting subcontractor may bid the work with the assumption that the two finish coats will be applied to the walls before casework or wall-mounted items are in place.

# 30 Ceramic Tile and Stone Flooring*

The ceramic tile and stone flooring (referred to here as ceramic tile) work is fairly straight forward. Although there are few problems with ceramic tile, a typical concern is whether the project schedule will be met. This is because restrooms involve more work that most other rooms and getting the work completed in the restrooms is often a key driver on the project schedule.

## SCOPE OF WORK ISSUES RELATED TO CERAMIC TILE AND STONE FLOORING

1. There are multiple ways to allocate interior stone work, each of which requires an educated and project-specific decision to be made by the general contractor and clearly relayed to the bidders via the bid instructions. Stone flooring is most commonly completed by the ceramic tile subcontractor, but masonry subcontractors are also capable of performing this work and often do so on projects with elaborate, high-end stone finishes. Generally, for a stone installation that is relatively simple, adhesively set, and primarily on the floor, the ceramic tile subcontractor will perform the work. This includes stone installations in shower surrounds, tub surrounds, and shower seats. For extensive stone cladding at the walls, the work is more commonly allocated to the masonry subcontractor. With stone cladding at walls and on floors, the stone and grout need to match perfectly, so it is best for the entire stone installation to be performed by a single subcontractor, most appropriately the masonry subcontractor.
   (a) If the wall stone is hung with mechanical anchors and not set with adhesives, the masonry subcontractor will invariably perform the work. Ceramic tile subcontractors are not commonly skilled in performing non-adhesive installations.
2. Exterior stone work to be set by an adhesive method can be allocated to either the ceramic tile or masonry subcontractors. It is usually preferable for the masonry subcontractor to complete this work, but if there is only a small amount of adhesively set exterior stone work it is most conveniently allocated to the ceramic tile subcontractor. If this small amount of work is assigned to the masonry subcontractor, they will not be able to perform it with their crew who may be on site setting CMU and clay brick, so a second

---

* MasterFormat Specifications Division 9

crew with specialized skills will need to be sent to the project for one or two days. This means there is another crew to coordinate with the overall construction operation. The ceramic tile subcontractors, who will surely have a crew on the project skilled in the adhesive set stone trade because of its similarity to ceramic tile setting, are invariably willing to include the exterior adhesive set stone in their scope of work.

3. The waterproofing at a shower, service sink, or other basin which is constructed of ceramic tile over cement board will have a waterproofing membrane below the tile, commonly referred to as a shower pan. Although it is often thought that the word pan implies a metal component, it is most often not metal. Metal components may be used as part of the system, but shower pans are predominantly a flexible membrane similar to a self-adhered sheet membrane. This shower pan will be installed by the ceramic tile subcontractor, not the waterproofing subcontractor. This is a common task performed by ceramic tile subcontractors and they need little direction. Although waterproofing components are often shown with exhaustive detailing, a shower pan is the exception. All that is needed is for the project specifications to identify the manufacturer and specific product for the membrane. The ceramic tile subcontractor can install it as per the manufacturer's standard details without any further direction from the design team.

4. Thresholds in doorways that lead to rooms with a ceramic tile floor are often made of marble or another stone. These stone thresholds are always provided and installed by the ceramic tile subcontractor. Though quite rare, stone thresholds may also be specified in locations where there is no ceramic tile or stone flooring. It is most convenient for this seemingly trivial work to also be assigned to the ceramic tile subcontractor.

5. Ceramic tile bidders regularly include caulking and sealants within and at the perimeter of their work. Conversely, the masonry subcontractors will commonly exclude the caulking and sealants work. The general contractor must clearly allocate the caulking and sealants for stone work to the masonry subcontractors. If unsuccessful in getting the masonry subcontractor to include this work, find another source, such as the caulking subcontractor, to complete it.

6. The specifications may call for a sealer at the tile or stone. It is best to include this in the ceramic tile subcontractor's scope of work. Some general contractors prefer to include this sealer in the final cleaning scope of work or to self-perform it. There is a particular advantage in having this sealer in the ceramic tile subcontract. The ceramic tile subcontractor will apply this sealer as soon as the grout has dried sufficiently after completing a room, thus eliminating an additional coordination effort by the general contractor. This approach is good for quality control. It should be noted that while the ceramic tile subcontractor will commonly list the sealer as an exclusion in their bid proposal, they can readily perform this work and can generally be persuaded to accept this responsibility.

(a)     There are exceptions to this rule. For instance, interior wall stone set by the masonry subcontractor rarely calls for a sealer, but when a sealer is required the masonry subcontractor should be held responsible. Also, for exterior stone the sealer specified will assuredly be water repellant, and as such, applied by the painting subcontractor.

# 31 Terrazzo and Epoxy Flooring*

High caliber buildings commonly use terrazzo flooring (a mixture of marble chips and colored cement) in the lobbies, grand stairs, porticos, and other signature locations. Epoxy flooring (Figure 31.1) is most common in rooms where the primary objective is to contain spilled or leaking water or other liquid.

Epoxy flooring comes in many forms, which differ primarily in the mode of application. The forms of application range from a simple roll-applied system, similar to a paint application, to a heavy duty, multi-layered, trowel-applied system. The type of epoxy flooring used will be based on several factors. Key factors would include the amount of traffic in the room, the likelihood of a spill or leak, and the type of liquids contained in the room. Suppose a mechanical room is located above a parking garage. A simple roll-applied epoxy floor would be considered adequate because a water leak, resulting in water accumulation in the parking garage, is not a big concern. Suppose a room is a heavily used laboratory where many harsh chemicals are kept. This floor should probably require a heavy duty trowel-applied epoxy floor system.

The terrazzo and epoxy flooring work will routinely be performed by different subcontractors. This occurs despite the fact that the work consists of quite similar fluid-applied flooring systems. It is because of the similarity of the work that these flooring treatments are discussed in the same chapter.

## SCOPE OF WORK ISSUES RELATED TO TERRAZZO FLOORING AND EPOXY FLOORING

1. The terrazzo work on a project often requires a considerable duration and can have a significant impact on the project schedule. A common error in bidding and planning is a failure to recognize the amount of time that will actually be required. Terrazzo is a high-end finish that is commonly specified for building lobbies or other primary entrances. Coincidentally, these will also be the primary construction entrances. Since terrazzo is commonly a large, time-consuming, and tremendously messy operation, when the terrazzo installation is in progress access through the area (likely the primary construction entrance) will be restricted. To avoid hampering access of other trades the terrazzo work is often phased such that the installation of the terrazzo is completed on one side of the entrance

---

* MasterFormat Specifications Division 9

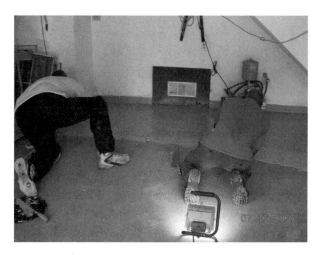

**FIGURE 31.1**    Epoxy flooring application. (Photo provided by Hi-Tech Flooring.)

while construction traffic remains uninterrupted on the other side. After the terrazzo work is completed on the first side, the terrazzo work shifts to the unfinished side while traffic reverts to the completed side. This phasing is costly, and thus needs to be clearly relayed to the terrazzo bidders during the bidding phase. Also, the project schedule issued with the general contractor's bid instructions must clearly indicate the duration allotted to perform the terrazzo work, the terrazzo completion date, and the impact this work will have on the other trades. Subcontractors representing other trades need to be informed if the terrazzo work will compromise their efficiency so that any inefficiency costs can be included in their base bids.

A similar mistake commonly made is not allowing sufficient time in the schedule for the epoxy flooring work. This occurs when it is errantly assumed the epoxy flooring is a simple roll-applied process instead of a multiple-coat trowel-applied system. Since the roll-applied epoxy systems are by far the most common, schedulers often fail to recognize that some flooring applications are complex and require a longer duration. A thorough review of the specifications prior to completing the project schedule will avoid such errors.

2. All caulking and sealants within and at the perimeter of the terrazzo and epoxy flooring should be the responsibility of the terrazzo and epoxy flooring subcontractors, respectively.

3. Terrazzo and epoxy flooring systems are both slippery when wet. These flooring systems are generally smooth or polished, but occasionally they are to be textured when greater slip resistance is desired. Review the drawings and specifications to verify the correct floor finish. In some cases, the epoxy flooring in one portion of a room must be textured while in another portion of the room it is to be smooth. This may occur in food service areas where a smooth finish is specified underneath fixed appliances and tables

because a smooth finish is easier to clean. A textured finish may then be specified at all walking areas to provide superior slip resistance.

4. Since many different subcontractors work hectically in close proximity to each other, minor construction damage to all finished systems should be expected. Each subcontractor installing finishes should maintain a small allowance for touch-up work, the value of which should be established by the general contractor in their bid instructions. Keep in mind that minor touch-up on terrazzo will involve grinding and buffing, so the allowance should be more substantial for this work. The terrazzo subcontractor's allowance must also allow for consequential work such as protection and cleanup (the dust generated will be widespread).

5. Transition strips from the terrazzo or epoxy flooring to an adjacent carpet, tile, resilient, wood, or any other type of flooring is best provided by the adjacent subcontractor. Conditions where terrazzo meets epoxy are rare. If this does occur, a project-specific decision regarding the responsible sub-contractor must be conveyed to the bidders in the bid instructions.

   If a full threshold (Figure 31.2) is specified in lieu of a transition strip (Figure 31.3), it should be furnished and installed by the door subcon-tractor. This is an easy allocation because the threshold will be specifi-cally noted in the door hardware schedule, whereas the transition strip will not.

6. It is best for all flooring subcontractors to furnish and install their own waxes and sealers. Although some general contractors prefer to self-perform this work, and still others assign this work to the final cleaning subcontrac-tor. All are acceptable and common practices, while the aforementioned approach of having the flooring subcontractors perform this work is most commonly preferred.

7. Terrazzo and epoxy flooring is protected during construction with simple kraft paper, which the terrazzo and/or epoxy flooring subcontractors should generally install. Since the terrazzo and epoxy flooring subcontractors will not be on site through the end of the project, the general contractor should maintain the protection and assume responsibility for subsequently remov-ing and disposing of the kraft paper.

   (a) At heavy traffic areas, such as the terrazzo entrance of a main build-ing, it is preferred to lay kraft paper aisles with Masonite (hard pressed board) on top of the kraft paper. The Masonite will provide a durable walking surface while the kraft paper will prevent minor abrasions from the rough underside of the Masonite board. It is further sug-gested to use delineators and caution tape along each side of the aisles. This is to keep workers from straying off the protected floor.

8. Terrazzo will commonly have odd-shaped divider strips creating a mosaic pattern or some type of lettering. Sometimes, these can be complex layouts. As with most other trades, the terrazzo subcontractor is to be responsible for their own layout in the field. Additionally, it should be clear whether the architect or the subcontractor is to provide a detailed and dimensioned layout plan. If this dimensioning is not clearly depicted in the architectural

**FIGURE 31.2**    Threshold. (Photo by author, courtesy of Hathaway Dinwiddie Construction Company and The University of Southern California.)

**FIGURE 31.3**    Transition strip. (Photo by author, courtesy of Hathaway Dinwiddie Construction Company and The University of Southern California.)

drawings, it is implied that the subcontractor will provide this information in their shop drawings.

9. Terrazzo and epoxy flooring subcontractors must perform moisture testing at the concrete slabs within their work. This test is to verify that there is an acceptably low moisture level to ensure that there will be no adhesion failures. (Moisture testing is discussed in greater detail in the chapter on carpeting and resilient flooring.)

10. The general contractor must provide the work areas for the terrazzo and epoxy flooring subcontractors in a broom-swept condition. The flooring surface is to be free and clear of any materials, equipment, or other items that might get in their way. The responsibility for any surface preparation beyond this broom-clean condition lies with the terrazzo and epoxy flooring subcontractors.

# 32 Acoustical Treatment*

For acoustical treatment work (referred to here as acoustical or ceiling work) it is relatively easy to define the scope of work. As a result, there are few change order that are associated with acoustical work. This work primarily consists of acoustical ceilings, while acoustical wall panels and ceiling clouds are also fairly common elements found on projects for which the acoustical subcontractor will be responsible. Ceiling clouds are horizontal panels of varying shapes constructed of acoustical materials and suspended below a ceiling. The actual assembly of acoustical panels and clouds vary greatly, but the basic assembly consists of a rigid acoustical board insulation wrapped with a fabric. For this basic assembly, the insulation board provides the acoustical properties and the fabric contributes to a decorative finished product.

## SCOPE OF WORK ISSUES RELATED TO ACOUSTICAL TREATMENT

1. Although few common change order issues are related to acoustical work, one issue seems to arise much more frequently than others. This relates to the responsibility for furnishing and installing seismic support wires for the lights and diffusers which are installed in the acoustical ceiling grid. These wires are regularly excluded by the acoustical, mechanical, and electrical subcontractors. The most practical approach is for the acoustical subcontractor to provide these wires in conjunction with the ceiling grid hanger wires. The mechanical and electrical subcontractors then maintain responsibility for connecting the wires to their lights, diffusers, and other ceiling mounted items. This seismic restraint is a common requirement, so all products designed to be mounted in a T-bar ceiling grid are manufactured with tie-off points at the corners.

2. Some acoustical subcontractors prefer to drop their ceiling wires through the metal decking prior to concrete placement. Unless this is absolutely unavoidable, for safety reasons this practice should not be allowed (Figure 32.1). The resultant grid of thick, pointed wires extending down from the ceiling is a serious safety hazard. These wires will pose a hazard to all workers installing overhead work. It is safer if these wires are left out until immediately prior to the ceiling installation. The wires will then be attached to the structural deck above by shot pins (Figure 32.2).

   (a) Dropping the suspending wires through the deck prior to the concrete pour is unquestionably the easiest approach for the acoustical ceiling subcontractor. They may even qualify their proposal with respect

---

* MasterFormat Specifications Division 9

**FIGURE 32.1**  Dropping ceiling wires through metal decking. (Photo by author, courtesy of Hathaway Dinwiddie Construction Company and Tishman Speyer.)

**FIGURE 32.2**  Acoustical ceiling wire hung with shot pin. (Photo by author, courtesy of Hathaway Dinwiddie Construction Company and The California Institute of Technology.)

to this method, so the bid proposals should be reviewed carefully to weed out this qualification. If the ceiling subcontractor is not informed about the restriction on this practice before contract award, a change order request will be imminent.

(b)  Installing the ceiling wires with shot pins is a widely used and well-accepted method of installation. Despite this, it can be argued that dropping the wires through the metal decking is a stronger attachment method. For this reason, some specifications will be encountered that specifically require wires to be dropped through the deck, specifically prohibiting the use of shot pins. The California Division of the State Architect is one of the governing bodies with this requirement. Thus, every new public building in California has ceiling wires anchored through the concrete deck above.

(c)  When the ceiling wires are dropped through the metal decking, the ceiling subcontractor must be responsible for bundling the wires. This is a good safety practice.

3. When acoustical ceiling wires are dropped through the decking ahead of the concrete, the wires will be suspended for a considerable time and many other activities will take place while these wires are dangling loose. Although the practice is discouraged, the MEP and other subcontractors with above ceiling work frequently cut the ceiling wires as their work progresses. They regard this cutting of wires as a necessity for the efficient execution of the above ceiling work, but this will obviously irritate the ceiling subcontractor who will bear the added cost of reinstalling the wires. If this issue is not properly addressed prior to bid, the ceiling subcontractor can be expected to submit a change order request to replace the cut ceiling wires, which depending on specific project conditions may or may not be a valid request. The acoustical ceiling subcontractor must be held responsible for coordinating their work with the other trades, but the other trades must make an effort to avoid needless cutting of the ceiling subcontractors support wires. This contractual language is a difficult aspect to address in the bid instructions and will vary from project to project depending primarily on the degree of difficulty of the above ceiling work. For instance, MEP systems above the ceiling in a hospital are extensive and cannot feasibly be coordinated such that all of this work avoids the ceiling wires. In such a case, the acoustical ceiling subcontractor must plan to replace many wires. In an office building the above ceiling MEP work is comparatively small, therefore very few of the ceiling wires should be cut. With few wires likely to be cut by diligent MEP subcontractors, a change order debate is highly unlikely.

4. Many subcontractors, particularly the mechanical and electrical subcontractors, will require above ceiling access for various reasons after the acoustical ceiling tiles are in place. They need to access air balancing dampers, air terminal boxes, pull boxes, and a myriad of other items. It is best to hold the subcontractor requiring access responsible for removing and replacing the ceiling tiles. This must be addressed in the bid instructions or the MEP subcontractors will remove countless tiles without replacing them. Because the edges are easily chipped, ceiling tiles are somewhat difficult to install, which is why the MEP subcontractors are reluctant to replace ceiling tiles they remove.

5. For rooms under positive air pressure, the acoustical ceiling system will be specified to have all tiles clipped to the ceiling grid. When this is specified, all but one of the tiles in a room will be clipped. To clip the tiles, a worker needs to be above the grid, but they cannot get above the grid with all the tiles in place. As a result, the tiles will be installed by starting in

one corner of the room and clipped one at a time all the way back to the opposite corner. The last tile will be set in place with a weight on top of it. Positive air pressure is common on projects where the level of cleanliness in a room must be very high. Examples include hospital operating rooms, semi-conductor clean rooms, and biotech manufacturing facilities. When a door to one of these positively pressured rooms is opened, the clean air from within the room will rush out of the room. Conversely, a room under balanced or negative air pressure would allow or draw the dirty exterior air into the room.

6. Ceiling wires have a code-required minimum spacing. When ductwork or equipment has a width or length exceeding this minimum spacing, the ceiling subcontractor must bridge under the obstructions to maintain the minimum spacing. The first inclination of the ceiling subcontractor will be to hang wires off the duct or equipment supports. This is not a commonly accepted or good construction practice. The ceiling subcontractor should be required to provide independent supports for this bridging work.

7. The construction documents will normally call for ceiling grids to be centered in a room. This requirement may only be observed by means of drawings that graphically depict a centered grid on the reflected ceiling drawings. It is a commonly accepted construction practice to center the grid on projects and this does not pose an issue for most ceiling subcontractors. It must be recognized that centering the grid in a room will double the number of ceiling tiles that need to be cut around the perimeter, thus, adding to the subcontractors cost. There are some ceiling subcontractors who will not perform this additional work unless the requirement is specifically provided in writing, making this an important requirement to address in the bid instructions.

8. If a wood ceiling is a mass-manufactured product constructed typical to a lay-in acoustical grid system, it will be the responsibility of the acoustical ceiling subcontractor. If a wood ceiling is a custom wood application that does not require a T-bar grid, the millwork subcontractor will furnish and install this system.

9. Light troughs commonly require an egg crate lens (Figure 32.3). This is a light lens, not a ceiling, so the ceiling subcontractor will exclude this work. At the same time, this lens is not part of the light fixture itself, so the electrical subcontractor will exclude this work as well. Since this egg crate lens is installed typical to an acoustical ceiling and lays into an acoustical ceiling L metal, it is most appropriate to allocate this work to the acoustical ceiling subcontractor.

10. If insulation is designed for installation simply lain on top of a grid ceiling, it is commonly installed by the ceiling subcontractor; though in some cases, at the general contractor's discretion, the insulation subcontractor may perform this work. There is no union jurisdictional problem between the ceiling and insulation subcontractors for this work.

11. The ceiling subcontractor will maintain responsibility for any cut-outs in the ceiling tiles. These cut-outs might be required to receive recessed can

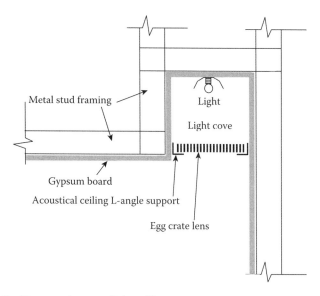

**FIGURE 32.3** Egg crate lens at a light soffit.

lighting or other items. The electrical, or other, subcontractor will then install their items in the pre-cut openings. An exception arises when drilling small holes in a ceiling panel for wiring. Such wiring might be for a smoke detector or motion detector. These holes are commonly made by the subcontractor installing the device. The rule of thumb is that if a hole in an acoustical ceiling panel can be made with a small drill bit, the subcontractor needing the hole will drill it. If the hole in the ceiling is made with a knife or blade, the ceiling subcontractor will provide the cut-out.

12. Acoustical wall panels will normally be installed on cleats which are attached to studs. With this method of installation, wall backing provided by the framing subcontractor is not required because the cleats run the full width of the panels and can be attached directly to the studs through the gypsum board. It is important to clarify this in the bid instructions, because a framing bidder may increase their bid with the assumption that backing will be required for the acoustical panels.

13. The acoustical paneling, like other finish items, should be installed after all dirty work on a project is completed. General contractors sometimes schedule these panels to be installed as soon as the walls are ready for them. This may be before the dirty work is complete. If work proceeds without dust protection, extensive cleaning of each panel may be necessary. It is normally acceptable to install these panels somewhat early if they are immediately covered with visqueen.

(a) The cleats will attach directly to the studs. Unfortunately, the studs are difficult to locate once a wall is painted. As a result, it is suggested that the cleats be installed prior to painting, and hang the panels later. With this approach, the acoustical subcontractor will be able to easily locate

the studs via the strips of taping compound and the panels will not be
exposed to dusty conditions. This will require multiple mobilizations
by the acoustical subcontractor. Since this approach is a method of
providing an efficient installation for the subcontractor, a change order
issue is not likely to occur. If a change order request is made, the general contractor can address the request by simply asking the acoustical
subcontractor to install the cleats after the walls are painted. Another
method is to simply avoid painting the walls behind acoustical panels,
but this is not always allowed by architects or owners.

14. In a performing arts theater or arena, acoustical "clouds" hanging from the
ceiling are common. These are to address acoustical concerns and to serve
as ornamental architectural elements. The method of attachment and lateral
(seismic) bracing is an important scope of work to define. The work will be
allocated to either the acoustical subcontractor or the miscellaneous metals
subcontractor. Each project is considerably different in how these elements
are hung. Projects differ by (1) roof/ceiling structure, (2) configuration
of the panels to the structural steel, trusses, or concrete deck, (3) length
they are suspended from the ceiling, (4) size and weight of the panels, and
(5) architectural design. The general contractor must make project-specific
decisions with regards to the allocation of this work, as ceiling subcontractors will not have the ability to provide the steel structures for large, heavy
panel systems suspended a significant distance from the underside of the
deck above.

# 33  Carpeting and Resilient Flooring*

The carpeting and resilient flooring (referred to here as flooring) work is the favorite work of many general contractors. This is because flooring is one of the last finishes to be installed. By the time flooring is installed, the walls have been painted, ceilings are complete, casework is set, and MEP systems are functioning. At this point the project is close to substantial completion and owner occupancy. Thus, for the general contractor, the flooring work is a sign that project completion is imminent. Other good aspects of flooring work include few change order issues and the duration is rarely a cause for concern. Despite these comments, the flooring work must still be run efficiently and it must be properly planned.

Separate bids may be submitted for carpeting and resilient flooring on lump sum competitively bid public sector projects, but it is more common for carpeting and resilient flooring to be bid as a package. On private projects, a general contractor will rarely award these scopes separately. Though these two types of floor coverings differ significantly, the installations are virtually identical. Regardless of the material, this work essentially consists of covering the floor with either a sheet or tiled product.

## SCOPE OF WORK ISSUES RELATED TO CARPETING AND RESILIENT FLOORING

1. The most common scope item missed by flooring subcontractors is probably the contrasting stripes at the top and bottom of carpeted stair treads. These contrasting stripes are not always shown on the contract drawings, but are required by the Americans with Disabilities Act. To avoid any confusion, this work should be clearly communicated in the instructions to bidders. The means of installation can be very difficult and must also be agreed upon. Contrasting stripes are commonly one inch back from the front edge of the tread. With thick carpeting, this 90°+ turn is difficult to make with a one inch sliver strip of carpet to hold it down on the tread. Most skilled flooring subcontractors can make this work, but it is not easy. This stair edge is exposed to heavy foot traffic and needs to be durable. If not properly constructed, the carpet strip may begin to pop up a short time after installation.

---

\* MasterFormat Specifications Division 9

2. The most common cause for flooring adhesion failure is when the flooring is adhered to a concrete slab with high moisture content. This moisture presence can be attributed to several causes:
   (a)  A new concrete slab that has not cured sufficiently.
   (b)  A new or existing concrete slab on grade constructed without a proper vapor barrier below it.
   (c)  A new concrete slab on grade constructed with a waterproofing membrane or vapor barrier, but the concrete was left exposed during rain storms.

   The following is an example of a new concrete slab over a vapor barrier that was left exposed to the rain. On this project, rainwater that ponded on the concrete slab, began to seep through the concrete and eventually collected on top of the vapor barrier. Since the water could not migrate past the vapor barrier, it remained trapped above the vapor barrier until it evaporated. This evaporation took place through the porous concrete slab. This process took a considerable amount of time. Before the water had all evaporated, the sheet vinyl flooring was installed. The flooring installation then acted as a trap to prevent the water from escaping through the top of the slab. As time went on, the water vaporized and these vapors seeped up through the concrete causing the adhesive to fail. The gaseous water vapors also caused bubbling in the sheet vinyl.

   To address this problem, the flooring subcontractor should not place the flooring until there is minimal moisture in the concrete. This drying process can obviously have a tremendous schedule impact. A calcium chloride test or in-situ relative humidity test, appropriately performed by the respective flooring subcontractor (Figure 33.1), can determine if the moisture content is low enough to meet the respective manufacturer's recommendations. If the moisture readings are higher than allowed by the manufacturer, one

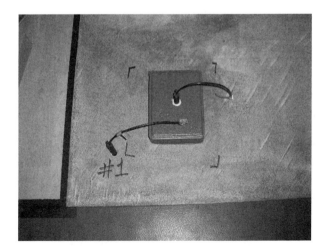

**FIGURE 33.1**   Slab moisture testing. (Photo provided by George Donnelly Testing and Inspections.)

remedy which does not significantly impact the schedule will be to seal the concrete; this is a costly solution.

The responsibility for the cost of sealing the concrete will vary from project to project and situation to situation. Different allocations of this responsibility may be as follows:

(a) High moisture content in an existing slab on grade is most commonly the owner's responsibility as an existing condition. In the case of existing slabs, it is recommended that the moisture testing be conducted before bidding the project. The areas that fail the test can be identified. Applying the concrete sealer can be included in the bid package to take advantage of the competitive bidding process. Otherwise, the sealing costs will need to be subsequently addressed through the costly change order process.

(b) On a well run project, a new slab on grade will be the owner's responsibility as a force majeure. This will apply if the general contractor has enclosed the building and got the HVAC air moving through the building within a reasonable timeframe as established in the project schedule. Dealing with moisture content in concrete is not an exact science or something that can be well managed even by the most diligent of general contractors. It has become an industry standard to consider excess moisture in concrete as a force majeure or act of God. On the other hand, suppose the general contractor was contractually bound to have the building enclosed and air moving six months before flooring installation. In actuality, the enclosure and air movement did not begin until one month before flooring installation. The general contractor could be held liable for problems arising from excess moisture in the concrete because if they had gotten the building enclosed and air moving through the building on schedule it could be speculated that moisture problems would have been averted.

3. The elevator subcontractor will provide the wall and ceiling finishes within the elevator cabs, but the floor finish is typically excluded. Flooring subcontractors should always examine the elevator details and finish schedule to verify the type of flooring required. This should also be discussed with the flooring subcontractor before awarding the subcontract to verify that this has been included. It is an industry standard practice for the elevator subcontractors to exclude the flooring work in the elevator cabs. This practice has evolved because architects routinely want the elevator flooring to match the flooring in the elevator lobbies. This can be best assured if the flooring materials are provided by the same subcontractor. The elevator cab walls and ceilings do not commonly need to match any finishes outside the elevator cab. As a result, the finishes for walls and ceilings have remained within the realm of the elevator subcontractor's scope of work.

4. The flooring subcontractor will provide all flooring base at their work, except for wood base. The wood base will most commonly be provided by the millwork subcontractor. Under some circumstances as described in the chapter on casework and millwork, the wood flooring subcontractor will

provide the wood base at carpet or resilient flooring locations. An exception to this rule that is important to address in the bid instructions occurs when a small amount of 1″ × 1/4″ wood base is specified (as on small residential projects). The flooring subcontractors will provide and install this small wood base trim.

5. The specifications may require pre-molded inside and outside corners for the rubber base, as opposed to simply bending the straight rubber base around the corners. The flooring subcontractor will assuredly provide these rubber base fabrications in their base bid because their estimators are known to diligently review the specifications. In addition, both the flooring subcontractor and general contractor, as part of their quality control effort, must ensure that the crew actually installs the pre-molded fabrications in the field. These fabrications might be omitted by the crew in the field because they do not often look at the specifications. Both the general contractor and flooring subcontractor need to keep the field crew informed about these types of requirements.

6. The construction tolerances for flatness of a structural deck and flatness of a finish flooring system differ significantly. It is best to hold the flooring sub-contractor responsible for floating out the structural deck to meet the industry standard tolerances for the flooring system. The extent of this work can be quite variable. It is recommended that the general contractor establish an estimated quantity for this work. The flooring subcontractor should then provide a base bid for this estimated work and a unit cost for any additive or deductive floating that varies from the estimate. This is discussed in greater detail in the chapter on wood flooring.

7. It is preferable for the flooring subcontractor to install protection for the flooring surface after the work is completed. This protection should be provided according to an estimated quantity or rough plan provided by the general contractor. Then the general contractor should maintain the protection as self-performed work and remove it when the protection is no longer needed. In this way, the flooring subcontractor will not need to include additional costs for policing the protection and making a special trip to roll up and dispose of the kraft paper protection.

8. Many interior floor expansion joints are designed with a recess on top of them to receive a carpet or resilient flooring inlay. This flooring inlay detail may only be identified in the contract documents with the single respective expansion joint detail or it may only be found by looking up the product data for the make/model of the expansion joint identified in the specifications. The flooring bidders will seldom look for this information. A diligent general contractor will specifically look for this information during the bidding phase and address it in the bid instructions. Ultimately, the flooring subcontractor will be responsible for this work. Prompt communication to the flooring subcontractor(s) about this expansion joint inlay will avoid disputes that might otherwise occur.

# 34 Wood Flooring*

Wood flooring is a comparatively expensive flooring system. There are two primary wood flooring types (Figure 34.1), and each of these types has many variations. One type is the traditional hardwood floor that is installed on a system of wood sleepers. The other type is manufactured laminate wood flooring laid on top of resilient acoustical matting, which is in turn directly on top of the wood or concrete subfloor. The latter type is gaining in popularity due to its lower cost, high acoustical value, and ease of installation. Wood flooring, even a manufactured wood flooring system, is a specialty trade item. It is always provided and installed by a wood flooring subcontractor, not the millwork or another flooring subcontractor.

## SCOPE OF WORK ISSUES RELATED TO WOOD FLOORING

1. Just as the carpeting and resilient flooring subcontractor is generally responsible for conducting moisture tests prior to installing the flooring, the wood flooring subcontractor also needs to perform such tests. High moisture content can cause the wood flooring to swell and buckle. In the case of a laminate floor, moisture in the subflooring can even cause delamination. Moisture testing is discussed in greater detail in the chapter on carpeting and resilient flooring.
2. The wood flooring subcontractor should be responsible for everything from the top of the structural deck to the final finish (wax) of the wood flooring. This will include resilient underlayment, leveling compounds, sleepers, moisture barriers, stains, varnishes, wax, or any other required treatment. To understand the exact composition of the wood flooring system, conduct a careful review of the manufacturer's product data. This often provides more information than is provided in the architectural drawings.
3. Wood flooring systems of differing manufacturers will vary in thickness. Be sure to verify that the thickness of the system qualified by each bidder closely matches the thickness of the system described in the construction drawings. The specifications will typically identify five or six acceptable manufacturers of wood flooring for use on the project. Only one of these types of flooring (usually identified by being bolded or by being listed first) is used by the architect as a basis of design. Each of these systems will have a slightly different total thickness, which will affect the finish floor elevation and how the wood flooring transitions to adjacent flooring systems.

---

* MasterFormat Specifications Division 9

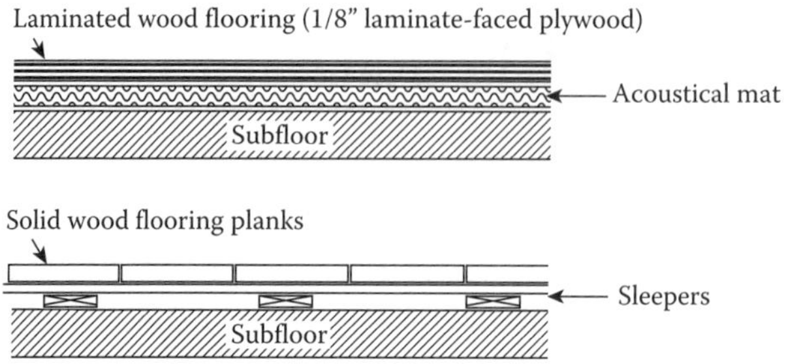

Laminated wood flooring (1/8" laminate-faced plywood)

Acoustical mat

Subfloor

Solid wood flooring planks

Sleepers

Subfloor

**FIGURE 34.1**   Wood flooring.

(a)   The most severe wood flooring transitions are generally those next to exposed concrete or sheet vinyl. The major cause of concern exists when the wood flooring system proposed by the bidder is too thick for a code compliant transition to the adjacent flooring type. The ridge between the flooring types might be too steep and violate the Americans with Disabilities Act (ADA) mandated maximum height gradient.

(b)   In many cases, gypcrete might be shown as a means to raise the adjacent flooring level to the height of the wood flooring system. This is necessary for the thicker wood flooring systems and is most common when the wood flooring system is placed on a sleeper system or on a thick resilient underlayment. Laminate wood flooring systems on top of a thin underlayment are generally thin enough to maintain a code compliant transition to adjacent surfaces. The wood flooring system being used by the flooring subcontractor may differ slightly in thickness from the type of flooring that was used as the basis for the design on the drawings. This will change the scope of work for the gypcrete subcontractor by the same amount that the flooring system thickness differs from the basis of design system. Thus, the actual required total thickness of the gypcrete will probably be a little thicker or thinner than specified. If gypcrete bids arrive at the same time as the wood flooring bids, there will generally be insufficient time to properly address this issue prior to bidding. Fortunately, there will still be sufficient time to analyze and address this issue prior to actually awarding the subcontracts.

4. The wood flooring subcontractor will furnish all wood base trim that is associated with the wood flooring. Conversely, the millwork subcontractor will provide all wood base trim which occurs with all other flooring types. The allocation of the wood base trim is relatively straightforward and this is now a well-accepted industry standard to which the various subcontractors are well accustomed. As might be expected, an exception still arises. Consider rooms or a continuation of rooms and hallways that are not separated by a door or other dividing element. Under such conditions, the wood base trim in these areas should be furnished and installed in its entirety by a single subcontractor for continuity. This will ensure that there are no

size variations in the trim and that there are no color matching issues. The driving factor in this scope of work allocation is that it is most important to have a good match between the wood base trim and the wood floor. This can be best assured if the entire wood base trim is furnished and installed by the wood flooring subcontractor, rather than the millwork subcontractor. Typically, the wood flooring subcontractors do not mind doing this additional work. The general contractor just needs to inform the wood flooring subcontractor about this additional scope of work prior to bidding.

5. The wood flooring subcontractor should be held responsible for all staining, varnishing, sealing, waxing, or any other finish specified for the wood flooring and base. These subcontractors are accustomed to this allocation of work and will almost always include it for mass-manufactured laminate flooring. This is almost automatic because the manufactured systems routinely come pre-finished from the factory. For a traditional solid wood flooring system, the subcontractor will often exclude the finishing. This exclusion is commonly made because it involves finishing the flooring on site. The painting subcontractor is another viable candidate for this finishing work. As a general practice, the wood flooring subcontractor should be responsible for finishing their own work, even when it requires the employment of a lower-tier subcontractor. As a result, the allocation of this finish work to the painting subcontractor should be made with care.

   (a)  If the wood flooring is designed to match the millwork, cabinetry, or other wood elements, be sure to closely coordinate the finishes among the appropriate subcontractors. This matching can be a difficult task, especially when dealing with mass-manufactured laminate flooring. An exception to the rule discussed above would be for traditional solid wood flooring that is scheduled to match the casework and millwork. It would be beneficial for the painting subcontractor to finish the wood flooring, in addition to the millwork, to provide for a consistent finished product.

6. The direction of the grain in wood finishes is often an important architectural feature. Despite this, the direction of the grain is not commonly shown on the architectural drawings. Instead, there is typically a single sentence in the specifications that states that all grains are to run in the same direction, but a problem persists in that crews who install the wood flooring in the field do not always review the specifications before they begin to work. They might decide to install the flooring in the easiest direction, which is the direction in which the longest pieces can be used. This might result in the wood flooring segments being installed with the grain running perpendicular to an adjacent segment (Figure 34.2). This is an important issue to watch to maintain the desired quality.

7. The wood flooring subcontractor should have full responsibility for all cut-outs in the wood flooring system. Such cut-outs might be made for electrical floor boxes or for floor diffusers.

8. If the thresholds from the wood flooring to an adjacent floor finish are a wood product, the wood flooring subcontractor must furnish and install them. This

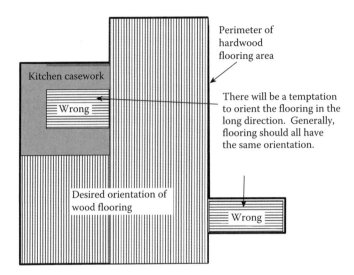

**FIGURE 34.2**   Wood flooring grain direction.

will help ensure a proper color match with the wood flooring. If the thresholds or transition strips are a material other than wood, the adjacent subcontractor should provide and install them, with one exception. The exception is if the transition is made with a threshold that is identified in the hardware specifications (or hardware schedule), in which case the door subcontractor will naturally provide it.

9. A common problem occurs from project to project in the definition of an accepted tolerance for flatness. A structural deck is accepted as being within industry standard tolerances when it is no more than 5/16″ out of plane in 10 feet. The industry standard tolerance for a wood flooring system is commonly 1/8″ out of plane in 10 feet. If the wood flooring is installed on top of a concrete deck it might be out of tolerance by 3/16″. (A wood floor installed directly on top of a concrete deck that is 5/16″ out of plane will cause the wood floor to be an equivalent 5/16″ out of plane. This is 3/16″ beyond the allowable tolerance for a wood flooring installation.) This assumes the structural deck in question has been constructed near the maximum deviation allowed for structural decks, which is out of plane beyond the accepted industry standard for wood flooring. To accommodate this difference, the concrete deck will need the low spots floated out with a leveling compound, and potentially the high spots ground down. This is done to bring the substrate to within the allowable level standard for wood flooring. The extent to which the structural deck is out of plane will not be known until the deck has been completed. Sufficient time must also pass to allow the deck to settle into place. Also, it must be recognized that an elevated structural deck will deflect continuously, though slightly, over time. With these variables, it is not possible for a bidder to put a firm lump sum value on this scope of work. It is suggested that prior to bidding the project an educated estimate

be made about the amount of preparatory work that is anticipated. The sub-contractors should be instructed to include this estimated amount in their base bids. The bid should also be provided with a unit cost for additive or deductive preparatory work. This typically comes in the form of a unit cost per square foot to be used for making final adjustments in the allowance amount.

(a)  The responsibility for preparatory work beyond the estimated quantity or the beneficiary for a deductive change order when the quantity is reduced is quite debatable. There is no standard industry practice regarding this. Since the deflection and settlement are largely unpredictable, it is most appropriate to consider this as a force majeure or act of God, and as such, the owner maintains responsibility. Since workmanship is a major contributor to the deflection and since deflection is inevitable, some believe the general contractor should anticipate the preparatory work in their base bid. The important point is to address this issue during prime contract negotiations and make sure the costs and risks are covered by one of the parties.

　(i)  Since the wood flooring subcontractor has no control over the concrete or structural steel subcontractors, it is not fair for the wood flooring subcontractor to be held responsible for this variable quantity of work.

(b)  It must be recognized that the preparatory work is quite expensive and the naked eye will not easily notice a 1/4″ deviation in floor flatness. This is especially true once a room is filled with furniture. As a result, some owners will reduce the flatness standard for their project as a value-engineered saving.

(c)  The issue of tolerance in flatness deviation is applicable to all flooring systems and many other finish systems. Similar issues arise in various other trades when structural components of a building meet finish components. The accepted industry standards for flatness or plumbness of structural components are significantly more lenient than those of finish components. This is a very important point and a common cause for change order debates among the various building trades.

10. As with other flooring subcontractors, it is preferable for the wood flooring subcontractor to furnish and install protection for their finished work. The general contractor should then maintain this protection as self-performed work and also remove it when it is no longer needed. Further, the general contractor needs to quantify the extent of required protection for bidding purposes, either by a square foot quantity or by providing a protection plan.

# Questions—Module Five (Chapters 24–34)

1. Identify five different tasks or types of work commonly performed by the framing subcontractor.
2. Which subcontractor typically provides the gypsum board product, such as DensDeck, between rigid roof insulation and PVC roof membrane?
3. Which subcontractor has sole responsibility for identifying and pricing in-wall backing?
4. Identify which of the following subcontractors (framing, casework, miscellaneous metals, or structural steel) will never have a part in furnishing or installing the low wall posts or steel counter support angles.
5. Which subcontractor should furnish access doors in gypsum board walls for domestic water valves and which subcontractor should install them?
6. Explain why allowances for undefined scopes of work, such as patching trade damage, should be in the form of a stipulated quantity of labor hours and dollar value for materials.
7. Which subcontractor will furnish and install the mesh tape at the cement board corners and flat joints backing a ceramic tile installation?
8. When a project has minimal or no insulation work outside of the stud-framed walls and ceilings, which subcontractor is generally responsible for this work? When there is a great deal of insulation occurring in locations other than within the framed walls and ceilings, which party generally takes the responsibility for getting this work done?
9. Discuss the statement that for simplistic scopes of work, such as building insulation, a quick cursory review of the proposal is all that is needed to ensure the scope of work is accurately covered.
10. Which subcontractor performs the fire safing at the heads of CMU walls?
11. Which subcontractor performs the spray-applied insulation work?
12. Identify four different subcontractors that may perform work for the doors, frames, and hardware scope of work on a public works project.
13. Which subcontractor furnishes the door hardware for steel gates and chain link gates? Which subcontractor installs them?
14. Which subcontractor will caulk the gaps from the hollow metal door and window frames to the gypsum board walls?
15. Fill in the blanks for the following statements:
    For casework, a stained finish will be completed by the _____ subcontractor, a painted finish will be primed by the _____ subcontractor, and a painted finish will be finish painted by the _____ subcontractor.

    For millwork, a stained finish will be completed by the _____ subcontractor, a painted finish will be primed by the _____

subcontractor, and a painted finish will be finish painted by the
_____ subcontractor.

   For doors, a stained finish will be completed by the _____
subcontractor, a painted finish will be primed by the _____
subcontractor, and a painted finish will be finish painted by the
_____ subcontractor.

16. Identify three different parties that may enter into direct contracts with the handicap door operator subcontractor.

17. Which subcontractor will furnish and install the power booster, with an integral low-voltage transformer, for the handicap door operator?

18. How is the push button for a handicap door operator at the exterior side of a storefront door commonly mounted if the mullion is too narrow for the button and its back box?

19. Which subcontractors will furnish the request to exit devices? (State the conditions for which a particular subcontractor will be responsible.)

20. Which subcontractor will install the electrified hinges for a wood door that are supplied by the security subcontractor?

21. Which party is preferred for the preparation of the keying tree?

22. Explain the purpose of the interlock between the coiling door motor and the manual coiling door lock.

23. What embedded items will need to be furnished and installed on large coiling doors located in a concrete wall? Which party will furnish this material?

24. What must wide coiling grilles always include to prevent the security issue of someone being able to pull up the center of the grille while another person slides underneath?

25. In what type of project is leaving out walls to maintain a clear path of travel for a large interior coiling door most common?

26. Which party should take the field measurements for the casework scope of work?

27. Which subcontractor provides the wood base when it occurs in conjunction with carpeting? Which subcontractor provides the wood base when it occurs in conjunction with wood flooring? Which subcontractor provides the wood base when it occurs in conjunction with sealed concrete? Which subcontractor provides the wood base when it occurs in an area with both wood flooring and sealed concrete?

28. Which of the following items (granite countertops, solid surface countertops, or plastic laminate countertops) should not be provided by the casework subcontractor (or a lower-tier subcontractor of theirs)?

29. Discuss the statement that on union projects the casework subcontractor cannot install the sinks, even in their shop, due to union jurisdictions.

30. Discuss the statement that if the casework subcontractor is not installing the sinks there is no need to rush the sink procurement, as they will not be needed until late in the project right before the plumber installs them.

31. If a painting bidder is not capable of completing wall covering work with their own forces, what should they do?

32. How extensive will the effort of the painting subcontractor be when reviewing construction documents for bidding?

33. Which party will provide the water repellants for a project? Which party will provide the graffiti coatings?

34. Discuss the statement that a Kynar painted finish on a hollow metal door will be completed in the factory, not on site by the painter.

35. Which subcontractor is responsible for filling and sanding nail holes in the millwork trim?

36. Which subcontractor will perform caulking around the door frames to the gypsum board walls?

37. Although caulking work at door frames is a variable at bid time, how likely is it that the general contractor will receive a change order request for this work?

38. Is it recommended that the general contractor allocate the cost of repairing and touching up minor trade damage by equally distributing the costs among all subcontractors on a project via unilateral backcharges because this is the easiest method of assigning blame?

39. Which party should be responsible for painting over graffiti at the pedestrian barricades?

40. Is it necessary to prime paint the gypsum board surface behind wall coverings?

41. The following activities occur at a gypsum board wall, put them in order:
    a. Second coat finish paint
    b. Install casework
    c. Touch-up taping imperfections
    d. Prime paint
    e. First coat finish paint

42. Explain why restrooms tend to drive project schedules more frequently than common offices, conference rooms, or bedrooms.

43. When there is an extensive amount of adhesive-set stone in a high-end building lobby, which party is most appropriate to perform this work?

44. Which subcontractor should provide shower pans below a ceramic tile surface?

45. Which subcontractors should perform the terrazzo and trowel-applied epoxy flooring work?

46. What type of finish (smooth or textured) should be provided under the counters in a commercial kitchen with an epoxy floor? What type of finish is best suited on walking surfaces in such a facility?

47. Which party should apply the floor sealer?

48. Which party should perform moisture testing of concrete slabs prior to flooring installation?

49. Which one of the following (acoustical panels, grid ceilings, ceiling clouds, acoustical wall insulation) will not be provided by the acoustical subcontractor?

50. Which party should furnish and install the ceiling wires used for seismic restraint of light fixtures in a T-bar ceiling grid? Which party should tie the wires to the lighting?

51. Discuss the statement that when an HVAC subcontractor removes an acoustical ceiling tile, they are also responsible for carefully replacing the ceiling tile such that the edges do not chip.

52. What is done when a large duct interferes with the acoustical ceiling wire spacing?

53. Which subcontractor is responsible for a mass-manufactured wood ceiling system designed to be laid into a T-bar ceiling grid?

54. Which subcontractor is responsible for cut-outs in an acoustical ceiling tile for a 6″ diameter recessed light?

55. Which subcontractor typically provides the supporting structure for large, heavy ceiling clouds suspended a significant distance from the underside of the structural deck above?

56. What is the most common cause for flooring adhesion failure on a concrete slab?

57. Which party should perform the moisture testing of the existing concrete decks on a renovation project?

58. With regards to elevator cabs, which subcontractor will provide the wall finishes? Which subcontractor will provide the flooring?

59. Discuss the statement that wood flooring will always be provided by the millwork subcontractor?

60. Which subcontractor should provide the leveling compounds under a wood flooring system?

61. Which subcontractor should provide the wood transition strip between a wood floor and adjacent non-wood floor? Which subcontractor should provide the threshold if it is on the hardware schedule? Which subcontractor should provide a plastic transition from a wood floor to a carpeted floor?

62. To provide a wood flooring installation within the acceptable tolerances for flatness on top of a concrete deck, which subcontractor should be required to provide a leveling compound to fill the low spots of the structural concrete deck?

63. Discuss the statement that the wood flooring subcontractor should be held fully responsible for all costs associated with filling low spots in a structural concrete deck in order to provide a wood flooring system within the appropriate industry standard tolerances for flatness.

# Module Six

# 35 Miscellaneous Specialty Work*

A construction project will have many miscellaneous components that do not fit within a major subcontractor's scope of work. This work is generally small in scope, inexpensive, or just plain easy to perform. This chapter will discuss these miscellaneous project elements.

## TOILET PARTITIONS AND ACCESSORIES

Toilet partitions and toilet accessories will typically be provided and installed by a single subcontractor. An exception is fairly frequent on medium to large hard-bid public works projects, where separate bids are common for each of these two scopes simply due to the nature of the public bidding environment and the large number of subcontractors trying to get even a small piece of work on the project.

1. If separate subcontractors are to provide and install the partitions and accessories, one important consideration will be to account for the cut-outs in the partitions for the accessories, as both of these subcontractors will probably exclude this work. The toilet partition subcontractor should always make the cut-outs because they are solely responsible for the structural integrity and final appearance of their partitions.
    (a) For solid plastic, phenolic, or wood partitions, making the cut-outs will be quite simple, but it is still advisable that the partition subcontractor make the cut-outs. It is best to avoid having the accessories subcontractor make the cut-outs in the field before partition installation, as this involves additional coordination. Similarly, avoid having the accessories subcontractor try to make these cut-outs with the partitions in place because workmanship quality can become an issue. Additionally, the toilet accessories subcontractor may not have skilled craftsman with the capability of satisfactorily modifying the partition panels. This capability may be a reason why a subcontractor only bids the accessories, and not also the partitions.
    (b) For metal partitions with a hollow core, the cut-outs will need to be reinforced. The most efficient means will be for the partition subcontractor to cut and reinforce the panels in the shop before shipping them to the project for Installation.

---

* MasterFormat Specifications Divisions 10, 11, and 12

2. All in-wall backing and blockouts for the partitions and accessories will be provided by the framing subcontractor. As a result, the accessories work should not entail any preliminary rough-in work prior to mobilizing for the final installation.

3. The restroom signage required by the Americans with Disabilities Act (ADA) should be provided by the toilet accessories subcontractor, not the signage subcontractor. This is an inexpensive scope of work generally consisting of the entry door signs and a sticker at the handicap stall doors.

4. Toilet partition fabrication will not commence until the subcontractor has verified the dimensions of the partitions by taking field measurements. Since ceramic tile, gypsum sheathing, and taping thicknesses vary little, most toilet partition subcontractors will take these field measurements when the rough framing is complete. The final dimensions are computed by simply subtracting the ceramic tile and gypsum board thickness from this overall dimension. This is an extremely important procurement issue, as field measurements for toilet partitions are not commonly possible until late in the project. Obviously, any procurement delays or fabrication errors in the toilet partitions could jeopardize the substantial completion date.

   (a) The toilet partitions subcontractor may prefer to take field measurements when the gypsum or ceramic tile work is complete. Unless there is the unlikely occurrence that time is not an issue in the scheduling path of this work, the general contractor should direct the toilet partitions subcontractor in writing to promptly take the field measurements based upon the rough framing dimensions. The gypsum and ceramic tile work will not commonly be complete until near the end of the project. Since the partitions can have a long procurement duration (12 weeks or longer have been noted), it is important to expedite these important project components.

   (b) The toilet partition subcontractor will not be on site prior to the installation of the partitions. As a result, they will not have firsthand knowledge of when the restrooms are ready for field measurements. The general contractor must manage this work by giving the partition subcontractor a courtesy call a couple weeks before the restrooms are ready for measurements. Another call must be made a couple of days ahead of time to confirm the restroom wall completion date. This will ensure that the partition subcontractor is on board and prepared to take the field measurements immediately upon framing completion. With a conscientious management effort, no time is wasted.

   (c) On a large project with many restrooms, field measurements should be taken in phases throughout the project in order for the partitions to be procured in phases. This phased work needs to be clearly shown in the master schedule issued prior to the bid date.

**FIGURE 35.1** Ceiling mounted toilet partition. (Photo by author, courtesy of Hathaway Dinwiddie Construction Company and The California Institute of Technology.)

(d) The general contractor must convey to all of the related trades the importance of adhering to the specified dimensions of the partitions for the gypsum board, taping, ceramic tile, and other wall finish work. This will help to ensure that the provisions for these systems calculated into the field measurements are accurate and that the accuracy is carried through with the subsequent construction activities. The general contractor should carefully monitor this work to help ensure the installed quality.

5. A common dispute between general contractors and architects relates to the support steel for ceiling-hung partitions (Figure 35.1). Unlike floor-mounted partitions, ceiling-hung partitions have a significant dead weight hanging from the ceiling which is too heavy for proper support by gauge metal ceiling joists and necessitates structural steel supports. Architects habitually deem the supporting steel a means and methods item similar to in-wall backing and, as with in-wall backing, will not regularly depict it on the drawings. Though general contractors typically disagree with this practice it is still an issue that must be addressed. The miscellaneous metals subcontractor is the appropriate party to design and install this support steel, so this work must be clearly included in their scope of work. If the general contractor addresses this item in the bid instructions, the routine debate mentioned above will not occur. This is truly the best way to solve a debate, by avoiding the debate altogether.

## TRASH CHUTES

Trash chutes are seemingly simple and are often disregarded by general contractors unfamiliar with them as not requiring any specific planning. In fact, work associated with trash chutes is quite convoluted and must be properly bid, planned, and executed to avoid costly change orders requested by subcontractors due to scope gaps (Figure 35.2).

Trash chutes are traditionally purchased by the general contractor and delivered FOB (which has two generally recognized definitions of freight on board or, more commonly, free on board) jobsite by the manufacturer. The HVAC subcontractor will typically install the chutes themselves, but there is a surprising amount of supplemental work also completed by various other subcontractors.

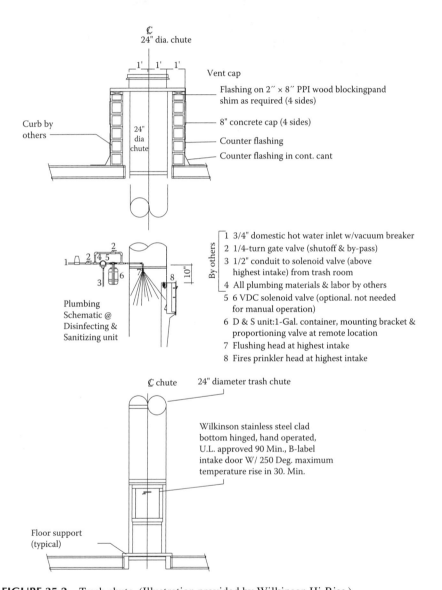

**FIGURE 35.2**    Trash chute. (Illustration provided by Wilkinson Hi-Rise.)

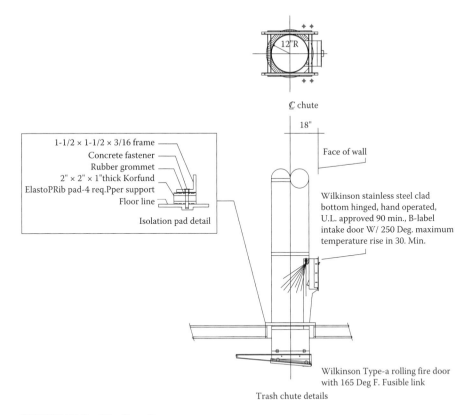

12"R

Ȼ chute

18"

Face of wall

1-1/2 × 1-1/2 × 3/16 frame
Concrete fastener
Rubber grommet
2" × 2" × 1"thick Korfund
ElastoPRib pad-4 req.Pper support
Floor line

Isolation pad detail

Wilkinson stainless steel clad
bottom hinged, hand operated,
U.L. approved 90 min., B-label
intake door W/ 250 Deg. maximum
temperature rise in 30. Min.

Wilkinson Type-a rolling fire door
with 165 Deg F. Fusible link

Trash chute details

**FIGURE 35.2**   (Continued)

1. It is important to address the issue of how trash chute installation work is
   properly allocated. This is not necessarily a simple task.
   (a) The trash chute supplier will furnish FOB jobsite the chute, doors,
       sanitizing spray heads, fire sprinkler heads, heat detectors, smoke
       detectors, and electrical door interlocks.
   (b) The HVAC subcontractor will install the chute itself, including the
       rain cap at the roof. While the HVAC subcontractor is generally capa-
       ble of also furnishing this rain cap, it should be furnished by the chute
       manufacturer. It is important to qualify this requirement, as chute
       manufacturers will sometimes exclude it.
   (c) When the design documents specify a wood roof curb it will be con-
       structed by the general contractor. If a concrete curb is specified, it
       should be installed by the concrete trades (formwork, rebar, and plac-
       ing and finishing). Some roofing systems, such as a PVC system, can
       be roofed right up the chute itself, but a built-up asphalt roofing system
       will typically require a curb. Although the necessity and details of
       the curb will vary from project to project, they will be depicted in
       the architectural drawings and the general contractor will be able to

properly and easily allocate the work based upon the information provided by the design team.

(d) The fire sprinkler subcontractor will provide branch piping and install the fire sprinkler heads in the chute. Cut-outs and offsets for the heads will already be prefabricated within the chute and, as mentioned above, the heads themselves will be furnished by the chute manufacturer.

(e) The plumbing subcontractor will provide branch piping and install the sanitizing heads, which, like the fire sprinkler heads, will be furnished by the chute manufacturer. The plumbing subcontractor will also install the sanitizing system, which could be floor or wall-mounted.

(f) For wall-mounted sanitizing systems the framing subcontractor will need to provide in-wall backing.

(g) The framing subcontractor will install the chute doors in conjunction with the framing and gypsum board work. This activity is similar to access doors for MEP or other in-wall items. In both cases, the doors are furnished by the subcontractor most closely aligned with the work and installed by the framing subcontractor. Access doors are discussed further in the chapter on framing.

(h) The electrical subcontractor will furnish and install all conduit, wire, and wiring terminations for the electrical door interlocks. The electrical subcontractor will also commission the electrical door interlocks. Only the interlock devices themselves and the maintenance key switch will be furnished by the chute supplier.

(i) The smoke and heat detectors will be furnished by the chute manufacturer. The installation, wiring, and commissioning will be by the fire alarm subcontractor. Further, the conduit for these devices will be provided by the electrical subcontractor.

As many as eleven different trades may be involved in the installation of trash chutes. This reveals the complexity of what would appear to be a simple activity and exemplifies the level of detail general contractors must get into during the bidding phase of a project. Underestimating the complexity of this or any other scope of work can create significant coordination problems.

2. The first scope gap with trash chutes often occurs when they arrive on site. This entails deciding who is responsible for taking them off the truck, uncrating them, and loading the chute sections and doors into the building. Since the HVAC subcontractor will be installing the sections, it is most appropriate for them to off-load the truck, uncrate the pallets, and place the chute sections at their respective locations of installation.

(a) After uncrating the pallets, the HVAC subcontractor will deliver the doors, sanitizing heads, sprinkler heads, door interlocks, and other peripherals to the general contractor. The general contractor will then notify the appropriate subcontractors to pick up these items.

3. Most jurisdictions require trash chute vents to extend three feet above the roof and to maintain the same chute diameter. Thus, a 24″ diameter trash chute will have a 24″ diameter vent extending 36″ above the roof.

(a) Some project conditions and local jurisdictions require trash chutes to extend 10 feet above the roof. Such extensions of the trash chutes will require guying. When a chute vent is required to extend 10 feet above the roof, the architectural documents should show a minimum requirement of three guy cables. The HVAC subcontractor will be responsible for completing the guying work including the means of mounting the guys through the roofing material. Since details will differ with different roofing systems, the architectural drawings should be reviewed to determine how to penetrate the specific roofing system used for the project. One effective and inexpensive method of penetrating the roofing material is with a small steel post, which the HVAC subcontractor should not have a problem with providing as long as they are aware of the requirement. This particular method is commonly accepted by roofing manufacturers.

4. Architects routinely design the blockouts for trash chutes to be the full width and length of the shaft. This is especially true for side-by-side chutes which occur in multi-family residential projects where one chute will be for trash and the other for recycling. Trash chute manufacturers typically furnish floor support brackets that are only capable of supporting the chute sections through a slightly oversized deck opening. For instance, a 24″ diameter chute will commonly be supplied with floor support brackets long enough to span across a 27″ wide deck penetration. If the deck penetration is not coordinated with the chute brackets, the chute brackets will be too short to span the deck penetration. When the floor support brackets are too short to reach the edge of the deck, the HVAC subcontractor will need to provide supplemental supports (or extensions) at every floor. The general contractor must verify that the brackets supplied by the chute manufacturer are the correct length to span the designed deck opening. This coordination problem needs to be addressed early in the project in one of three ways. The general contractor will either need to order longer supports for the chute, pay the HVAC subcontractor to furnish supplemental supports, or change the dimensions of the deck opening. Changing the size of the deck opening might seem to be a quick and simple solution, but if the trash chute chase is also to be utilized for piping or conduit risers, the general contractor will need to keep the larger deck opening. The only viable options might be to order larger chute support brackets or pay the HVAC subcontractor to provide supplemental supports.

## FIRE PROTECTION SPECIALTIES

1. The fire extinguishers and cabinets (FECs) can be furnished FOB jobsite by the supplier for installation by the general contractor or they can just as easily be furnished and installed by a specialty subcontractor. The task is very simple, consisting of little more than screwing them into place. Some general contractors will even have the framing subcontractor perform this work, which is another perfectly acceptable and efficient route. As always,

ensuring this work is covered, but not double covered, is the important thing.

2. The fire hose cabinets are usually furnished and installed by the fire sprinkler subcontractor, since they need to run the piping to them anyway. This is the more common and preferred route, but a general contractor might also decide to allocate this installation to the FEC specialty subcontractor.

In whatever manner this scope is handled by the general contractor, it is important to order the hose cabinets early because they will need to be installed during the rough-in phase so the piping connection can be made before the walls are covered with gypsum board. Failure to make a prompt connection will cause a delay in closing up the walls. This delay happens frequently.

## RESIDENTIAL APPLIANCES

Even for projects with a kitchen, cafe, or other food service component that necessitates the need for a specialty food service subcontractor, the residential appliances will traditionally still be furnished and set in place by the general contractor. Residential appliances are easy to distinguish from the food service appliances as they will not occur in the same rooms as the food service work and will have their own specification section. Residential appliances are common household grade and will be located in areas that are not under the jurisdiction of the health department. Examples of areas that are not monitored by health departments include break rooms, kitchenettes, lounges, and other areas where food is not prepared for sale. On the contrary, food service equipment will be located in commercial kitchen areas, which are under the jurisdiction of the health department.

The residential appliances commonly found on projects are as follows:

1. Refrigerators will be off-loaded, uncrated, and set in place by the general contractor. It is necessary for the plumber to connect the ice makers.
2. Dishwashers will be off-loaded, uncrated, and set adjacent to their point of installation by the general contractor. The plumber will then connect the dishwashers and push them into place. Since dishwashers are simply plugged in electrically, not hardwired, the electrician does not need to become involved with the dishwashers themselves. This is typical for most residential appliances.
3. Garbage disposals will be off-loaded by the general contractor and delivered to the plumber for installation. Until they are installed, the garbage disposals should be stored in the site lock-up area. These are highly susceptible to theft if left out in the open.
4. Electric ranges will be off-loaded, uncrated, set in place, and plugged in by the general contractor. An important item to remember when procuring the electric ranges is that manufacturers sell the electrical cords for ranges separately from the units. Make sure that the electrical cords are procured, as this is a common omission.

5. Gas ranges will be off-loaded, uncrated, set in position, and plugged in (electrically) by the general contractor. These will often have a 110 V plug for the display panel, igniter, and other electrical features. The plumbing subcontractor will then make the gas connection.

   (a) On the contrary, gas or electric cook tops integral to the casework should be set in position and mounted by the casework subcontractor. The gas connections will still need to be made by the plumbing subcontractor.

6. Recirculating exhaust hoods will be off-loaded, uncrated, and set adjacent to the point of installation by the general contractor. The electrical subcontractor will then install and wire the hoods. Contrary to most residential appliances, exhaust hoods are commonly hardwired, not simply plugged into an electrical receptacle.

7. Ducted exhaust hoods will be off-loaded, uncrated, and set adjacent to the point of installation by the general contractor. In this case, the HVAC subcontractor will install the hoods and the electrical subcontractor will complete the wiring terminations.

8. Microwaves will be installed by the general contractor complete when they are simply set in place. When mounted in casework, the casework subcontractor will be responsible for installing them.

9. Washing machines and dryers will be off-loaded, uncrated, set in place and plugged in by the general contractor. The HVAC subcontractor will furnish and install the dryer ducting. If the dryer is gas fired, this connection will be made by the plumbing subcontractor.

   (a) If the dryer duct requires a booster fan, this fan will be furnished and installed by the HVAC subcontractor. The electrical subcontractor will provide the often-missed interlock wiring with the dryer. This interlock wiring between the booster fan and the dryer is the signal from the dryer to the booster fan that turns the fan on while the dryer is running, then off when the dryer is off. This interlock circuit is commonly 120V and also powers the fan.

## MANUFACTURED NATURAL GAS FIREPLACE UNITS

Fireplaces are much more complex than would be apparent by their appearance. Design considerations for a fireplace are even more complex, but this discussion will focus solely on the divisions of work.

1. Fireplace units will be furnished and installed by a specialty fireplace subcontractor. It can be difficult to locate a union subcontractor for this work, which is critical when running a union project.

2. The gas connection to the fireplace will be completed by the plumber.

3. The flue, including rain cap, will be furnished and installed by the HVAC subcontractor. This particular division of work is typical of the HVAC subcontractor's responsibilities for a domestic water heater flue, and a generator exhaust stack.

4. The electrician will provide power to the fireplace unit when required. This is commonly necessary for the electronic ignition, accent lighting, digital control panel, or other electrical features.

5. The electrician will provide and install conduit, mounting of the control box, and the wiring and terminations from the fireplace components to the control box. Commissioning of the controls will be by the fireplace subcontractor.

6. The HVAC subcontractor will interlock the make-up air terminal box to the fireplace. The flame depletes oxygen in the room, so when the flame is burning the HVAC system must provide make-up air to supplement the air lost to combustion and routed up the flue. This make-up air will be shown on the mechanical drawings, but the interlock controls wiring may not be clearly indicated.

   (a)  The fireplace subcontractor will collaborate with the HVAC subcontractor in the commissioning of the make-up air.

   (b)  The work allocation for the interlock controls conduit, wiring, and terminations should be identical to how the HVAC controls scope has been allocated. This will ensure project consistency. This division of work will vary based upon individual project conditions and is described in the chapter on HVAC.

## LOADING DOCK LEVELERS AND EQUIPMENT

Dock levelers (Figure 35.3) are a relatively straightforward item of work but, like most other miscellaneous specialty items, they require coordination with, and contributions from, various subcontractors. Dock levelers and other dock equipment are not set in place until near the end of a project. The coordination of this work begins at project start to ensure that the proper rough-in work is completed during

**FIGURE 35.3**  Dock leveler. (Photo by author, courtesy of The University of Southern California.)

the structural phase. Proper and complete rough-in work will make way for a very quick and easy installation of the dock equipment.

1. The architectural drawings will dimension the leveler pit based upon the make and model used for the basis of design. Different manufacturers and models will require slightly different pit dimensions. Before constructing the pits, the general contractor must coordinate the pit dimensions with the specific make and model of dock leveler(s) proposed by the subcontractor. This coordination is crucial. If those pits are not constructed correctly the first time, the remedial effort involved to demolish and reconstruct this concrete section will be quite costly.
2. The miscellaneous metals subcontractor will furnish the embedded edge angles for the leveler pit FOB jobsite for installation by the formwork subcontractor.
3. The electrician will provide a single point of connection for power to the dock levelers and dock lights. Dock lights are spotlights mounted on flexible arms next to the truck bays. The dock lights are used to shine into the back of a truck because semi-trailers seldom have their own interior lighting.
4. The electrician will also provide conduit, wire, and terminations for the dock leveler controls. Commissioning of the controls will be performed by the dock leveler subcontractor.
5. The dock leveler subcontractor will also furnish and install the dock equipment such as dock lights, wheel chocks, dock bumpers, and dock seals (Figure 35.4). It is easy to differentiate the dock lights from lighting provided by the electrical subcontractor because they will be a manufactured product listed in the dock leveler specification section, not in the electrical

**FIGURE 35.4**   Dock seals and bumpers.

specifications. Also, the dock lights will not be reflected on the lighting schedule. These accessories often require embeds which must be procured at the onset of the project and delivered to the formwork subcontractor for installation well ahead of concrete placement. This is often a missed and expensive coordination task in the early stages of a project.

6. When there is no wall adjacent to the dock leveler, the controls will be mounted on a post. When needed, this post with a mounting plate for the controls box will be furnished and installed by the miscellaneous metals subcontractor. The size and shape of the mounting plate needs to be properly coordinated. A hole(s) must be drilled in the back of the mounting plate by the miscellaneous metals subcontractor in the proper location for the conduit to enter the back of the control box.

7. Dock lights will sometimes come with their own post, in which case they will be installed by the dock equipment subcontractor. At other times, these lights will require a post furnished by others, in which case the miscellaneous metals subcontractor will furnish and install it. This division of work can be verified by checking the construction drawings, dock light specification section, and product data sheets. If and when the need arises, the post can serve a dual purpose for both the dock light and the dock controller.

8. When a dock leveler is installed at the edge of a building (rather than in a recessed, interior, truck bay) with a coiling door (or grille) landing at the front edge, an interlock between the dock leveler and the coiling door will be identified in the design documents. This interlock will prevent the dock leveler from lifting up if the coiling door is in the down position. Conversely, it will not let the door come down unless the leveler is in a level position (level with the finish floor). This protects both the leveler and the door from damage. The conduit, wire, and terminations for this interlock will be provided by the electrician, while the respective sensing device will be furnished by the coiling door subcontractor. Commissioning of the interlock will be a joint effort between the coiling door and dock leveler subcontractors.

9. The majority of dock levelers have the hydraulics built into the leveler itself, but some dock levelers will have an auxiliary hydraulic pump. In the case of the latter, the electrical subcontractor will need to install conduit sleeves for routing the hydraulic lines through the concrete. There is no reasonable way to avoid routing these lines through the concrete slab, so this will be a very costly repair if these sleeves are omitted.

## METAL LOCKERS

Several contrasts in responsibility may occur within the scope of work of lockers. Wood lockers will be furnished and installed as part of the casework scope of work, but metal lockers will be furnished and installed by a specialty metal locker (abbreviated to locker) subcontractor. This division of work is never confusing. Aside from the obvious difference in materials, metal lockers also have their own specification section and will always be a mass-manufactured product. Wood lockers, on the other hand, will almost always be a custom product and described in the casework specifications.

1. The key differentiator between the metal and wood lockers is that one is a mass-manufactured product and the other a custom product. If by chance a mass-manufactured wood locker is scheduled for installation, it will be furnished and installed by the locker subcontractor. This exception to the rule is quite rare.
2. Metal lockers are fabricated and sent to the project in banks of multiple lockers, not individually. The size of these sections must be coordinated with the path of travel through the building (Figure 35.5) to the point of installation. Directing the lockers to be delivered to the site in smaller sections is far less expensive than leaving out walls and other finishes in order for longer sections to be delivered. Therefore, the responsible party to assure the lockers will fit through the path of travel in the building should be the locker subcontractor.
   (a) During the bidding process many general contractors send the locker bidders a fax of the locker specification, elevation of the lockers, and a floor plan of the rooms where the lockers will be installed. Without a drawing of the path of travel, the locker subcontractor will not know the limitations in size of sections to ship. Without careful planning, the locker subcontractor may show up on the job only to find the banks of lockers need to be disassembled before they can be installed. Since the lockers are factory-painted in the same banks as they are shipped, this disassembly and reassembly will reveal lines that require touch-up paint. This touch-up paint will undoubtedly look patchy because the color and texture of the field finish will not perfectly match the factory applied baked enamel finish. This coordination error will result in diminished quality product and probably a change order request.

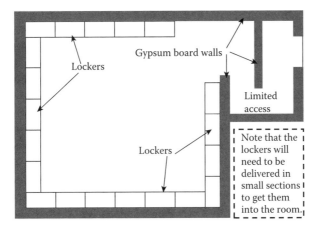

**FIGURE 35.5**   Path of travel for large items must be considered.

**FIGURE 35.6**   Locker with base and sloped top. (Photo by author, courtesy of Hathaway Dinwiddie Construction Company and The University of Southern California.)

3. Metal lockers sometimes have a concrete base, which is completed by the formwork, rebar, and placing and finishing subcontractors. When the lockers have a metal base that matches the lockers, the base will be provided by the locker subcontractor. The locker base is a common oversight. When lockers are installed in wet areas, such as locker rooms and restrooms, it is more common to have the lockers set on concrete curbs for water containment. Otherwise, the water will run under the metal bases and create unsanitary conditions. Metal bases are most commonly located in corridors, break rooms, and other dry areas.

4. Banks of lockers are typically six feet tall with a 6-inch base, rarely extending to the ceiling. The tops of the lockers are flat, but architects and owners do not like to see assorted items piled on top of the lockers by the building occupants. To prevent this, they will add a sloped top to make it physically impossible to set anything on top of the lockers. This sloped top must match the lockers and will be furnished and installed by the locker subcontractor (Figure 35.6).

## PROJECTION SCREENS

Projections screens have many different installation configurations (Figure 35.7), each with their own coordination issues and various subcontractors that will be

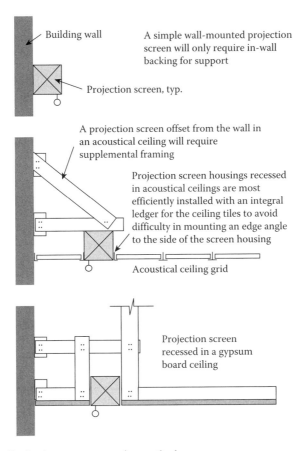

Building wall

A simple wall-mounted projection screen will only require in-wall backing for support

Projection screen, typ.

A projection screen offset from the wall in an acoustical ceiling will require supplemental framing

Projection screen housings recessed in acoustical ceilings are most efficiently installed with an integral ledger for the ceiling tiles to avoid difficulty in mounting an edge angle to the side of the screen housing

Acoustical ceiling grid

Projection screen recessed in a gypsum board ceiling

**FIGURE 35.7**   Projection screen mounting methods.

involved. A few of the primary installation methods will be described. Other methods are simply hybrids of these examples.

1. Simple wall-mounted manual (not motorized) projection screens are the easiest to coordinate and install. The framer will provide in-wall backing for each screen and the general contractor will determine who will furnish and install these screens. The installation for this screen entails no more than screwing the projection screen housing through the mounting holes to the in-wall backing.

2. For motorized projection screens that are recessed in a gypsum board ceiling, the framing subcontractor will be involved by boxing out their framing to accept the recessed screen. In addition to furnishing electrical power, the electrician will also be responsible for providing the controls conduit, wire, terminations, and commissioning. The switch itself will be furnished with the screen for installation by the electrical subcontractor.

   (a)   There is a dichotomy of work that should be pointed out in that permanently mounted projection screens are hardwired, not plugged in.

On the contrary (as described in the audio visual section of the chapter on electrical) permanently mounted projectors (unlike the screens) have simple cords that will plug into an above ceiling mounted receptacle.

3. Motorized projection screens recessed in an acoustical ceiling, but flush with the wall, will require work similar to that discussed above. This includes backing from the framing subcontractor, a hardwired connection from the electrician, and controls from the electrician. This application will also have a very important coordination issue with the acoustical ceiling subcontractor. The ceiling edge support angle will need to attach around the perimeter of the projection screen housing, but if screws are used the screws may stick into the screen housing and conflict with the screen. This difficult condition requires the ceiling subcontractor to attach these edge supports with adhesive in lieu of screws and add wires on the perpendicular runners as close to the screen housing as possible. Note that this installation method is typical of the means of installation described in the operable partitions section of this chapter. It is important for the general contractor to follow up diligently on the quality of work to ensure that the crew in the field understands this coordination issue and does not accidentally put a screw through the screen or insufficiently mount the grid edge angle.

4. Motorized screens recessed in an acoustical grid ceiling, but set off from the wall, will be installed similar to the screens which are flush with the wall. There is one exception. In lieu of furnishing backing in this case, the framer will need to provide stud framing supports for the screens since they are too heavy to be supported by the ceiling grid itself (Figure 35.7). This is typically considered a contractor's means and methods issue; therefore, it is not commonly shown on the architectural drawings. The framing subcontractor will not realize this work needs to be included unless they are specifically directed in the general contractor's bid instructions. It is important to address this support framing prior to bid or it will become a scope of work gap. If omitted, the work will become the financial burden of the general contractor.

There are many different approaches to the allocation of furnishing and installing the projection screens themselves. The general contractor could furnish and install these screens, or furnish them for installation by others. Other options would be to have the electrical, audio visual, or a miscellaneous specialties subcontractor furnish and/or install them. Normally the projection screens will be identified on the architectural drawings and have their own specification. There are also instances where the screens are actually shown on the audio visual drawings and specified within the audio visual specifications. In such instances, to avoid confusing the subcontractors or complicating the bidding process, it is suggested that the audio visual subcontractor furnish and install the screens.

## OPERABLE PARTITIONS

Operable partitions will be furnished and installed by a specialty operable partitions subcontractor, whether they are motorized or manually rolled into position. Since manually operated partitions are essentially identical to motorized partitions, except for the motor, the description of this work will focus on the installation of a motorized partition.

1. The operable partition track (Figure 35.8) is designed and fabricated to be suspended a few inches below a structural support. Consequently, a structural support for this track will need to be furnished and installed by the miscellaneous metals subcontractor. This support steel commonly consists of steel tubes and angles, but can also be Unistrut for the lighter variety of partitions. This steel supporting structure will be shown graphically on the architectural drawings, but the details will require a design-build effort by the miscellaneous metals subcontractor.

   (a)  In addition to the structural support for the track, the miscellaneous metals subcontractor will need to provide supports, or possibly a platform, for mounting the motor.

   (b)  At the general contractor's discretion, the supporting steel work may also be allocated to the structural steel subcontractor. This is the preferred route for large operable partitions that will require very large supporting steel, which is more efficiently swung into position with a crane in conjunction with the structural steel operation than specially rigged into position with the miscellaneous metals operation later in the project.

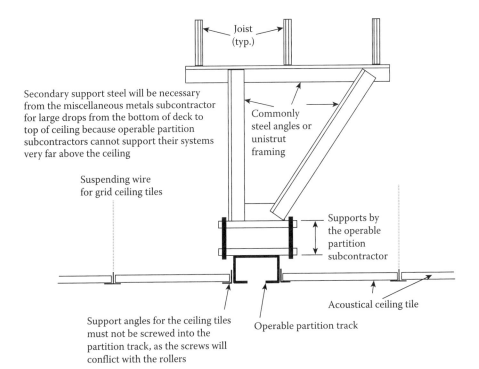

**FIGURE 35.8**   Operable partition track.

2. The electrical subcontractor will provide a single point of connection of medium-voltage power for each motor location. They will also provide conduit, back boxes, wiring, and terminations for the controls. The operable partitions subcontractor will provide the step-down transformer and commissioning for the system.

3. Most operable partitions will be acoustical folding partitions, and as such need a very tight seal at the floor and walls. The allowable construction tolerances for the partition seal is often closer than the allowable construction tolerances for the walls and floors, especially if the floor decks are elevated and expected to deflect slightly, but continuously, over time. The acoustical values and allowable construction tolerances for acoustical folding partitions vary from model to model. For the selected model, the general contractor must review the product data of the panel system and coordinate heightened requirements with the framing and flooring subcontractors. This may mean that the framing subcontractor needs to pay special attention to the related walls and construct them to a higher standard. This may also mean that the floor deck needs to be floated out with a topping compound to bring the floor flatness within the appropriate tolerance range. Since the necessity and extent of this preparatory work will not be fully realized until the decks have had sufficient time to settle and the walls are constructed (well after bidding has been completed), it is recommended that the general contractor set aside adequate funds as an allowance to cover this anticipated work. If the project has a great deal of operable partitions and the leveling work is expected to be a significant quantity of work, it would be more prudent to make an educated estimate as to the quantity of work and have the flooring subcontractor include this leveling work in their base bid with a unit cost for additive or deductive leveling work quantified as a cost per square foot.

4. The ceiling track will most often be recessed in a stud-framed soffit, which is quite simple for the framing subcontractor to box out. At times a track will be recessed at an acoustical grid ceiling. In the case of a grid ceiling, a significant coordination issue will arise in how the ceiling grid edge angle will be mounted to the track. Screws cannot be used because they will screw right through the track and conflict with the rollers. The most effective means of accomplishing this is to support the grid with wires at each perpendicular runner as close to the partition track as possible. The use of adhesives will help support and align the edge angle.

5. For large partitions, such as high auditoriums, the path of travel through the nearly completed building to the point of installation may not accommodate these large panels. It will be necessary to review the conditions of the project. It may be determined that windows or walls will need to be left out to accommodate passage of these panels. This is an important coordination task of the general contractor.

    (a) The first instinct to solve this problem may be to procure the panels early, then store and protect them in the auditorium until their installation. Remember that the panels cannot be procured until field dimensions are taken. These field dimensions cannot be taken until the auditorium rough framing is complete, so it is not likely that getting the panels early and storing them in the auditorium will be feasible.

6. Operable partitions will have the finish applied in the shop. This finish is sometimes a painted finish and other times a fabric identical to the acoustical wall panels in the room. In the case of the fabric finish, the best management practice to ensure a perfect material match is to have the acoustical paneling subcontractor purchase all of the panel fabric materials and furnish it FOB at the operable panel factory. This will ensure that the materials are derived from the same manufacturer and from the same production run to ensure a perfect match. Because the operable panel procurement does not begin until field measurements can be taken, there should not be a problem obtaining this fabric in time.

## ROOF HATCHES AND PREFABRICATED SKYLIGHTS

Roof hatches and prefabricated skylights will often be provided by different suppliers, but they should both be installed either by the general contractor or the same designated subcontractor. The installation methods for these items are commonly identical and simple. The skylights being described are the small, one-piece, prefabricated products that are simply set in place on a roof. This discussion would not apply to a large custom skylight similar to a curtain wall system.

1. The hatches and skylights are self-flashing and will sit on the top of curbs (Figure 35.9) provided by others. These curbs will usually be made of pressure treated wood and constructed by either the framing subcontractor or the general contractor.

**FIGURE 35.9**   Roof hatch. (Photo by author, courtesy of Hathaway Dinwiddie Construction Company and The California Institute of Technology.)

2. Though the hatches and skylights are defined as self-flashing, the reality is that this self-flashing leg is typically manufactured as only 1-1/2″ long. A three-inch down leg is preferable to provide greater assurance of no water intrusion. Compounding this minimal effort by the manufacturers is the construction tolerance for the roofing systems, especially a built-up asphalt roofing system. The roofing will be run up the curb and be in place before the hatch or skylight is installed. In reality, the roofing will seldom extend all the way to the top of the curb for the entire periphery of the down leg. After taking the construction tolerances into account it is not uncommon to find that there is only a 3/4″ or 1″ overlap with the down leg. When the roofing inspector notes this deficiency, it will be deemed that the contractor did not provide the necessary overlap. Thus, the contractor may be held responsible for corrective measures. This may initiate a change order debate between the general contractor, architect, owner, and roofing subcontractor.

   (a)  Arguably, the only time this debate will not occur on a project is when the roofing inspector does not catch the deficiency. Note that these deficiencies frequently do go undetected. Nevertheless, this is still a deficiency and a potential location for a water leak.

   (b)  To combat this issue, it is recommended to add a three to four inch galvanized sheet metal counterflashing that is designed to go in after the hatches and skylights are set in place. This counterflashing will be tucked all the way up to the top of the 1-1/2″ self flashing leg, thus ensuring the 1-1/2″ overlap is attained. This counterflashing should be the responsibility of the flashing subcontractor.

      (i)  Many architects are beginning to show this counterflashing on the design drawings, which is an excellent practice. Coincidentally, when this overlap problem is noted as a deficiency, the counterflashing described above is a common remedy. It is recommended that plans be made ahead of time to ensure that this flashing work is included in the flashing subcontractor's base bid.

   (c)  There are some products on the market, such as Bilco's Bilclip (Figure 35.10), that are designed to mitigate this lapping problem.

3. Unless there is a tower crane on the project, subcontractors that are responsible for installing items need to provide their own hoisting. Subcontractors often expect the general contractor to coordinate the shared use of a crane. When this expectation exists, the subcontractors will often exclude hoisting from their base bid.

4. It is important to order roof hatches and prefabricated skylights early and have them ready to install as soon as the roofing is complete to close off these holes in the roof from the weather. If these items are not on hand, the openings will need to be temporarily covered. Experience has shown that this temporary protection, commonly constructed with visqueen, is a maintenance hassle and often blows off in heavy storms. This leaves the interior of the building exposed.

5. Wherever hatches and skylights are to be installed, there is inherently a floor opening that exposes workers to fall hazards. Numerous worker fatalities

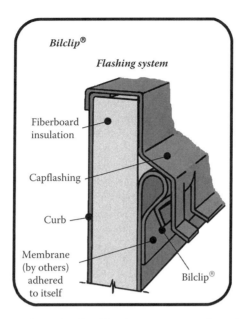

**FIGURE 35.10**   The Bilclip product is utilized to ensure the roof membrane extends fully up and behind the integral flashing element. (Illustration provided by The Bilco Company.)

and serious injuries have occurred because appropriate safeguards were not in place as fall protection. These openings can be protected by such measures as rails or by shielding them with a structurally sound cover. Such safety measures are to be put in place as soon as the hazard is created.

## SIGNAGE

The scope of work for the signage subcontractor will consist of all signage not otherwise furnished by another subcontractor. Signage subcontractors are generally knowledgeable about the signage elements furnished and installed by others, so the potential for gaps in this scope of work is comparatively low. Following are some of the more common divisions of signage work.

1. Signage for large, especially complicated, projects can become a scheduling problem. The signage plan requires shop drawing preparation, architectural submittal review, fire marshal review, and a lengthy procurement process (as long as four months in some cases). Consequently, it is important for the entire sign procurement process to begin early in the project. A general contractor needs to avoid getting so caught up in the day-to-day construction activities that they fail to think about signage a year before completion. It is important that early actions are taken to ensure that the signage arrives on time. The building cannot be occupied until the signage is installed. If the signage is not ready at project completion, the general contractor will typically incur additional expenses by installing temporary signage in order for a temporary certificate of occupancy to be issued.

(a)  Temporary certificates of occupancy are commonly issued by inspectors when a project still has minor deficiencies, provided that the facility can be safely occupied. The inspector would in this case issue a six to twelve month temporary certificate of occupancy contingent upon the permanent signage installation.

2. Exterior parking signage around an asphalt parking lot may include stop signs, handicap parking signage, and vehicle directional signage. This signage will be furnished and installed by the asphaltic-concrete paving subcontractor turnkey, including post-hole drilling and concrete footings. Parking signage that is not placed in the asphalt paved area, such as a parking garage, dirt parking lot, or concrete parking lot, should be provided by the signage subcontractor.

3. A project may specify exterior convenience signage for pedestrian traffic, such as directional signs and any other signage not specifically associated with vehicular traffic. This signage should be furnished and installed by the signage subcontractor turnkey, including post-hole drilling, concrete footing, and backfilling over the top of the footing.

4. Code-required signage for the fire sprinkler system will be provided by the fire sprinkler subcontractor.

5. Mechanical, electrical, and plumbing identification signs and tags should be provided by the MEP subcontractors.

6. The exterior building sign and building address will be furnished and installed by the signage subcontractor. Note that the structural and other support work for large signs will be provided by others.

(a)  If a large sign is mounted to the building exterior, the miscellaneous metals subcontractor will be responsible for providing and installing any structural steel supports within the wall.

(b)  If a large sign is pedestal-mounted in front of the building, a concrete foundation will be provided by the site concrete subcontractor. The general rule of thumb is that if a footing has rebar the concrete subcontractors will provide it; if there is no rebar the signage subcontractor will provide it. (Only small footings will be designed without reinforcing steel.)

(c)  If a sign of any size is illuminated, the electrical subcontractor will provide a single point of connection for 110 V power. If the lighting is low voltage the step-down transformer will be provided by the signage subcontractor.

7. Illuminated exit signage will be furnished and installed by the electrical subcontractor. This includes illuminated exit signage that is not connected to a power source, also known as nuclear signs.

8. Restroom signage, required by the Americans with Disabilities Act (ADA), will be provided by the toilet accessories subcontractor.

9. Elevator signage required by code will be provided by the elevator subcontractor.

10. All interior room numbers, convenience signage, and directional signage will be furnished and installed by the signage subcontractor. Larger signs will be secured to in-wall backing provided by the framing subcontractor.

## WINDOW TREATMENTS

Window treatment (sometimes termed window coverings) for a project includes window blinds, draperies, roller shades, black-out curtains, or any combination of these items. Major commercial window treatment subcontractors are capable of providing and installing all of these items. There will be cases, such as on large public works projects, where a small window blind subcontractor may not be capable of providing and installing draperies, roller shades, or other window coverings. This is when one or more additional specialty subcontractors will be required.

There is little coordination that needs to occur between the window treatment subcontractor and the other trades. Two specific issues are worth mentioning:

1. Motorized curtains and roller shades (Figure 35.11) are more complex than most window treatments. The electrical subcontractor will be required to provide a single point connection of 110 V power to each motor, as well as conduit, wire, and terminations for the controls. This controls work will normally consist of a 3/4″ conduit from the motor to the wall switch, mounting of the wall switch, and pulling three or four wires in the conduit. Commissioning of the roller shades will be completed by the window treatment subcontractor.

2. For curtain rods or other surface-mounted items that do not reside in front of the window header or other solid stud backing, the framing subcontractor will be required to provide backing.

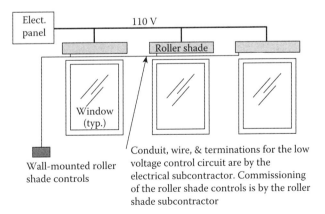

**FIGURE 35.11**   Roller shades.

## MARKER, CHALK, AND TACK BOARDS

Many general contractors prefer to perform the marker, chalk, and tack board (referred to here as marker boards because this scope consists primarily of marker boards) installation themselves. Others prefer to subcontract this work to a specialty subcontractor that may also provide the toilet partitions, FECs, and other appurtenances. Each of these approaches is an efficient and effective means of completing this work.

1. Marker boards are routinely hung on z-clips, which require in-wall backing for support. The in-wall backing must be included in the framing subcontractor's scope and the general contractor needs to verify its installation through their quality control program.

## CORNER AND WALL GUARDS

Corner and wall guards, whether they are stainless steel or plastic, will be handled similarly to the marker boards. They are routinely furnished by the same supplier and installed by the same crew, whether it is the general contractor's crew or a subcontractor's crew. Wall guards will require backing, but because a stud will invariably be found at a wall corner or end, the corner guards do not need to be added to the backing schedule.

## POSTAL SPECIALTIES

Postal specialties can be the sole responsibility of the general contractor or this responsibility could be assigned to a subcontractor. The postal specialties will be furnished FOB jobsite from the materials supplier. The responsible party will off-load and install these items. Installing the mailboxes and package lockers is a simple task, as they are simply set in the blockout constructed by the framing subcontractor and secured with screws. If the general contractor does not plan to perform this installation, the framing, casework, and miscellaneous specialties subcontractors are all capable of installing these items as well.

## RECESSED FLOOR MATS

There are several general approaches for the installation of the recessed floor mats (Figure 35.12). The first is to procure the frames and pans early so they can be set in the initial concrete pour. A 1-1/8″ sheet of plywood can then be set into the frame for construction purposes so the mat itself can be stored and is not damaged during construction. The drawback with this method is that the top edge of the embedded aluminum frame remains exposed to damage. The second approach is to block out the concrete slab for the frame prior to placing the concrete and plate over these blockouts until nearing project completion (Figure 35.13). As project completion draws near, the frames will be set and the blockouts will be filled. When choosing the blockout method, ensure that the formwork subcontractor and the placing and

FIGURE 35.12   Recessed floor mat. (Photo by author, courtesy of Hathaway Dinwiddie Construction Company and The California Institute of Technology.)

FIGURE 35.13   Blockout for a recessed floor mat. (Photo by author, courtesy of Hathaway Dinwiddie Construction Company and The California Institute of Technology.)

finishing subcontractor have included the necessary return work, to avoid receiving change order requests for their respective inefficiencies.

The latter approach is generally preferred. First, this method eliminates the worry of damaging the top edge of the aluminum trench frame, in which case the frame and concrete would need to be demolished for replacement. Secondly, when setting the frames prior to pouring the slab there are no walls or other tangible frames of reference beyond the survey control points. As a result, this presents a relatively significant margin of error. Since the positioning of these frames must be held within strict dimensional tolerances, it is best to set them after the walls have been constructed and the frame can be physically aligned with them.

Recessed floor mats are traditionally purchased by the general contractor, who then furnishes the frames to the formwork subcontractor to embed in the concrete pour. Recessed mats (or gratings) are sometimes just set in place, while others are secured with four hold-down screws. Regardless, the general contractor should store these gratings until the end of the project when they are placed in position.

## FLAGPOLES

It is common for the miscellaneous metals subcontractor to furnish and install the steel flagpoles. Many flagpoles are made of aluminum. On a union project, this will be a concern as the miscellaneous metals subcontractor cannot perform aluminum work themselves due to union jurisdictional issues. In the chapter on miscellaneous metals it was noted that the miscellaneous metals subcontractors regularly bring in lower-tier subcontractors for aluminum fabrications such as railings. Although this issue is quite similar, industry standard practices are not to allocate aluminum flagpoles to the miscellaneous metals subcontractor. The general rule of thumb in this regard is whether the work in question is specified in Division 5, where miscellaneous metals and railings are identified, or if the work in question is specified in other specification divisions. For example, flagpoles are specified in Division 10.

The general contractor could employ a specialty subcontractor for the aluminum flagpoles or maintain this as self-performed work. It is not as difficult as it might appear to locate a specialty subcontractor to furnish and install aluminum flagpoles. Flagpole manufacturers maintain a list of local installers with whom they have worked in the past. If it is germane, they can even provide information on those firms with union labor agreements.

1. If a specialty subcontractor will perform this work, they need to include their own crane. Smaller companies often exclude the cost of a crane with the assumption that there will be a crane on site for their use. If such crane availability cannot be assured, the cost of hoisting the flagpole must be included in the bid quotation.
2. The foundation for each flagpole will be completed turnkey by the site concrete subcontractor.
3. Procuring the anchor bolts and the setting template for the poles is not typically difficult. The poles will generally be shipped with anchor bolts and templates, but if the shipment is not received in time to pour the pole foundation, they can generally be acquired as an inexpensive off-the-shelf purchase from a local hardware supplier. The setting pattern is commonly a simple four-bolt square pattern for which a template can be quickly fabricated on site.
4. If the lanyard, pulleys, spherical top, and other hardware are provided loose with the pole, be sure to attach them before hoisting the pole into position. Otherwise it will be difficult to attach these items once the flagpole is set.

## PAYPHONES AND ENCLOSURES

In this age of cellular phones, fewer payphones are being installed. They do occur on some facilities, especially on public works projects. Payphones will have their own specification section and it is best for the general contractor to furnish the phone and enclosure. After the general contractor's installation of the enclosure, the phone itself is presented to the tele/data subcontractor for installation in conjunction with their data (and depending on the payphone model sometimes power by the electrician as well) connections. Of course, allocating the payphones and enclosures to a miscellaneous specialty subcontractor is another perfectly viable option.

## SUMMARY

Every project will have its own variety of miscellaneous specialties. Some of the more common issues related to these specialties have been presented. Some other approaches will routinely be required to address the intricacies of individual and unique projects. The general contractor must stay alert and intelligently allocate the responsibility for each of the various project components. It is the general contractors task to make sure each element of work for each individual item is properly covered by the appropriate subcontractors, as well as assuring that nothing is double covered. This is a difficult task, but not one that is insurmountable. The general contractor must have the innate ability to identify obscure or uncommon items during their comprehensive estimating review to ensure all items are included in the scope of work of the appropriate subcontractors.

# 36 Food Service Equipment*

Food service equipment includes various items of equipment found in a commercial kitchen or servery. These include ovens, walk-in coolers, stainless steel countertops, coffee makers, exhaust hoods, stainless steel wainscot, drink dispensers, sneeze guards, cash register stations, toasters, salad bars, rolling stainless steel racks, rubber trash cans, reach-in refrigerators, industrial size kettles, industrial size dishwashers, storage room shelving, and the list goes on. This scope of work entails a myriad of different materials, fixtures and equipment, but devising the scope of work of the various subcontractors is a relatively simple process (Figure 36.1).

The design and construction of food service facilities has evolved differently than the mainstream of general construction. Most of the food service projects are simple restaurant renovations and such projects are completed by only three primary parties: an owner, a food service designer, and a food service contractor. The food service contractor will typically employ about a half dozen subcontractors to complete the Ansul systems, walk-in coolers, and a few other scopes of work. The food service contractor even has the capability of employing the mechanical, electrical, and plumbing subcontractors, though most often the owner will directly employ these trades. For this common scenario, the food service designer and food service contractor have each augmented their services to effectively fulfill the functions normally performed by the general contractor on a large scale project. This experience has proven to be a tremendous help to the general contractor when the food service contractor is in fact employed by them as a subcontractor on a project.

For the examples in this discussion, the food service contractor will actually be a subcontractor under direct contract with the general contractor.

A. The food service construction industry is a small one. The designers and food service subcontractors in a given region all know each other, work together regularly, and for the most part work collaboratively as a cohesive team. This collaborative spirit and teamwork is a tremendous help to the general contractor in effectively getting the job done.
B. Food service subcontractors have become proficient at organizing their lower-tier subcontractors and in dealing with owners. This expertise helps the food service subcontractor understand the needs and viewpoints of the general contractor and owner, which adds to their collaborative abilities and spirit of teamwork. The food service subcontractors are among the most self-sufficient and reliable subcontractors in the industry.
C. One role the food service designers have learned to fill, in the absence of a general contractor, is assisting the owner and food service contractor in

---

* MasterFormat Specifications Division 11

**FIGURE 36.1**   Food service equipment. (Photo by author, courtesy of The University of Southern California.)

properly allocating the scopes of work. The food service drawings, unlike any other design discipline, commonly have extensive sheet notes itemizing which specific elements each and every subcontractor is to perform. They do this by clearly identifying the specific responsibilities each individual subcontractor has for each individual piece of equipment in either written notes or in a matrix for the mechanical, electrical, and plumbing scopes of work. When a general contractor is controlling the project, they need to review how the food service designer has distributed the work, identify any scope allocations they wish to modify, and qualify those modifications in their bid instructions. For the most part, there is not much the general contractor will want or need to revise.

(a) The extensive level of detail that food service designers put into scoping the work of different subcontractors is an excellent example of what this book is based upon—a thorough review and analysis of the contract documents focused on making sure each subcontractor knows exactly what they are supposed to do. Food service designers are industry leaders in this regard.

(b) The written scope notes, or responsibility matrix, will provide information that can be clearly obtained from the contract drawings. Some general contractors may think that providing a narrative and/or matrix outlining exactly what is expected from each subcontractor is providing too much detailed information and the subcontractors should review the documents themselves. The rationale for this view is that if the food service designer forgets to put something from the drawings into the matrix the subcontractor will have the basis for a change order request. There may be some validity to this view, but the reality is that these scope notes and matrixes do tremendously more good than harm and constitute a tremendous aid in executing the food service work. It is important that the general contractor makes it clear to the bidders that these written notes and matrixes are to be used as complimentary documents for their use in checking to see whether they have missed anything in their review of the drawings. This may not eliminate all respective change order requests, but the more clearly this is conveyed to the bidders, the more compelling the argument will be that subcontractors are responsible for making independent evaluations of the construction documents.

## SCOPE OF WORK ISSUES RELATED TO FOOD SERVICE EQUIPMENT

1. Health department permit responsibility is an important issue to address. The permit drawings are prepared, submitted, and followed through until all plan check comments are answered and the permit is ready to pull by the food service designer. The food service subcontractor then pays for and obtains the permit.

   (a) At times the food service designer will allocate the submission and all other direct dealings with the health department to the food service subcontractor. This is not the most efficient means of procuring a permit, as it is best that the party with the questions (the health department) and the party with the answers (the food service designer) maintain direct contact and open communications.

   The real big-picture point here is to be sure that the responsibility for procuring permits is covered for all trades. If the general contractor does not allocate permit procurement to the subcontractors, they may have a dozen supplemental permits to pull on every job themselves where they will essentially be acting as an intermediary. This is an extremely inefficient method of permit procurement, as the most efficient method of permit procurement is to promote direct and open communication between the designer (the food service designer in this case) and the city plan checker. The general contractor could end up designating a staff member to spend several hours a week dealing with permits as an intermediary between the various designers and the city, when their time could be more effectively spent planning and managing the project.

2. One of the most convoluted, also most important, divisions of work with respect to the food service equipment scope of work involves coordination with the casework subcontractor. The primary division of work is quite simple. That is, any cabinetry or counters made of metal will be provided by the food service subcontractor and any made of wood will be provided by the casework subcontractor. This general rule will adequately address most of the allocations of the work between the food service and casework subcontractors.

   (a) A fixture such as a salad bar may have base cabinets and countertops constructed by the casework subcontractor. Then the refrigerated tray, compressor, and sneeze guard would be mounted to the casework by the food service subcontractor.

(b)  Service counters will commonly be a casework product, but the casework subcontractors must make cut-outs and other provisions in the casework for soda lines, plumbing to coffee makers, heated soup inserts, cabling to cash registers, electrical lines to toasters, mounting points for sneeze guards, and any other food service item associated with the casework.

(c)  When display cases or other food service equipment occur in line with the cabinetry, the casework subcontractor may need to provide wood trim around the food service equipment.

Each installation will have some unique features. The key issue of this discussion is to make a conscious effort to review this division of work and properly/logically allocate the work for the project.

3. Another significant coordination effort occurs with the walk-in coolers. The food service subcontractor will provide the vast majority of this work, but other trades will play key roles as well (Figure 36.2).

(a)  The food service subcontractor will furnish and install the compressor (cooling unit), whether it is mounted at the building roof or directly on top of the walk-in itself. They will also furnish and install all refrigeration lines between the compressor and the built-in fan coil within the unit.

(b)  The roof curbs for the compressors will be of the same type and furnished by the same party(s) that provide the other roof curbs for the project. These curbs may be wood, concrete, or prefabricated metal components.

(c)  If the compressor is mounted on the roof, the sleeves for the refrigerant lines will not commonly be provided by the food service subcontractor. These sleeves are best allocated to the HVAC scope of work simply because refrigeration lines for other equipment are commonly an HVAC responsibility. The general contractor or electrical, plumbing, or formwork subcontractors could also perform the quick and easy task of setting these sleeves.

(d)  The electrical subcontractor will provide single points of connection for medium-voltage power separately to the compressor and electrical panel built into the walk-in cooler. They will also provide a conduit raceway, wiring, and terminations for the controls circuit from the compressor (cooling unit) to the fan coil inside the cooler. Commissioning the controls circuit will be performed by the food service subcontractor. The electrical subcontractor will seal all conduit penetrations made in the cooler panels.

(e)  An industry standard procedure which has developed due to union jurisdictional issues is that the lighting inside the coolers is furnished by the food service subcontractor FOB jobsite for installation by the electrical subcontractor. The electrical subcontractor will install the lighting, including all conduits, wiring, and switching.

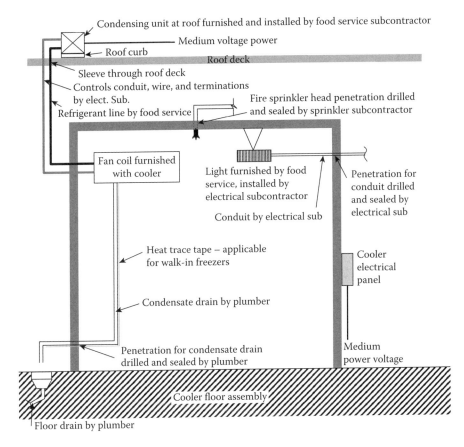

Condensing unit at roof furnished and installed by food service subcontractor

Medium voltage power

Roof curb

Roof deck

Sleeve through roof deck

Controls conduit, wire, and terminations by elect. Sub.

Refrigerant line by food service

Fire sprinkler head penetration drilled and sealed by sprinkler subcontractor

Fan coil furnished with cooler

Light furnished by food service, installed by electrical subcontractor

Penetration for conduit drilled and sealed by electrical sub

Conduit by electrical sub

Heat trace tape – applicable for walk-in freezers

Cooler electrical panel

Condensate drain by plumber

Penetration for condensate drain drilled and sealed by plumber

Medium power voltage

Cooler floor assembly

Floor drain by plumber

**FIGURE 36.2**   Walk-in cooler.

(f)   The plumbing subcontractor will provide a condensate drain from the fan coil inside the cooler to the nearest available floor drain. The plumber will also be held responsible for drilling and sealing the condensate line penetration in the cooler wall panel.

(g)   For walk-in freezers the electrical subcontractor will provide heat trace tape on the fan coil condensate drain to keep it from freezing.

(h)   The fire sprinkler subcontractor will furnish and install sprinklers in the coolers and freezers where required by code. They will be required to seal their penetrations through the panels as well.

4.   Walk-in coolers do not come standard with plastic curtains at the doorways for energy conservation, but these are commonly required by the documents and local jurisdictions. It is important to ensure that the food service subcontractor provides these energy saving components when specified.

5.   The floor assemblies for walk-in coolers are commonly prefabricated panelized flooring systems provided by food service subcontractors on renovation

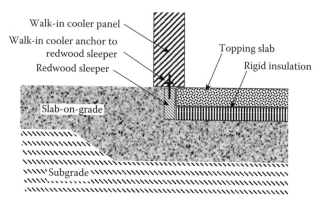

**FIGURE 36.3**   Walk-in cooler floor assembly.

projects. On new construction projects, custom floating concrete floor assemblies are more common (Figure 36.3). The reason designers choose to use a more expensive floor assembly for new construction is to keep the cooler floor at the same level as the adjacent floor, thus eliminating the need for a ramp. Since the redwood sleepers and rigid insulation of this custom floor assembly can be provided by the formwork subcontractor, this custom floor assembly will only include the formwork, rebar, and placing and finishing subcontractors.

(a)   A good, and common, question is: Why is the redwood sleeper necessary? Can't we just anchor the wall panels directly to the concrete with expansion anchors? While this is a good question, the answer lies in the fact that the redwood sleeper actually serves a triple purpose. In addition to acting as a substrate for anchoring the base of the panels and a form for the floor depression, this redwood sleeper serves as a thermal barrier between the cooler slab and the slab immediately outside the cooler. Without this thermal break it would result in a cold concrete slab outside the cooler. This will then develop condensation and become a slippery condition. Since the flooring in a food service area is commonly a smooth tile, a slippery safety problem would result.

(b)   A custom-built floating concrete floor will need to be sealed. The walk-in coolers are not always shown on the room finish schedule. The subcontractor performing the concrete floor sealing work may not realize this work is required unless informed by the general contractor in the bid instructions. This requirement may not be called out anywhere except in the food service documents and the subcontractor performing the concrete sealing work may not know to look there. (Refer to the chapter on concrete placing and finishing for further discussion on how the concrete sealer work should be allocated.)

6. The health department requires all surfaces within a food service area be easily cleaned with a common rag. Tight spaces behind and on top of fixed equipment are not allowed, nor are tight gaps in which bacteria can gather, such as in the gap between a counter and the wall. The objective is

to construct a kitchen area with all surfaces easily cleaned. For all permanently installed furniture, fixtures, and equipment, there are several tasks that must be completed by the food service subcontractor.

(a)    Walk-in coolers will require stainless steel closure plates to seal off the cooler panels to the building walls and ceiling. It would be thought that if there is a four foot space between the top of the cooler and underside of the ceiling that this space can be cleaned, but the reality is that this space traditionally is not cleaned and becomes caked with grease, grime, dust, and the resultant bacteria. Because this has become a common health code deficiency, the health departments in most jurisdictions have begun requiring stainless steel closure panels to keep this area from becoming a problem. Health departments enforce this requirement within reason (on larger fixed items) and do not apply it to smaller fixed equipment such as reach-in refrigerators.

(b)    Countertops will need to be caulked to the wall. Otherwise, cleaning this tight gap would require flossing, which is not realistic. When this gap is filled with a caulk joint, the surface can effectively be cleaned and maintained with a cloth rag.

Pipes and conduits within the food service area cannot be exposed. They must either be concealed in the wall or covered with a stainless steel enclosure. If left exposed, gunk will build up between and behind the lines. It may be possible to get away with exposed piping and conduits if tight difficult-to-clean spaces are avoided by separating and holding them off the wall to attain a one-inch or more clearance all the way around. The acceptability of this approach must be approved by the local health department.

(i)    If stainless steel enclosures are used they should be performed by the food service subcontractor. Since the enclosures are commonly a means and methods of construction, the subcontractor must be informed that these enclosures are necessary prior to bidding.

(ii)   Note that pipe insulation does not have a surface that can be cleaned, consequently the plumber will likely need to cover this insulation with a PVC jacket. Even with the PVC jacket, permission must be obtained from the health department if this is to be left exposed.

These are a few examples of the health department requirements for closure panels and caulking for which the food service subcontractor is responsible. The scope of work must be written such that the food service subcontractor maintains full responsibility for providing easily cleaned surfaces throughout the entire food service area. Food service subcontractors are well accustomed to providing this work and will likely include it in their base bid without specific direction, but the general contractor should provide this direction anyway to leave no doubt.

7. The exhaust hoods above hot areas, with a high level of grease, require special fire suppression systems, such as the systems manufactured by Ansul.

These systems will be furnished and installed by the food service subcontractor, not the fire sprinkler subcontractor. The fire sprinkler subcontractor will provide only a single point of connection for supply water from the sprinkler branch piping to this system.

Of all the different municipal authorities, health departments are arguably the most approachable, easy to access, and eager to help. When unsure of a health code requirement, builders are encouraged to give them a call. Health inspectors are very appreciative when getting a call from a general contractor during the planning stages of a project. This demonstrates to the inspectors that the general contractor is on top of things and doing what they can to avoid problems in the future. The inspectors would rather field a question from a contractor early in the project than deal with problems that could have been avoided with a diligent planning effort.

# 37 Elevators*

Elevators are arguably the most scrutinized trade on a project. They have their own state agency and also their own federal code. There is also a jobsite protocol prohibiting other trades from working in the elevator shafts and machine rooms once the elevator construction has begun. The only exception would be when others are supervised by the elevator subcontractor for tie-ins such as the electrical and telephone lines. Conduits, piping, ductwork, and any other work which is not directly associated with the elevators cannot be routed through the elevator shafts or machine rooms. As soon as the elevator subcontractor begins working on site, the elevator machine rooms are locked. These strict rules, regulations, and protocols are unique to the elevator trade, which puts this trade in a division of its own when it comes to managing them (Figure 37.1 and 37.2).

There are two primary types of elevators. Traction elevators operate via a hoist at the top of the elevator shaft and hydraulic elevators operate via a hydraulic plunger buried beneath the elevator pit.

Elevator subcontractors require support from other trades, but what they require are somewhat small items that are virtually identical from project to project (Figure 37.3). The elevator subcontractor will essentially complete all significant work within the elevator shafts and elevator machine rooms. The following elements will be completed by others in coordination with the elevator subcontractor:

A. The electrical subcontractor will provide several interfaces with the elevator work:
   (a) The electrical subcontractor will provide power to the elevator, which is first brought to a disconnect switch in the elevator machine room, then routed from the disconnect switch to the controller. The electrical subcontractor will also provide a shunt trip breaker. This is a safety device tied to the fire alarm system which shuts off power to the elevator when the sprinkler system goes off. The reason for this is that if the sprinkler head at the top of the elevator shaft is activated it will shower water over all of the electrical components above the elevator car, and then rain down into the car, potentially electrocuting the occupants of the elevator cab. The circuit from the disconnect switch to the controller and the shunt trip breaker, which is located in line between the disconnect switch and the controller, are sometimes excluded in electrical bids when the electrical subcontractor assumes they will be provided by the elevator subcontractor,

---

* MasterFormat Specifications Division 14

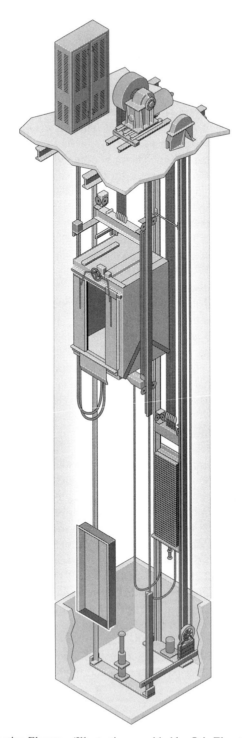

**FIGURE 37.1**   Traction Elevator. (Illustration provided by Otis Elevator Company.)

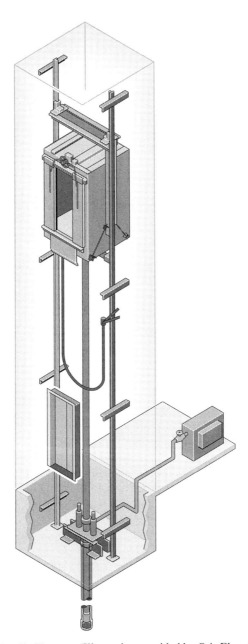

**FIGURE 37.2**   Hydraulic Elevator. (Illustration provided by Otis Elevator Company.)

  while the elevator subcontractors routinely exclude these items. The general contractor must verify the coverage of this work during the bidding phase.

(b) For most elevators, the electrical subcontractor will provide a second point of connection for 110V power from an emergency circuit for

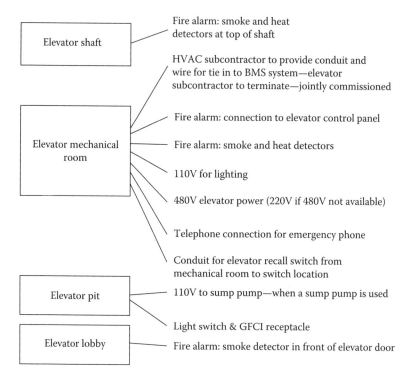

**FIGURE 37.3** Elevator responsibilities.

the elevator cab lighting, which is required by code to ensure that the elevator cab lighting remains functional in the event of a power outage, even when the elevator itself is not tied to an emergency circuit and will be stuck in the event of a power outage. The electrical subcontractor will simply bring this conduit and wire to the elevator machine room and end the wiring at a wall mounted disconnect switch. The elevator subcontractor will provide conduit and wire downstream of the disconnect switch.

(c)  The electrical subcontractor will install a duplex convenience receptacle at the machine room and a second duplex convenience receptacle at the elevator pit. Each of these receptacles will be equipped with a ground fault circuit interrupter (GFCI).

(d)  The electrical subcontractor will furnish and install a light and switch at both the elevator pit and machine room.

B.  The fire alarm subcontractor will provide a circuit to the elevator machine room, which will be terminated by the elevator subcontractor in the elevator control panel. This circuit will be commissioned via a collaborative effort between the elevator and fire alarm subcontractors. The purpose of this connection is to signal the elevator when the fire alarm has been activated anywhere in the building. After the alarm is given, the elevator controller will switch to emergency recall mode and automatically bring the elevator to the main building level, termed the primary recall floor. The elevator will then open the doors to allow anyone who is in the elevator at the time of the alarm out of the building in the most expedient manner possible. (If the alarm happens to have occurred in the elevator lobby on the primary recall floor the fire alarm system will signal the elevator to ferry the passengers to a secondary recall floor, usually the second floor.)

C. The telecom subcontractor will provide a dedicated telephone line to the elevator machine room for the emergency phone in the elevator cab. Similar to the fire alarm connection, this will be terminated in the elevator control panel by the elevator subcontractor and then commissioned in a joint effort between the telecom and elevator subcontractors. The elevator subcontractor will perform all work from the control panel to the elevator cab, including furnishing of the phone itself.

    (a)    Older generations of elevators had the emergency phone behind an access panel in the cab. This phone was provided by the telecom subcontractor or owner and installed by the elevator subcontractor. With new construction, the phone in the elevator cab will always be furnished and installed by the elevator subcontractor, as they are now built into the control panel with a simple push-button operation.

D. The fire alarm subcontractor will furnish and install heat and smoke detectors at the top of the elevator shaft and on the machine room ceiling. They will also provide a circuit to the shunt trip breaker discussed above.

E. The fire sprinkler subcontractor will provide sprinkler heads in the elevator machine room, at the top of the elevator hoistway, and at the elevator pit. Each of these sprinkler heads will have a protective guard, also termed a cage, and will be an isolated branch line. An isolated branch line means the piping to these sprinkler heads will terminate at each head and cannot continue through the shaft, machine room, or pit.

F. The HVAC subcontractor will always provide cooling at the machine room. For larger elevators they will provide ventilation at the shaft as well.

G. The miscellaneous metals subcontractor will furnish and install the elevator pit ladder. They will also furnish embedded sill support angles FOB jobsite for installation at each floor by the formwork subcontractor.

H. An elevator shaft with multiple elevators side by side will require a divider screen between each elevator for the safety of the elevator maintenance workers. This divider screen allows the crew to work on one elevator within the pit while being protected from the travel of adjacent elevators that remain in operation. This divider screen is most commonly chain link (other construction types may also be used) and is provided by the chain link fencing subcontractor.

I. The elevator flooring will be provided by the respective flooring subcontractor, depending on the type of flooring specified. The elevator subcontractor will provide all other elevator finishes, but because architects commonly prefer to have the elevator cab flooring match the elevator lobby flooring, this item of work will be furnished and installed by the flooring

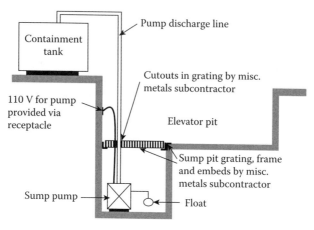

**FIGURE 37.4**   Elevator pit.

subcontractor. It is now common for the elevator subcontractor to exclude the flooring on every project, as it is assumed that the respective flooring subcontractor will provide it.

J. Sump pits were a staple in elevator shafts just a few years ago, but they are rarely specified on new construction projects. These sump pits are no longer required by federal or most local elevator codes. When a sump is required, there will be multiple subcontractors contributing to its construction (Figure 37.4).

   (a)   The plumbing subcontractor will provide the sump pump, commissioning of the pump controls, containment tank, and all associated piping.

   (b)   The electrical contractor will provide power to the sump pump via a non-GFCI receptacle and a disconnect switch in the elevator pit. The elevator sump pumps are small and activated by an integral float, therefore these pumps do not typically have auxiliary control panels.

   (c)   The miscellaneous metals subcontractor will furnish the embedded sump pit edge angles and grating FOB jobsite for installation by the formwork subcontractor. The grating itself simply sets in place, but the miscellaneous metals subcontractor will be responsible for cut-outs in the grating for the pump discharge line and electrical cord.

K. A wall-hung fire extinguisher in the machine room will be provided by the general contractor or the subcontractor that is completing the FECs throughout the project. Refer to the chapter on miscellaneous specialties for a more detailed discussion on allocating the FEC work.

## SCOPE OF WORK ISSUES RELATED TO HYDRAULIC AND TRACTION ELEVATORS

1. The elevator subcontractor will prepare engineered drawings, prepare a submittal for the elevator permit, complete all interactions with the plan

reviewer directly, and pay for the elevator permit. This is a standard procedure in the industry and is rarely noted as excluded in the elevator bids.

2. The elevator bidders will very clearly and firmly qualify that their warranties commence the day the elevators are put into use. This presents a problem in that the elevators will be complete three months or more before the project reaches substantial completion. The general contractor will be using the elevators, or at least one of them, for construction purposes. It is standard scheduling practice to begin dismantling the material hoist (Figure 37.5) the day after the first elevator is operational. With this in mind, the general contractor needs to advise the elevator bidders which elevator(s) will be put to use for construction purposes and for what duration. The elevator bidders will then include three additional items in their base bid. The first inclusion will be additional costs associated with an extended warranty. Secondly, the elevator subcontractor will refurbish the elevators after completion of construction. This essentially entails running through the commissioning check list one last time, cleaning and addressing any minor concerns. Lastly, because elevators cannot be operated by anyone other than a certified elevator technician until such time as the building receives at least a temporary certificate of occupancy the elevator subcontractor will include the cost of an elevator operator for construction purposes.

3. The smoke and heat detectors are simple to install at the machine room ceiling, but the top of the shaft presents a much greater degree of difficulty. Heat and smoke detectors must be accessible for maintenance purposes to a fire alarm technician per code. Since a fire alarm technician is not able to access the top of an elevator shaft after construction, a dilemma is obvious. The solution (Figure 37.6) is to put an access door at the side of the elevator shaft near the top. Then, these detectors can be mounted to the inside face of the access door with enough flexible conduit to account for the full door swing. This will be a rated access door, so the means of

**FIGURE 37.5** Material hoist.

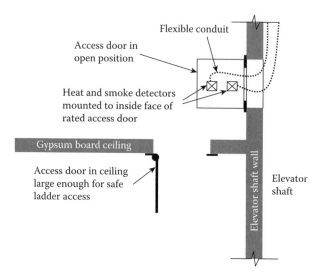

**FIGURE 37.6**   Smoke and heat detectors at elevator shaft.

mounting to the door cannot compromise the integrity of the door rating. Then, when these devices need to be accessed, a fire alarm technician simply opens the access door and the devices swing right out for easy access. Further, the position of this access door must be safely accessible from a standard ladder. This means it cannot be very high above an acoustical tile ceiling. There must also be a second, ladder-sized, large access door if the respective ceiling is a gypsum board finish.

4. An important coordination issue occurs with the flooring installation. The elevator subcontractor must recess the cab floor to accommodate the scheduled flooring thickness. For example, stone tile flooring will require the elevator cab floor to be depressed about 1″, whereas only a 1/4″ depression may be necessary for carpet tiles. This required depression depth must be confirmed by the respective flooring subcontractor. The depression depth for carpet and resilient flooring will be easy to determine, but there is considerable variability in the thicknesses of stone or tile flooring which can range in thickness from 1/2″ to 2″.

(a) There is a common problem with thickset stone flooring applications. Some elevator manufacturers can only provide a maximum floor depression depth of 1″, while a thickset tile system may require a 2″ depression. The stone flooring subcontractors can typically accommodate these thicknesses by simply decreasing the thickness of the grout setting bed. Generally, architects do not mind a slight deviation from the contract documents in this small area and the decreased grout thickness is rarely a future problem. In the case of this common problem, the general contractor bears responsibility for coordinating the solution with the flooring subcontractor, elevator subcontractor, and architect.

5. Hydraulic elevators will require the ram (plunger) hole to be drilled and the casing (for the plunger) set during the excavation phase of construction (Figure 37.7). The spoils for this operation will be stockpiled by the elevator subcontractor for removal by the earthwork subcontractor.

(a)  A very common and expensive scope gap occurs on projects with more than two basement levels when access for the drilling rig in and out of the excavation is not properly planned. Excavations with more than two levels below grade will not commonly have temporary trucking ramps; whereas one or two-level excavations commonly will have a temporary trucking ramp. This leaves two options for getting the elevator ram drilling rig into, and back out of, the excavation when there are more than two basement levels. The first option is hoisting the rig, but it is extremely heavy and if the shoring is not designed for a tremendous surcharge (Figure 37.8) this approach will be unsafe. This is often not considered to be a viable option due to safety concerns. The second option, assuming the excavation is to accommodate a parking garage as is typical, is to leave a blockout in the pit base. The hole can

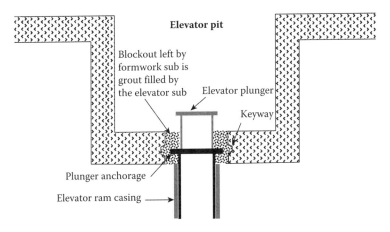

**FIGURE 37.7**   Blockout for elevator ram.

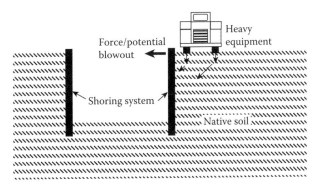

**FIGURE 37.8**   Equipment adjacent to shoring.

then be drilled after the parking garage is complete by a drilling rig that can be driven down the parking ramps. At face value, this second option may sound simple. In reality, the method is usually successful only after major challenges have been overcome. For example, a normal drilling rig is too tall to fit into the garage, so a myriad of costly revisions to the scope become necessary. This begins with the use of a special smaller and much slower drilling rig. Also, removing the drilling spoils from the basement level must be completed by hand and with small trucks. This removal is done with a couple of simple scoops of an excavator bucket if the casing is drilled during the mass excavation phase, but responsibility for spoils removal in this case will be best allocated to the elevator subcontractor since the excavation subcontractor will have long since demobilized by this time.

6. Elevator plungers cannot regularly be procured in time for installation before the pit bases are poured. Therefore, the formwork subcontractor must blockout an opening in the bottom of the pit for this plunger that will be set at a later date. If this blockout is anticipated, it must be included in the formwork scope of work and a qualification noted in the elevator scope that they will be responsible for filling this blockout after installing the plunger. This is a common division of work for which the elevator and formwork subcontractors rarely have a problem. If it is not addressed prior to bidding, this coordination issue is likely to generate a change order debate.

7. Many general contractors are so busy getting the project off the ground at the onset of construction that they fail to think about elements, such as the elevator, that do not begin installation for twelve months or longer. There are three significant reasons to award the elevator subcontract as early in the project as possible. First, as discussed, for hydraulic elevators the ram hole should be drilled during the excavation phase. Secondly, preparing the engineered drawings, going through the submittal process with the architect, and then procuring the elevator permit commonly takes a year to complete. This process should not take a year, but the reality is that it regularly takes this long. Lastly, the elevator fabrication itself can take six months or longer. This fabrication will not begin until the engineered drawings are complete and the architect has approved their submittal. So, if the engineering and submittal processes take six months and fabrication takes six months, the elevators will not arrive on site until twelve months after issuance of the elevator subcontract. Elevators have one of the longest lead times in the construction industry and are a common cause of schedule delays.

8. In recent years elevator subcontractors have begun providing systems and controls that can be maintained by any licensed elevator contractor. This is in contrast to the past when they were known to provide proprietary systems and controls that could only be maintained by the subcontractor installing the elevators. In this case, when elevators with proprietary systems and controls are provided, the owner is locked into using the elevator subcontractor for elevator maintenance throughout the life of the building. Elevator subcontractors have actually been known to use this as leverage in

their bidding strategies. They would submit a very low bid for construction of the elevators, knowing the owner would have to hire them for all elevator maintenance throughout the life of the elevator. They would then plan on using inflated labor rates for elevator maintenance in future years because they have a monopoly on that work. Elevator subcontractors are now taking this proprietary approach much less frequently, but some instances still occur. Be cautious when reviewing the stipulations (such as labor rates for maintenance work) of elevator bids. Be sure to include appropriate language prohibiting proprietary systems in the elevator subcontract.

9. The structural engineer will design support steel for the elevators, but this is done before the elevator subcontractor has been selected. Since each brand of elevator will have slightly different structural requirements, the support steel may subsequently require modifications. If additional steel or significant revisions are required, it must be addressed during the bidding phase to reduce the potential of future change order requests. The elevator subcontractor must be alerted in the bid instructions that, unless otherwise agreed during the bidding phase, they will be financially responsible for any additional structural work beyond that shown on the contract drawings that is necessary for their elevator system.

10. During construction within the shaft, the elevator crew will be required to wear safety harnesses with lanyards tied off above them. Since elevator shafts are tall with flat walls, there are often no suitable anchor points for the lanyards. Therefore, tie-off points or anchors may need to be added to ensure the safety of the workers in the elevator shafts. The general contractor must verify the quantity of anchor points required and include them in the miscellaneous metals scope of work. Since these tie-off points are only a means and methods of construction (not a required part of the finished building), they will not be shown on the contract drawings. Thus, this information must be addressed in the bid instructions.

11. The elevator subcontractor must be consulted to verify that the dimensions of the elevator pit(s) shown on the architectural drawings match the dimensions required for the selected elevator system. Additionally, be sure the elevator pit ladder will fit between the elevator cab and the wall. Of course, the ladder must be sufficiently far off the wall to provide the code-required toe clearance. Most jurisdictions require a 4″ space between the wall and the ladder for an adequate and safe toe-hold on the ladder. If a ladder is too close to the wall a worker using the ladder cannot get a safe toe-hold onto the ladder rungs. This toe clearance is frequently overlooked. Ladders too close to the wall violate code requirements. Inspectors may show leniency if the pit is shallow and the ladder is not more than an inch or so too close to the wall. If this occurs, the general contractor will be spared the cost of demolishing and reconstructing the pits. It is not prudent to count on obtaining the inspectors leniency. Code compliance should be the objective as demolishing an elevator pit after the elevators have been constructed and the project is nearly completed will be catastrophic to the substantial completion date.

12. Freight elevators require an embedded steel edge angle in the concrete deck, or a steel member of another sort for wood-framed construction. This is located immediately in front of the elevator doors, and is termed the elevator sill angle. Passenger elevators may require this sill angle, but this is not always the case. The necessity of the angles at passenger elevators will be determined by the elevator subcontractor, as only some elevator manufacturers require these sill angles for their systems. Architects routinely show these edge angles for all elevators, so the miscellaneous metals and formwork (for concrete slabs the formwork subcontractor will install the angles, but for wood-framed construction the miscellaneous metals subcontractor will furnish and install the angles) subcontractors will account for this work in their bids. If a particular elevator model does not need these angles, the architect may remove them from the documents in exchange for a small reduction in the formwork and miscellaneous metals bids, but this needs addressing during the bidding and subcontract negotiation period to ensure that the pricing is completed during the competitive bid environment.

13. Most jurisdictions require a remote elevator recall switch at or near the building entrance. This is a key switch for use by the fire department in the event of an emergency to summon the elevator down to the ground floor. This remote switch will require a conduit raceway from the switch to the elevator machine room to be provided by the electrical subcontractor. The wiring, terminations, and commissioning will be completed by the elevator subcontractor.

14. The elevator subcontractor will provide all code-required signage for the elevators, including floor numbers (located at each floor level opening) and machine room identification.

15. The construction of glass elevators must be thoroughly reviewed by the architect and owner. It is difficult to make the shaft components of elevators aesthetically pleasing and each different elevator manufacturer will have a slightly different system to expose. The general contractor, owner and architect are encouraged to visit a similar glass elevator constructed by the low bidding elevator subcontractor to verify that the quality of work meets the expectations of the architect and the owner.

    (a)  This type of verification by examining past work of subcontractors is a good practice. On one construction project, the owner examined a glass elevator in progress and concluded that the end result would be entirely unacceptable for the proposed project. In that instance, the owner had a solid brick wall constructed in place of the glass. This was a bad experience for all involved, as the glass elevators are often key architectural features of new buildings. The elimination of this feature was an embarrassment that resulted from poor quality workmanship, which is the last thing for which a general contractor wants to be known. Checking on the quality standards achieved by elevator subcontractors before the elevator subcontractor is selected can avoid such problems.

# Questions—Module Six (Chapters 35–37)

1. When separate subcontracts are awarded for the toilet partitions and toilet accessories work, which subcontractor should maintain responsibility for cut-outs in the toilet partitions for the toilet accessories?

2. Which subcontractor will furnish and install the steel supporting structure located above the ceiling for ceiling-hung toilet partitions?

3. Which party will furnish the trash chute sanitizing system? Which subcontractor will install it?

4. Name eight different parties that might be involved in the trash chute installation.

5. What is the diameter of a trash chute vent for a 30″ diameter trash chute?

6. Which party will install all fire extinguisher cabinets?

7. Which parties might supply the fire hose cabinets? (Name two.)

8. Explain how residential appliances are differentiated from food service equipment.

9. Identify four residential appliances for which the plumbing subcontractor will be involved.

10. Which subcontractor should install the microwaves mounted integrally to the upper casework cabinets?

11. Describe the availability of union fireplace subcontractors?

12. Which subcontractor will furnish the embedded steel edge angles for a dock leveler pit? Which subcontractor will install them?

13. Discuss the statement that the formwork subcontractor will always form the dock leveler pits per the exact dimensions on the architectural drawings?

14. Identify which two of the following are optional metal locker components that will be the responsibility of the metal locker subcontractor: concrete base, metal base, sloped top, and in-wall backing.

15. State whether the following items are hard-wired or plugged into an electrical receptacle: motorized projection screens and ceiling-mounted projectors.

16. A project has a considerable number of fabric acoustical panels scheduled at gypsum board walls and a few operable partitions which are scheduled to receive the same fabric. Which party should furnish the fabric for the acoustical panels and which party should perform the installation? Which party should furnish the fabric for the operable partitions and who should install it?

17. For a project without a tower crane, which party should take responsibility for hoisting the roof hatches?

18. Discuss the statement that the signage scope of work requires virtually no management and is a relief for general contractors, and that signage subcontractors need not be contacted until about a week prior to installation.

19. Which subcontractor should provide the handicap parking signage at an asphalt pavement? Which party should provide a sign directing people to the building entrance that is located at an asphalt pavement?

20. A large building sign is to be furnished and installed with integral low-voltage lighting in the landscaped area at the front of a building. Which party should provide each of the following: the concrete foundation, the electrical connection, and the step-down transformer?

21. Which subcontractor should have the responsibility for the aluminum railings, and which subcontractor has responsibility for the aluminum flagpoles?

22. Discuss the statement that food service subcontractors are among the most collaborative and self-sufficient trades on a project.

23. For optimal efficiency, which party should submit and procure the health department permit for food service equipment?

24. In general, the division of work between the casework subcontractor and the food service subcontractor will be determined by the materials constituting the cabinetry and countertops. What general rule distinguishes the work performed by each party?

25. (Fill in the blanks) For a plastic laminate service counter for soups, the cabinet will be provided by the _____ subcontractor, the soup inserts will be provided by the _____ subcontractor, and the cut-outs in the service counter for the soup inserts will be performed by the _____ subcontractor.

26. Which subcontractor will furnish the lighting inside a walk-in cooler? Which subcontractor will install it?

27. What must be provided for a condensate line from the fan coil inside a walk-in freezer to keep the water from freezing? Which subcontractor is responsible for drilling and sealing the penetration through the cooler panel for the condensate line to exit the walk-in freezer?

28. Which party will furnish and install the redwood sleeper and rigid insulation components of a walk-in cooler floor assembly?

29. Describe the functions of the redwood sleeper at the perimeter of a walk-in cooler floor assembly.

30. Which subcontractor will furnish the fire suppression system in the exhaust hood over a commercial kitchen range?

31. Describe the purpose of a shunt trip breaker in an elevator machine room.

32. Identify for which location (elevator shaft, pit, cab, or machine room) the fire sprinkler subcontractor will not provide a sprinkler head.

33. What safeguard is required when multiple elevators are installed side by side in a shaft so that one elevator can be shut down and work can be safely

performed in the respective pit while the adjacent elevator(s) remains in operation?

34. What are three additional costs the elevator subcontractor will need to include in their base bid with respect to the construction use of the elevator(s)?

35. Give three reasons for awarding and issuing the hydraulic elevator subcontract as early in the project as possible.

36. Discuss the statement that permanent elevator pit ladders may be omitted if there is not enough room in the pit to attain the required toe-clearance for the ladder rungs.

37. Which subcontractor will furnish and which subcontractor will install embedded elevator sill angles at a concrete deck?

# Module Seven

# 38 Plumbing*

Plumbing work is the first of the technical MEPF (mechanical, electrical, plumbing, and fire sprinkler, more commonly referred to simply as MEP) trades discussed in this book. Buildings are not the inanimate objects they appear to be. Buildings can be viewed as living, breathing, man-made objects for which the MEP trades are the brains (electrical), intestines (plumbing), lungs (mechanical, or more commonly termed HVAC), and blood (fire sprinklers) systems (Figures 38.1a–c).

There are five primary subcontractors dealing with inflow and outflow of water on the project, and they all have distinct divisions of work:

a. The site utilities subcontractor will bring two individual water service lines, a sanitary sewer line, and a storm sewer line from the city main to within five feet of the face of the building.

b. One water service will be for the fire sprinkler subcontractor and the other for the plumbing subcontractor. Each of these subcontractors will bring their systems into the building from five feet outside the building.

c. All waste water originating from within the building will be tied into the sanitary sewer line. This will be done by the plumbing subcontractor who will complete all work to the point of connection five feet outside the building line.

d. All waste water originating from outside the building (including the roof, balconies, exterior plazas, and any other place where rain will collect) will be connected to the storm sewer. The plumbing subcontractor will also complete this work to the point of connection five feet outside the building line. The reason storm and sanitary systems are separated is because sewerage treatment is extremely expensive. For this reason, most cities will route the sanitary sewer waste water to a plant for extensive treatment, while the storm sewer receives minimal or no treatment before being discharged back to nature.

   (i) There are exceptions to this rule. For example, New York City and Sacramento do not separate the sanitary and storm sewer. Cities with such combined sanitary/storm systems will treat all the water as if it were sanitary waste. These combined systems can create serious sewerage treatment problems when a city has a heavy rainfall.

e. The fourth and fifth subcontractors (in addition to the site utilities, fire sprinkler, and plumbing subcontractors) that deal with water on a project are the landscaping and HVAC subcontractors. These subcontractors

---

* MasterFormat Specifications Division 22

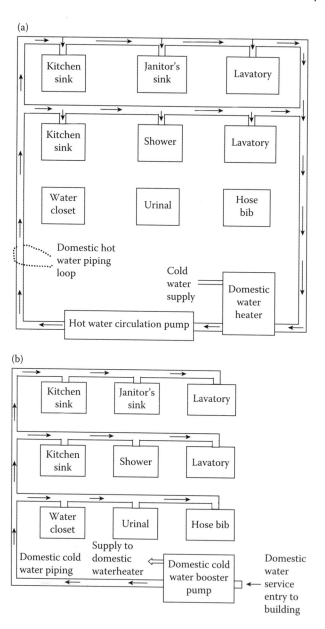

**FIGURE 38.1** Plumbing system.

will focus their efforts on the irrigation work and hydronic (heating and cooling that uses water as the medium) system, respectively. These subcontractors will begin their work at specific connection points identified on the drawings and provided by the plumbing subcontractor. Each of these systems will require a backflow preventer (BFP) to avoid contamination of the domestic water system from the irrigation or hydronic

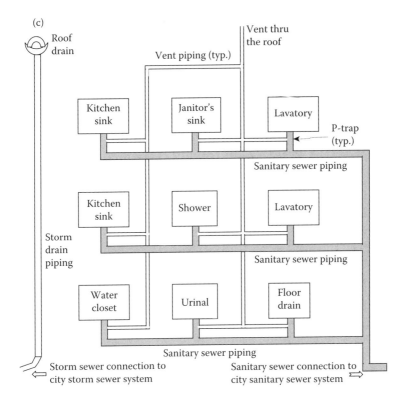

**FIGURE 38.1** (Continued)

systems. The division of work most commonly calls for the plumbing subcontractor to furnish and install the BFP, then the irrigation and HVAC subcontractors will begin their work at the downstream end of the backflow preventer.

Natural gas service is slightly different than the wet utilities in that it will be provided by a coordinated effort of the site utilities subcontractor and local gas company, whose joint efforts culminate at the gas meter. The plumbing subcontractor will then continue the natural gas piping work from a point of connection on the building side of the gas meter. Note that the gas meter is commonly immediately adjacent to the building, not five feet away.

Due to the technical nature of their jobs, MEP tradespeople undergo significant training on the job and in the classroom. In many cases, this training is the worthy equivalent of a traditional college education. This training is put to good use in the field, as the MEP work on a project can predominantly be considered a design-assist approach. That is, MEP design documents are conceptual in nature showing only the major system components and general design intent. The MEP subcontractors must design the detailed aspects of their systems. To accomplish this task, the MEP subcontractors will prepare extensive shop drawings and undergo a collaborative

MEP coordination effort. The MEP coordination process is an important aspect of every project and entails the mechanical, electrical, plumbing, and fire sprinkler subcontractors getting together to overlay their shop drawings and ensure that there are no conflicts in their systems, such as plumbing running through the same space the HVAC subcontractor has planned to place ductwork. There is typically very little space above ceilings and in walls for all of the MEP elements to fit, so this is an important process through which a tremendous number of conflicts are resolved.

## SCOPE OF WORK ISSUES RELATED TO PLUMBING

1. The general building permit will be pulled by either the owner or the general contractor, but the MEP permits for a project will be separate from the general building permit and will commonly be pulled by each individual subcontractor. When these permits are pulled by the respective subcontractors, it is a common practice for the MEP subcontractors to include the cost of the permits in their base bids. This is not always the case. For example, the owner may issue a check for the permit and deliver it via the general contractor to the plumbing subcontractor. The plumbing subcontractor will take the owners check to the building department and deliver it in exchange for the permit. Other times, in fairness to the subcontractors, because permit fees are a variable until such time as the city provides the final tabulation, the owner will reimburse the subcontractors, via the general contractor, for the exact cost of the permit fees via a change order. It is crucial to verify how permit fees will be handled on each specific project. This information should then be incorporated with any permit acquisitions responsibilities in the bid instructions and subsequent subcontract language. Permit costs are quite expensive and can place a significant financial burden on the responsible party if this is not properly accounted for prior to bidding a project.

2. The backflow preventers (BFPs) for both the domestic water supply and fire services (Figure 38.2) will, with few exceptions, be located outside the building. The site utilities subcontractor is most often in the best position to provide these items. If the BFPs are located within five feet of the building a problem may present itself. In such a case, the plumbing subcontractors will claim all piping work within five feet of the building. Plumbing subcontractors are capable of performing this work, so the BFP locations need to be verified on each individual project. An educated, project-specific decision must be made by the general contractor regarding the best allocation of this work.

3. Grease interceptors (Figure 38.3), which separate and trap grease before it is discharged into the sewer system, will be necessary for projects with commercial kitchens. The grease interceptors (also termed grease traps) will be located outside the building and, similar to the BFPs, are normally furnished and installed by the site utilities subcontractor. If the grease interceptors reside within five feet of the building line, allocating this work to the plumbing subcontractor may be necessary. This is particularly true on union projects where union jurisdictions must be recognized and honored.

**FIGURE 38.2**   Backflow preventer.

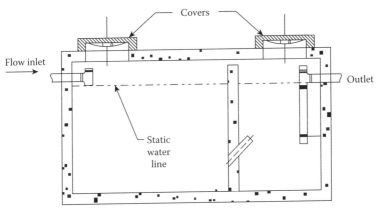

**FIGURE 38.3**   Grease Interceptor. (Illustration provided by The Plumbing and Driange Institute.)

4. The plumbing subcontractor will complete the subdrainage system at the perimeter of the building and, when applicable, directly below the building (Figure 38.4). Subdrainage systems are used to drain and pump ground water away from the building as an added below-grade waterproofing measure. This is essentially a secondary system that lessens the degree to which the building is relying on the below-grade waterproofing membrane.

   The plumbing subcontractor will need to remove the lagging in the lower two feet of the shoring system to access the native soil for the perimeter subdrainage work. In addition to the piping, drain rock, and filter fabric, the plumbing subcontractor needs to be held responsible for hand digging for this perimeter drain, even though they will regularly exclude this work. Considerable persuasion may be required before they accept this

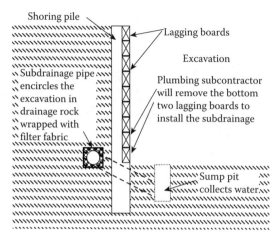

**FIGURE 38.4**   Subdrainage system.

responsibility, but it is rare when the plumbing subcontractor does not in the end accept responsibility for this tedious digging work. This is an often overlooked scope of work that can be costly for the general contractor.

5. For below-grade penetrations through the waterproofing membrane, the plumbing subcontractor should furnish and set in place a galvanized steel sleeve. All waterproofing work outside the sleeve, including sealing the membrane to the sleeve, will be performed by the waterproofing subcontractor. All work within the sleeve will be performed by the plumbing subcontractor, which commonly consists only of a pipe and a somewhat elaborate donut-shaped waterproofing seal.

6. For buildings with a great deal of plumbing (such as biotech, hospitals, and multi-family residential projects), the plumbing subcontractor will require a significant amount of time to layout and set the pipe sleeves in each concrete deck. These sleeves require precise placement to ensure perfect alignment within the walls. If this layout work is done in haste, it will only result in future problems when pipe sleeves are found to be out of alignment with the walls. Setting these sleeves must be performed in a window of time after the deck formwork is complete, but before the rebar placement can commence. This is often overlooked by the general contractor when preparing the project schedules. In contrast, a common office building has comparatively little plumbing, and the sleeves can be set with little more than a day's work. For a common office building, even if this activity is not specifically shown on the schedule, it will not likely become a cause of delay.

7. The only piping the fire sprinkler subcontractor will not perform with regard to the fire sprinkler system will be the sprinkler drains (Figure 38.5). This drain will have a special receptor and is used solely for the fire sprinkler system drainage, but is otherwise just another typical building drain tied to the sanitary waste system.

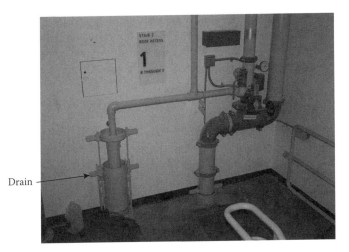

**FIGURE 38.5** Fire sprinkler drain. (Photo by author, courtesy of Hathaway Dinwiddie Construction Company and The California Institute of Technology.)

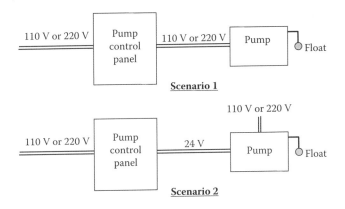

**FIGURE 38.6** Sump pump controls.

8. Medium to large plumbing pumps commonly require control panels and sensors. For a sump pump this sensor will be a float valve, which measures the depth of water in a sump pit (Figure 38.6). In either case, the electrical subcontractor will be required to provide conduit, wire, and terminations. The plumbing subcontractor will then commission the pump system. There are two common scope gaps concerning plumbing pump controls. First, both the electrical and plumbing subcontractors commonly exclude the conduit, wire, and terminations. Secondly, the controls panel may have a low-voltage circuit. This requires a step-down transformer that should be provided by the plumbing subcontractor as part of the controls panel. If the manufacturer of the panel does not include the step-down transformer as a standard part of the panel, the electrical and plumbing subcontractors will each think that the other is providing it. This will result in a scope gap.

(a)  The circuit from the controls panel to the pump is usually the 110V or 220V power supply for the pump. On occasion, power is brought directly to both the control panel and the pump, which will necessitate a low-voltage circuit between the controls panel and the pump. For each pump on each project it is important to review the electrical single line diagram in the pump product literature to verify this division of work.

(b)  It is worth noting in this discussion the disparity that smaller pumps, such as condensate pumps at a fan coil or small submersible pumps, are actually self-contained units that house the controls, pump, and float valve. Electrically, these pumps require only a 110V connection from the electrical subcontractor.

(c)  Make special note of the disparity in allocating the controls work between the plumbing and HVAC scopes. The HVAC controls conduit, wire, and terminations are generally completed via the HVAC subcontractor by a lower-tier electrical subcontractor that specializes in low-voltage controls work, not the project electrical subcontractor as is the case with plumbing controls.

9. Normally, all irrigation piping will be completed by the landscaping subcontractor. This piping work begins at the point of connection provided by the plumbing subcontractor. An exception to this division of work occurs when the irrigation piping actually runs within the building (Figure 38.7). PVC piping is not allowed by code in interior spaces, so the irrigation lines running within a building must be constructed with copper piping. Since landscape subcontractors are not skilled in hanging or soldering copper piping, the plumbing subcontractor will perform this work. On union projects, the union jurisdictions will also prohibit the irrigation subcontractors from performing any copper work.

(a)  Steps can be taken to reduce the amount of costly copper piping and to eliminate a number of risky penetrations through the waterproofing membrane at elevated courtyards. To do this, many general contractors

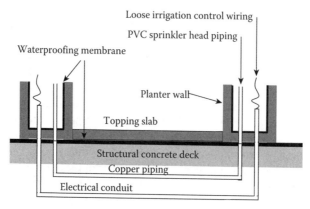

**FIGURE 38.7**  Irrigation piping and control wiring through the building.

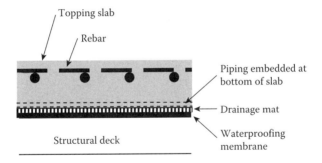

**FIGURE 38.8**    Irrigation piping below topping slab.

will elect to run PVC irrigation piping above the waterproofing membrane and at the bottom of the topping slab or paver system (Figure 38.8). Design drawings traditionally show only where irrigation needs to occur. The means and methods of how this piping is routed to the specified location is the general contractor's decision. Therefore, how this piping is to be routed must be identified in the bid instructions to clarify whether it is work to be performed by the plumbing subcontractor or the irrigation subcontractor. Otherwise, each subcontractor may assume that the other will perform this work. This scope gap needs to be avoided.

10. While the HVAC subcontractor will provide their own hydronic piping, they will not provide the condensate or other drains from the HVAC equipment to the nearest drain. The plumbing subcontractor will provide these drains from a point of connection on equipment that either produces condensate, namely anything with a compressor, or that is filled with water. Examples of items requiring drain lines include air handling units, fan coils, expansion tanks, hydronic boilers, and air separators (Figure 38.9). Plumbing subcontractors are quite accustomed to performing this work and rarely exclude it. A conscious effort still needs to be made by the general contractor to double check that the low bidder has appropriately included this work in their base bid.

(a) For concealed equipment, such as a fan coil in the ceiling of a condominium unit, a condensate pump may be required. This pump will be required if the condensate piping cannot achieve a proper slope due to obstructions within the ceiling framing or other obstacles. If a condensate pump is necessary, it is usually not an integral part of the equipment that will be furnished and installed by the plumbing subcontractor. This small pump comes as a complete self-contained unit, with the sensor, controls, and pump. Unlike the larger sump and water pumps, this small device will not require any controls wiring, only a 110V power connection to be provided by the electrical subcontractor. This power connection will be shown on the design documents when the pump is absolutely necessary as part of the plumbing design. If the condensate pumps are added by the plumbing subcontractor or general

**FIGURE 38.9**  Condensate drains. (Photo by author, courtesy of Hathaway Dinwiddie Construction Company and The California Institute of Technology.)

contractor as an alternative means and methods, the electrical subcontractor will not have this connection in their base bid unless specifically directed to do so in the general contractor's bid instructions.

   (i)  An example will clarify why the general contractor might decide to add a condensate pump. Assume that a condensate line is shown on the drawings to be sloped to a drain, say 40 feet away. The general contractor might decide to add a pump so this condensate line can simply be run to the floor above where a drain happens to be only a few feet away. Such a change to the design drawings will not alter the intent or final product described by the design documents. This should not be considered a material change. This is considered a means and methods issue of construction which does not require the architect's approval.

11. Floor mounted pumps commonly require the baseplates to be grouted to the floor. While the plumbing bidders commonly exclude grouting work from their bids, they are the best party to perform this work. It is generally an easy task to persuade the plumbing bidders to include grouting in their scope of work. If the general contractor is unsuccessful in allocating this work to the plumbing subcontractor, the placing and finishing subcontractor or general contractor's laborers will generally be capable of performing this work as well.

12. Trash chute manufacturers will provide all of the parts and pieces for the chutes FOB jobsite. (The trash chute division of work is discussed in more detail in the chapter on miscellaneous specialties.) The plumbing subcontractor will include installing and commissioning the trash chute sanitizing system. The sanitizing heads and detergent rack will be furnished by the chute manufacturer, but all piping and other accessories must be furnished by the plumbing subcontractor.

13. The plumbing subcontractor will complete all caulking, sealants, and fire stopping related to their work. This includes work such as caulking water closets to the wall, fire caulking a pipe penetration through a fire-rated wall, and sealing a hose bib penetration through an exterior wall.

14. Equipment and fixtures not furnished by the plumbing subcontractor, but which require plumbing connections, should have the connections completed by the plumbing subcontractor. Quite often the plumbing subcontractor will provide gas or water to a shut-off valve, also termed an angle stop, but the connection from the valve to the equipment is not addressed by any subcontractors. This creates a scope gap. Though the different trades furnishing and setting the equipment in place could complete these connections, it is preferable to hold the plumbing subcontractor responsible. The fixtures and equipment that need to be connected include gas-fired hydronic boilers, fireplaces, refrigerator ice makers, dishwashers, and washing machines.

15. Bathtubs and fiberglass showers are furnished and installed by the plumbing subcontractor, who is also capable of providing the shower doors. Regardless, they regularly exclude the shower doors in their bid proposals, assuming the general contractor will employ a specialty shower door subcontractor for this work. As always, the important thing is to make sure the shower doors are covered, but not double covered. Since plumbing subcontractors do not install many shower doors, their employees are generally not highly skilled at this task; therefore, it is preferable to employ a specialty shower door subcontractor for this work. Another problem may surface when attempting to directly employ a specialty shower door subcontractor in that very few of them are signatory to union agreements, which can yield a limited number of bidders for this work depending on how many union shower door subcontractors are available in the region. The general contractor must make a project- and regional-specific decision concerning this work allocation.

# 39 Fire Sprinklers*

Essentially all of the MEP work will be performed in what can be considered a design-assist manner. The fire sprinkler (abbreviated to sprinkler) subcontractor shows considerably more independence than the HVAC, plumbing, and electrical trades by routinely performing their work in a design-build role. The basic bid criteria a fire sprinkler subcontractor will be provided in the bid documents are as follows:

A. A specification which describes the intricacies of the fire sprinkler system, including the types of heads that will be acceptable, whether the heads must be concealed (Figure 39.1) or exposed, the types of pipe fittings that will be acceptable, and any other system variables of concern.

B. A partial sprinkler head layout will be indicated on the architectural reflected ceiling plans. The architect will lay out the sprinkler heads in rooms where this layout holds an important aesthetic value, but they will not normally specify the locations of sprinkler heads in mechanical rooms, janitor's closets, or other non-public locations.

C. The location of the water source point of connection five feet outside the building will be indicated on the civil drawings.

D. Confirmation must be provided on whether or not a fire pump is required. If a pump is required, the capacity of the pump will be provided in the equipment schedule on the plumbing drawings. The design team will test the city water pressure serving the project site and perform a simplistic calculation based upon this pressure. By considering the height of the building and available water pressure, a determination will be made about the need for a fire pump. If a fire pump is required, the design team will size the pump and specify acceptable manufacturers and specific models. The necessary electrical connections, equipment pads, and pump location will also be incorporated into the project documents.

E. A blanket qualification that the sprinkler subcontractor must provide a sprinkler system that is compliant with all applicable building codes.

Throughout the industry, exceptions to these provisions are found. On smaller projects in particular, the design team sometimes partners with a fire sprinkler engineer who fully designs the system. The sprinkler subcontractor then bids this scope, similar to the plumbing, HVAC, and electrical work. For larger projects this exception rarely occurs.

---

* MasterFormat Specifications Division 21

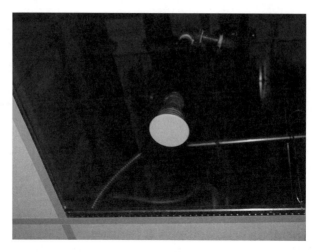

**FIGURE 39.1**    Concealed sprinkler head. (Photo by author, courtesy of Hathaway Dinwiddie Construction Company and The University of Southern California.)

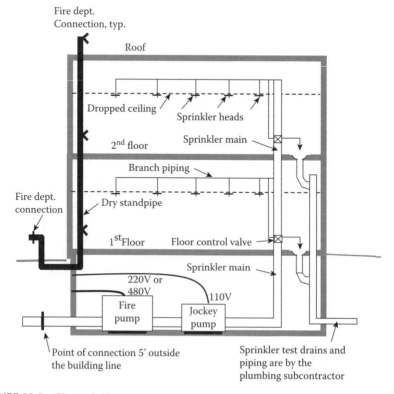

**FIGURE 39.2**    Fire sprinkler system.

Though common terminology for Divisions 21 through Division 28 trades is MEP (mechanical, electrical, and plumbing), not MEPF, inclusion of the fire sprinkler trade is always implied (Figure 39.2).

## SCOPE OF WORK ISSUES RELATED TO FIRE SPRINKLERS

1. The fire sprinkler permit, similar to the elevator and health permits, is a novelty in that it is not obtained from the building department. This permit is actually issued by the city fire marshal. The sprinkler subcontractor will be fully responsible for all work associated with obtaining the fire sprinkler permit, including the design, engineering, submission, interaction with the fire marshal, payment of fees, and pulling the permit. This method of permitting has become routine in the industry. Fire marshals and sprinkler subcontractors often have developed excellent working relationships and, although the general contractor will oversee this process, they rarely need to become directly involved in this permit procurement.

2. As discussed in the chapter on site utilities, the fire sprinkler subcontractor will begin their system five feet outside the building. The site utilities subcontractor will provide the water service to this connection point. The fire sprinkler subcontractor will then pipe the water service into the building and properly seal the penetration made through the waterproofing membrane or vapor barrier.

3. If a fire water booster pump is required, the building code dictates that it must be tied to emergency power. If the building has a generator, the fire pump will be an electric pump that is tied to the building generator. For buildings without generators, the fire pump will commonly have a diesel engine to generate its own power (Figure 39.3). A diesel fire pump is

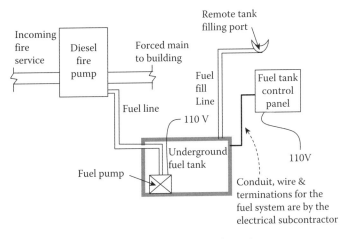

**FIGURE 39.3**  Fire pump with factory mounted control panel.

effectively a diesel generator, as it can essentially be considered a pump with a generator attached to it. The generator component of this pump has additional project requirements typical of those for a traditional diesel generator. This is discussed in greater detail in the chapter on electrical. The following divisions of work are based upon a diesel fire pump. The divisions of work for an electric pump are similar, except for the absence of the generator components.

(a)    The fire pump and all associated piping will be provided by the fire sprinkler subcontractor.

(b)    Fire pumps are normally located inside the building; therefore the equipment pad for the pump will be completed by the formwork, rebar, and placing and finishing subcontractors.

(c)    The separate electrical services to the fire pump and control panel will be completed by the electrical subcontractor. An exception would be made if this control panel was factory-mounted to the pump. This electrical work will be indicated on the electrical drawings.

(d)    The conduit, wire, and terminations between the fire pump control panel and the pump, as well as all other control circuits, will be completed by the electrical subcontractor. These controls will be commissioned by the fire sprinkler subcontractor. Naturally, for a control panel that is factory-mounted to the pump, this interconnection is not required. The general contractor must verify the necessity for this work by evaluating the specific model scheduled for the project. Unlike medium-voltage circuits, low-voltage circuits are not shown on the electrical drawings.

(e)    If the fire pump is located outdoors, has a sufficiently sized integral fuel tank, and is located where a fuel truck can simply pull up next to it for refueling, the plumber will not be required to get involved. If the fire pump has an auxiliary fuel tank, usually located underground, this fuel tank installation and associated excavation, backfill, and piping will be performed by the plumbing subcontractor. Additionally, if the integral and/or auxiliary fuel tank is not located in an accessible location for refueling, such as in a building basement, piping from the tank to an accessible remote filling location will be provided by the plumbing subcontractor (Figure 39.3). The controls conduit, wire, and terminations related to the fuel tank will be completed by the electrical subcontractor. These controls will be commissioned by the plumbing subcontractor.

(f)    The exhaust routing for the diesel engine will be completed, on union projects, by the HVAC subcontractor. Regardless of the installation location, the muffler will commonly be furnished with the fire pump FOB jobsite, where it will change hands to the HVAC subcontractor for installation. It is important to verify the responsibility for the specific fire pump specified and to properly allocate responsibility for supplying and installing the materials. For an interior fire pump, the exhaust stack will not be furnished with the pump, nor will the incidental materials required for installing this stack. Therefore, for an interior pump the muffler will be installed by the HVAC subcontractor, but all exhaust materials beyond the muffler will need to be both furnished and installed by the HVAC subcontractor.

4. A fire sprinkler system will have sensors, called tamper switches and flow switches, to alert the fire alarm system when the sprinklers have been activated. For example, a sprinkler system contains stagnant water unless a sprinkler head is activated, which is the only occurrence that causes the water to flow. A flow switch senses this water movement and alerts the fire alarm system that the sprinklers have been activated. This sounds the building alarm system for evacuation. These sensors are furnished and installed by the fire sprinkler subcontractor. The fire alarm subcontractor, in coordination with the electrical subcontractor, will provide all conduit, wire, and terminations. Commissioning of these sensors is a joint effort of the sprinkler and fire alarm subcontractors.

5. When it is not possible to construct a fire-rated wall as required by code requirements, such as a large fixed opening from a room to a rated corridor, or where a storefront is desired in lieu of a solid wall, the design team will commonly call for a water curtain. A water curtain is essentially a line of tightly spaced sprinkler heads that will create a water-wall in the event of a fire, thus providing the necessary fire rating. Water curtains will be identified by the architect on the architectural floor plans. This water curtain will be provided by the sprinkler subcontractor, but commonly requires sprinkler baffles to direct the water flow down in front of the opening (Figure 39.4). These baffles can be an infinite variety of materials and are always custom products. Because of this, they are common scope gaps when not specifically addressed by the general contractor. The most common baffle type is a glass, or plexiglass, plate set, and anchored in a U-shaped receiver. This common type of baffle is most appropriately provided by the glass

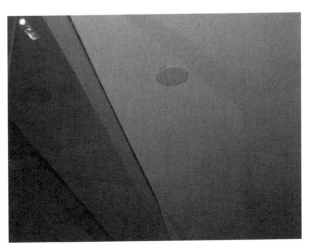

**FIGURE 39.4**   Sprinkler baffle.

subcontractor and though they will routinely exclude this work it does not often take much convincing to get them to pick it up. Also, note that these baffles are most commonly mounted to the underside of a gypsum board soffit and will require the framing subcontractor to provide backing.

6. During construction, the building dry standpipes must be installed floor by floor as the building is erected in compliance with the fire code for fire protection during construction. These standpipes are not filled with water. They have a connection on the exterior of the building called a fire department connection (FDC) at the ground level (Figure 39.5). During a fire, the fire department will connect either a fire hydrant or their truck to the FDC, which then pumps water into the dry standpipe. At each floor within the building there are fire hose connections that can be used for firefighting once water is in the standpipe. This standpipe eliminates the timely effort of dragging fire hoses throughout the building and thereby expedites the firefighting effort. This standpipe is required for the permanent building, so to avoid duplicating efforts the permanent standpipe is simply installed early for use during construction. Fire sprinkler subcontractors are aware of this requirement and perform this sequenced work on every project. Because the fire sprinkler subcontractor will not be on site full time during the structural phase of construction, the general contractor needs to keep the sprinkler subcontractor informed of when each standpipe will become ready to be extended vertically to the next building level. Once a building reaches a certain height, commonly 35 feet, no combustible materials may be loaded onto a floor unless this standpipe is in place. Thus, it is imperative that the fire sprinkler crew follow diligently behind the building erection.

7. Fire hose cabinets must be installed during the wall rough-in phase of construction and are many times allocated to the fire extinguisher subcontractor. Traditionally, fire extinguisher subcontractors do not come on site

**FIGURE 39.5**  Temporary fire department connection.

until the project is in the latter stages of interior finishes. Since a connection to the fire sprinkler system is required, it is more appropriate to have the fire sprinkler subcontractor furnish and install these items. If the hose cabinets serve a dual purpose by also housing a fire extinguisher, the extinguisher itself would be furnished with the fire hose cabinet, thus these extinguishers should be eliminated from the fire extinguisher subcontractor's scope of work.

8. Fire sprinklers will be required in many routine, and many obscure, code-required locations. Regardless, the fire sprinkler subcontractor must be held accountable for identifying and properly accommodating the requirements in their sprinkler system design. Some of the obscure locations may include the following:

   (a) Sprinkler heads at the elevator pit and top of the elevator shaft. Each of these heads will be required to have a wire guard.

   (b) Sprinkler heads inside large air handling units which have interior corridors. These large air handling units are commonly factory-equipped with lighting in these corridors, but they are not factory-equipped with sprinkler heads.

   (c) Jurisdictions will commonly require sprinkler heads inside walk-in refrigerated coolers and freezers.

   (d) Large void spaces above ceilings, in wall cavities, in mechanical shafts, and in other similar areas will generally require sprinklers. A very important void space commonly requiring sprinklers is above ceilings. The installation costs in these areas tend to be high. Large distances between the ceiling and deck above will require a sprinkler grid above the ceiling in addition to the grid below the ceiling. These grids will be installed with high-and-low heads off the same branch piping. The specific criteria for sprinklers in void spaces vary slightly in different jurisdictions. As a result, the sprinkler subcontractors are responsible for being knowledgeable of the fire code requirements in the local jurisdiction.

   (e) Large ductwork—eight feet wide, for example—will require sprinklers both above and below the duct due to the obstruction to the sprinkler spray pattern caused by the large duct.

   (f) Canopies at the exterior of the building will require sprinklers unless they are less than four feet from the innermost face of building. For example, if a door is set back from the face of building, this dimension is measured from the face of door, not the face of building. This minimum canopy dimension varies in different municipalities, but four feet is a common maximum. Make special note of these canopy sprinklers and ensure that they are designed and installed in an aesthetically pleasing manner. Exterior canopies are commonly regarded as key architectural elements. As a result, the installation of ugly, raw, sprinkler piping, and heads below the canopies is a source of concern for architects and owners. At a minimum this piping will need to be painted, which must be coordinated with the painters scope of work.

(g)   Trash chutes will have sprinklers within the chute that are protected by specially designed offsets to keep them out of the line of dropping trash. Trash chutes are a special circumstance when it comes to fire sprinklers in that the chute manufacturer will typically furnish the sprinkler heads FOB jobsite, shipped loose in the same shipment as the trash chute itself. These heads should be furnished to the sprinkler subcontractor for installation. The chute manufacturer will also pre-drill the mounting holes for the sprinkler heads in the specially designed chute offsets. All piping, fittings, and other components needed to install the fire sprinklers in the trash chutes will be the responsibility of the sprinkler subcontractor.

(h)   Large pallet racks require fire sprinklers at each rack level. This is a common omission from sprinkler bids when the storage racks are not clearly identified on the drawings. Since these large racks are commonly furnished and installed by the owner they are not always detailed on the construction drawings. For this reason, fire sprinklers in the pallet racks commonly become very large change order debates. The responsibility for this scope bust will differ between projects, but the general contractor may bear ultimate responsibility if they were aware that pallet racks were being provided by the owner, but failed to transmit this information to the fire sprinkler bidders.

9. Dry-type fire suppression systems (not to be confused with dry-pipe preaction systems, which are actually wet systems, as will be discussed below) are completed by specialty subcontractors with only minor support being provided by the fire sprinkler subcontractor. Dry-type systems have two common types and are respectively found in two different locations.

(a)   First, dry systems (such as an Ansul system) are required by code in commercial kitchens over the range and any other cooking locations susceptible to grease fires. Water suppression systems are not effective with grease fires or other flammable materials that are lighter than water, because in a fire the flames will float on top of the water. Dry-type systems in a commercial kitchen will be furnished and installed by a lower-tier subcontractor via the kitchen equipment subcontractor. Also, while these systems are termed dry, they are not completely dry and will require a water supply which most commonly comes from the sprinkler systems and is provided by the sprinkler subcontractor. Some jurisdictions allow water to be provided from the domestic plumbing, but this practice is not as common.

(b)   The second common location for a dry-type fire suppression system (such as a Novec 1230 system) is a server room or other location with a considerable amount of sensitive electronic equipment. An activated water sprinkler system may ruin expensive electronic equipment and destroy stored data. Owners will commonly pay the premium cost for a dry-type fire suppression system to avoid such losses. This dry-type system is a true dry system and does not require a water connection.

(i) The electrical subcontractor will provide power to the system control panel. The step-down transformer in the control panel and all controls, conduit, wire, terminations, devices, and commissioning will be the responsibility of the dry-type fire suppression subcontractor, though they often employ the on-site electrical subcontractor to perform this work for them.

(ii) These dry systems, installed at the discretion of an owner, are simply a luxury and do not commonly meet the local fire marshal's requirements for fire suppression. As a result, fire sprinklers are also required in these rooms, essentially double-protecting the rooms. These redundant systems require that the dry system will activate much sooner than the wet system. Conceptually, the dry system will activate and extinguish the fire before the wet system is activated.

10. A preaction fire sprinkler system (Figure 39.6) is a dry-pipe system that involves a two-step process before the heads will activate. First, when a smoke detector is activated, an electronic valve will open to fill the pipes with water. Once the pipes are filled with water, this sprinkler system is identical to a normal sprinkler system requiring only that a sprinkler head be heat activated.

Preaction type fire suppression systems are used in two different types of locations. First, these systems are used where sprinkler heads that are broken or falsely activated will cause tremendous damage. This activation might cause valuables to be ruined if showered with water, such as a

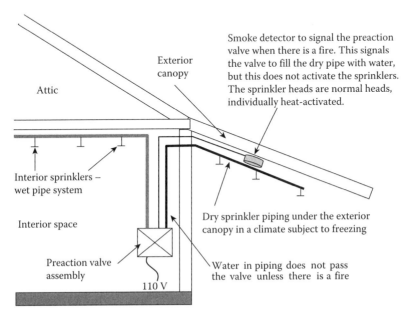

**FIGURE 39.6**   Preaction sprinkler system.

museum with priceless paintings. Secondly, preaction systems are placed in locations subject to freezing, such as walk-in freezers, water curtains or the exterior of a building located in a harsh climate.

These systems have an electronic valve furnished and installed by the fire sprinkler subcontractor. Conduit for this valve is provided by the electrical subcontractor. The wiring and terminations work tying this valve into the fire alarm system is completed by the fire alarm subcontractor (who is commonly under a lower-tier subcontract with the electrical subcontractor). Commissioning of this preaction system is a joint effort between the fire alarm and fire sprinkler subcontractors.

# 40 Mechanical (HVAC)*

The mechanical subcontractor (more commonly referred to as the heating, ventilation, and air conditioning, or HVAC, subcontractor) will furnish and install all equipment, ductwork, and piping for the systems that regulate air movement and modify the ambient temperature. As with all MEP systems, HVAC systems are complex, so this discussion will focus on the coordination of subcontractor scopes of work and not on the intricacies of each different type of HVAC system. Following are some basic examples of the most common of the myriad of HVAC systems that might be encountered.

A. Air handling units with variable air volume (VAN) terminal boxes are standard on many non-residential construction projects (Figure 40.1). This type of system includes a hydronic boiler system that is provided by the HVAC subcontractor. The plumbing subcontractor will provide only a single point of connection for supply water to this system, i.e., the HVAC subcontractors have their own pipe fitters (Figure 40.2).

   (a) In a union setting, their formal training is somewhat different, but pipe fitters for hydronic piping systems share the same union with plumbers. In fact, these workers can be employed in either trade. In practice, pipe fitters rarely perform plumbing work and plumbers rarely perform pipe fitting work. It would otherwise seem intuitive to someone new or unfamiliar with the construction industry that these trades are one and the same.

B. Residential projects, such as mixed use, condominiums, or apartment buildings, will most commonly have split systems for heating and cooling. These split systems consist of a fan coil within the ceiling or a closet and a condensing unit on the roof.

C. The HVAC subcontractor will provide all exhaust systems (Figure 40.3), including the following:

   (a) Restroom exhaust that must have its own dedicated duct system and exhaust fan, due to potential odors.

   (b) Laundry exhaust, due to lint collection, will also have its own ducting (with integrated lint traps) and exhaust fan.

   (c) Commercial kitchen exhaust from the range and other greasy cooking areas will be designed as an isolated system because of the accumulation of grease in the ductwork.

---

* MasterFormat Specifications Division 23

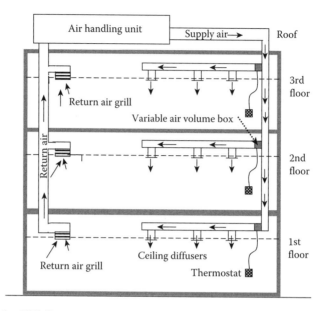

**FIGURE 40.1** HVAC system.

**FIGURE 40.2** Hydronic piping at an air handling unit (AHU). (Photo by author, courtesy of Hathaway Dinwiddie Construction Company and California State University Northridge.)

> (d) Parking garage exhaust is also routinely designed as a stand-alone system.
>
> D. Radiators and radiant heating will be provided by the HVAC subcontractor with their hydronic piping crew. Although this system involves a considerable amount of piping and does not move air, it does control the air temperature. Therefore, it is not the plumbing subcontractor's work as is often confused.

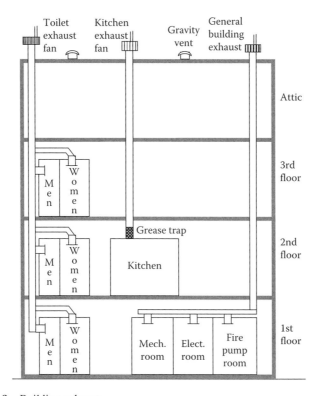

**FIGURE 40.3**    Building exhaust.

This is not a complete list of the systems an HVAC subcontractor can provide, but it is a good sampling of their responsibilities. When determining if work is to be performed by the HVAC subcontractor, a basic rule can provide guidance: if the system moves air, permits the movement of air, or has a primary purpose of conditioning the air temperature, it is most likely the HVAC subcontractor's work. There are few exceptions to this rule.

## SCOPE OF WORK ISSUES RELATED TO MECHANICAL (HVAC)

1. The HVAC subcontractor will commonly be responsible for obtaining the mechanical permit from the building department. As with the plumbing permit, the HVAC subcontractor will also be responsible for paying for the permit if stated in the contract documents.
2. HVAC work involves a significant amount of low-voltage controls work. The HVAC subcontractor normally provides the low-voltage controls work via their own specialized lower-tier low-voltage electrical subcontractor. These controls can be simple, such as in a condo tower where most controls are simply wall-mounted thermostats tied to their respective fan coil. They can also be quite complex, such as for a high-end commercial building where each piece of equipment, VAV terminal box, temperature

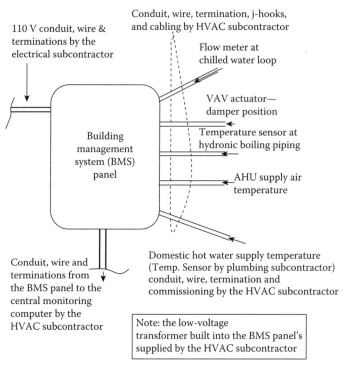

FIGURE 40.4   Building management system.

gauge, zone thermostat, air flow sensor, pressure sensor, and so on will be wired to a central computer called the building management system (BMS) (Figure 40.4). The BMS allows the building maintenance staff to monitor, and sometimes control, the HVAC systems remotely. This expensive management system is particularly common and useful for large campuses such as at universities and major industries.

(a)   Note that systems of other subcontractors, such as generators, domestic water pumps, and elevators, are also commonly monitored by the BMS system. The sensors and control points for these non-HVAC systems will be furnished and installed by the subcontractor responsible for the respective system. Then the HVAC subcontractor will provide conduit, wire, and terminations to tie these points into the BMS system, as well as program the BMS system to recognize these points. Naturally, commissioning of these points will be a joint effort.

3. In a high-end commercial building with significant controls work, various methods may be encountered for connecting 110V power to the VAV terminal boxes and fire smoke dampers (FSDs). Traditionally, the electrical engineer will indicate power to each VAV box and FSD, with the power being provided by the project's electrical subcontractor. This is the most common method of design and construction. The electrical engineer may also take advantage of the fact that the HVAC subcontractor will have their

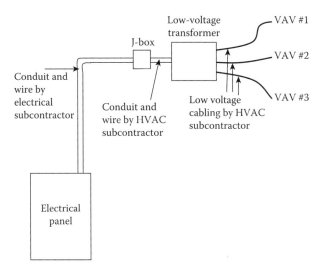

**FIGURE 40.5**   Power to VAVs and FSDs.

own lower-tier electrical subcontractor for controls. They may simply show an electrical circuit serving multiple components being brought to a central location and terminated in a junction box (Figure 40.5). The controls subcontractor would then be responsible for distributing the power from that point to each component. Electrical engineers may take this approach because MEP coordination is the general contractor's responsibility. In this coordination effort, VAV boxes and FSDs are commonly relocated. The electrical engineer will be cautious about indicating points of connection that the general contractor will later change in fear that these changes will be the genesis of change order requests. It is important to review how the power for VAVs and FSDs has been designed to ensure the divisions of work are properly covered by the bidders.

4. Commissioning of the HVAC system is described in most specifications as being performed by a third-party commissioning firm not affiliated with the HVAC subcontractor. Most large HVAC subcontractors have a wholly owned subsidiary firm that performs commissioning work. HVAC subcontractors commonly use this subsidiary as their third-party commissioning agent. The fact that this company is not a true, unbiased third party is rarely noticed. When this is discovered, the project team typically takes no action. It is recommended that the project team keep the subcontractors in full compliance with the contract documents. As such, they should disallow the use of subsidiary commissioning firms. Make note that subsidiary firms are often given a completely different name than the parent company, specifically so the association with the parent company is not readily apparent.

5. Wherever a low-voltage circuit begins (typically at independent control panels and control panels integral to equipment), a step-down transformer will be required. This will convert the 110V power provided by

the electrical subcontractor to a lower voltage, commonly 24V. These step-down transformers often come as part of the independent control panels and control panels integral to the equipment, but not always. If not specifically addressed in the subcontractor's scope of work, both the electrical subcontractor and HVAC subcontractor may exclude furnishing and mounting these transformers when they are not factory-installed in the control panels. It is best that the HVAC subcontractor provide these transformers for two reasons. First, the HVAC subcontractor is most informed about which step-down transformers are already included as integral parts of the equipment and which need to be furnished separately. Secondly, the HVAC subcontractor will at times custom fabricate the independent control panels. They must fabricate these panels with sufficient room to accommodate the step-down transformer inside; otherwise an auxiliary wall-mounted box will be necessary to house the small transformer. This eliminates the coordination issue and the necessity of the additional system component by requiring the HVAC subcontractor to mount these transformers in the shop.

6. Air handling units, exhaust fans, and other equipment are commonly designed to run at variable speeds by means of a device called a variable frequency drive (VFD) (Figure 40.6). Like control panels and step-down transformers, Variable frequency drivers (VFDs) are sometimes integral to the equipment and are factory-mounted, but sometimes they are furnished and installed separately. It is best to have the HVAC subcontractor furnish and mount the VFDs on or adjacent to the respective equipment. Since the electrical connection will be no less than 110V (quite often 460V), the electrical subcontractor must provide all conduit, wire, and terminations from the VFD to the equipment, in addition to their obvious work of bringing power to the VFD. The electrical subcontractor will assuredly pick up the

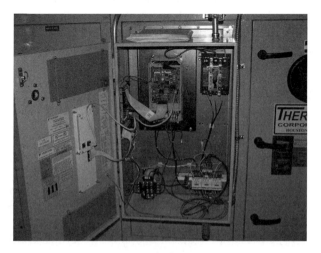

**FIGURE 40.6**  Variable frequency drive (VFD). (Photo by author, courtesy of Hathaway Dinwiddie Construction Company and The California Institute of Technology.)

conduit, wire, and terminations in their base bid, but furnishing and mounting the VFD is a common exclusion by both of these subcontractors. The general contractor must make sure that both furnishing and installing the VFDs is properly covered in the bids.

(a)   Mechanical equipment manufacturers do not commonly allow anything, including VFDs, to be screwed into their equipment in the field. To do this could diminish or completely void the warranty. If there is no wall immediately adjacent to the electrical point of connection, the HVAC subcontractor will need to provide a floor-mounted post or rack for the VFD. This is an important inclusion in the scope of work. For example, the HVAC subcontractor may argue that this mount is not in their scope and insist on simply mounting the VFD on the closest available wall, which could be twenty or thirty feet away. This would add considerably to the conduit and wire provided and routed by the unsuspecting electrical subcontractor. It is suggested that the HVAC subcontractor be contractually required to mount the VFDs within six feet of the equipment point of electrical connection.

7. Duct smoke detectors will be furnished by the fire alarm subcontractor FOB jobsite to the HVAC subcontractor. The HVAC subcontractor will mount the smoke detectors in the ductwork and provide an access door in the duct for access by the fire alarm subcontractor. The electrical subcontractor will provide the conduit and the fire alarm subcontractor will complete all wiring, terminations, and commissioning of the devices.

8. The HVAC subcontractor will provide all conduit, wiring, terminations, and commissioning for all controls circuits interconnecting the HVAC system to peripheral devices. An example of a peripheral device would be the make-up air for a fireplace as discussed in the chapter on miscellaneous specialties. When a fireplace is burning, the combustion depletes the oxygen in the room and sends air up the flue; therefore, make-up air must be provided. This make-up air is not necessary if the fireplace is off. Thus, a controls circuit from the fireplace to the HVAC system is necessary for the make-up air to run when the fireplace is on and to stop when the fireplace is off. This circuit will be completed by the HVAC subcontractor, but, naturally, the fireplace subcontractor will collaborate with the HVAC subcontractor in commissioning this circuit.

(a)   Fume hoods are another example of a peripheral device requiring interconnection. They will require variable levels of make-up air, depending on the height to which the sash is raised.

9. Parking garages regularly use a carbon monoxide (CO) detection system to control the garage exhaust fans. These detectors are spaced throughout the garage. When one detector senses that the CO level is above the preset limit, it signals the exhaust fan to turn on (Figure 40.7). Responsibility for the controls conduit, wire, and terminations for this work will vary by project. It is most typical for this scope of work to be allocated to the HVAC subcontractor. For a project with no significant HVAC controls it is more appropriate for the electrical subcontractor to provide the conduit, wire,

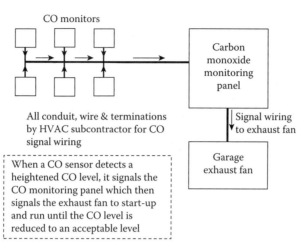

**FIGURE 40.7** Carbon monoxide detection system.

and terminations for this system because the HVAC subcontractor will not have a lower-tier low-voltage electrical subcontractor readily available on the project. The HVAC subcontractor will always be responsible for commissioning this system.

(a)  As a general rule, for a smaller project with few HVAC controls circuits the general contractor should consider allocating the HVAC controls circuits to the electrical subcontractor to avoid a lower-tier electrical subcontractor being employed by the HVAC subcontractor for only one or two days' worth of work.

10. The HVAC subcontractor will perform their own hydronic piping work. On a union project, the union jurisdiction will dictate that condensate piping is a drain and because plumbers claim all drains, it is to be performed by the plumbing subcontractor.

11. Large equipment (such as chillers, water storage tanks, and air handling units) located inside the building must be procured early in the project if they are too large to travel through the building once the walls are constructed. This equipment will require expedited procurement and significant protection while stored in place during heavy construction activities. Be sure this expedited procurement is clearly defined in the project schedule, bidding instructions, subcontract agreement, and expediting plan. Late delivery of such large equipment is a common cause of delay to interior framing, CMU, and exterior wall activities. The subcontractor providing the equipment (the HVAC subcontractor for this example) should always be responsible for providing protection of their own equipment, as well as the eventual removal and disposal of the protection materials. This protection should be sufficiently adequate to require no maintenance throughout the construction duration. (Note the disparity in responsibility between protection of large equipment versus protection of finishes, such as carpeting. In the case of building finishes the protection is provided by the respective

subcontractor, but maintenance and removal of the protection is actually performed by the general contractor.)

(a)   Equipment is only situated at inaccessible locations if there are no viable accessible locations. This is not a major concern for owners because large mechanical equipment is generally very reliable. Large equipment commonly has a 30-year or longer expected lifetime and, when necessary, malfunctioning parts can be repaired or replaced without replacing the entire unit. Owners often recognize that when it is time to replace major equipment in the distant future that the building will be in need of a major renovation. The equipment replacement can be timed to occur along with the renovation work.

12. The HVAC subcontractor should be held responsible for furnishing and installing all roof flashings for their own work. This includes roof jacks, flashing from equipment over curbs (Figure 40.8), and flashings required for duct penetrations (Figure 40.9). The HVAC subcontractor must also provide all sealants related to these flashings, maintaining full responsibility for waterproofing protection of their penetrations through the roofing membrane.

(a)   The flashing subcontractor is capable of completing flashings around ducts and equipment. Nonetheless, it is in the best interest of the project to keep responsibilities for the waterproofing integrity with the party that makes the roof penetrations. While there are exceptions to this general rule as discussed in various portions of this book, this approach greatly helps to avoid problems enforcing warranties in the event of a leak. The precise source of a leak is difficult to determine, so if the HVAC subcontractor installed a duct through a roof curb, the flashing subcontractor provided the sheet metal flashings from the duct over the curb, and the caulking subcontractor sealed the flashings, none would readily accept responsibility. Such warranty problems are common when multiple subcontractors are involved. If only one party is involved, the warranty is easily enforced. If more than one

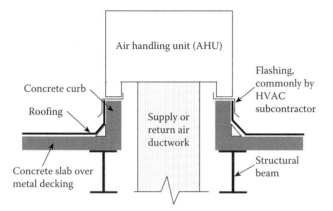

**FIGURE 40.8**   Cap flashing at equipment pad.

**FIGURE 40.9** Duct flashing at roof. (Photo by author, courtesy of The University of Southern California.)

party holds responsibility for a warranty, the reality is that there may be no enforceable warranty.

13. The HVAC subcontractor will install the trash chute and furnish all mounting hardware that is not otherwise provided by the chute manufacturer. The intricacies of the trash chute divisions of work are described further in the chapter on miscellaneous specialties, but there are additional items worth noting in this HVAC discussion:

   (a)  Assume that the trash chute manufacturer will provide mounting brackets for a 24″ diameter chute (24″ diameter is the most common size chute). These brackets are commonly only capable of spanning a 27″ × 27″ floor blockout. Trash chute installations typically consist of two chutes side-by-side (one for trash and one for recycling). A common problem occurs because this is often shown on the architectural drawings as a single large blockout, in lieu of two individual blockouts, and much wider than 27″. If the floor openings are too large for the manufacturer-supplied brackets, supplemental supports will be necessary and the HVAC subcontractor will assuredly submit a change order request for the additional work required to mount the chutes. The proper size of these floor openings may vary slightly from manufacturer to manufacturer, so the actual dimensions must be confirmed and coordinated by the general contractor.

   (b)  Trash chutes do not always come standard from the manufacturer with a rain cap. If this rain cap is excluded in the chute manufacturer's bid, it can easily be provided by the HVAC subcontractor, as long as they are notified during the bidding phase.

14. Secondary drain pans (Figure 40.10) will be required at locations shown on the drawings, such as below fan coils above gypsum board ceilings. Additional secondary drain pans may also be necessary due to the general

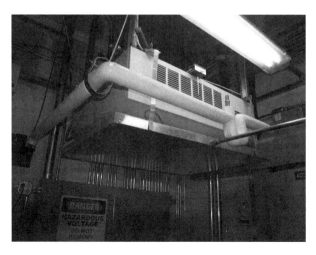

**FIGURE 40.10**   Secondary drain pan. (Photo by author, courtesy of Hathaway Dinwiddie Construction Company and The University of Southern California.)

contractor's means and methods, such as under piping in an electrical room. While the actual dimensions may vary from jurisdiction to jurisdiction, piping cannot be run above electrical gear within a specified lateral distance (stated in feet) of the gear. It is always best to route wet utilities around electrical rooms, but at times the only possible route for piping (whether it be plumbing, hydronic, or fire sprinkler) might be through the electrical room. A secondary drain pan will be required underneath this piping so that if a leak does occur it will be caught in the pan. These drain pans are sloped to drain in one corner at which point a screwed fitting is provided where the plumbing subcontractor will connect a drain line. While the secondary drain pans required as a result of the contractor's means and methods will not be shown on the construction drawings, a thorough pre-bid review of the MEP documents will provide the general contractor with a good estimate of the required number of these items. The general contractor must then inform the subcontractors of this additional work in the bid instructions.

(a)   Note that secondary drain pans can be furnished by either the flashing or HVAC subcontractors at the general contractor's discretion.

(b)   There may not be a building drain that can be reached by sloping the pipe to the drain. In that event, a small pump (similar to the fan coil condensate pump discussed in the chapter on plumbing) will be required. If a pump is required it should be furnished by the plumbing subcontractor, coordinated with the HVAC subcontractor, and powered by the electrical subcontractor. Again, for secondary drain pans added as a result of the general contractor's means and methods, this work must be identified and clearly conveyed to the bidders in the general contractor's bid instructions.

(c)   While all piping mains may be able to route around the electrical rooms, this will always be an issue for fire sprinkler piping. It is unavoidable that fire sprinklers will be located in the electrical room. Quite often sprinkler piping can be routed through the room while avoiding the prohibited space above and around the electrical gear. This may be done by entering the electrical room above the door (there will be no electrical gear in front of a door). Then route the sprinkler piping through the middle of the room, as far away from the electrical equipment as possible. Successfully routing the piping away from the electrical gear is largely dependent on the size, shape, and configuration of each individual electrical room. The pipe routing must be analyzed for each individual project.

15. Louvers on a building have a significant aesthetic value. These louvers, commonly termed architectural louvers, will most commonly be provided by the flashing subcontractor. At the general contractor's discretion, when there are only a few architectural louvers and they are all related to the HVAC work, they can be efficiently allocated to the HVAC subcontractor. Architectural louvers are usually located at the exterior walls, are easy to identify on the building elevation drawings, and have their own specification section in Division 10. This makes it easy to identify, quantify, and allocate this work.

16. The HVAC subcontractor will furnish and install flues from gas-fired equipment such as water heaters, boilers, and fireplaces complete.

# 41 Electrical*

The electrical scope of work is regularly the most technical and confusing work on a project. Electrical design documents are schematic in nature and require an educated, experienced subcontractor to complete the work in (similar to the other MEP trades) a basic design-assist approach (Figure 41.1).

Throughout this textbook, the electrical subcontractor has been mentioned as being responsible for various electrical points of connection, portions of low-voltage control circuit work, and numerous coordination issues. Many, but not all, of these responsibilities will be restated in this chapter. The objective of this chapter, and this book as a whole, is not for the reader to simply memorize a list of scope issues, but more broadly for the reader to learn how to look for, identify, and properly allocate responsibility for specific coordination issues. This will be done through a series of examples regarding electrical work.

Low-voltage electrical work is completed by a myriad of different specialty subcontractors, including the fire alarm (Figure 41.2), telecommunications (Figure 41.3), security (Figure 41.4), and audio/visual (Figure 41.5) subcontractors. On union projects, low-voltage electrical subcontractors cannot provide conduits, wiremold, or other raceways, due to union jurisdictions, so they rely on the electrical subcontractor for this work. Because of this, many low-voltage trade subcontractors are traditionally placed under a lower-tier subcontract agreement with the electrical subcontractor. Allocating these subcontractors to the electrical subcontractor is encouraged as coordination of the raceways for these design-build and design-assist low-voltage systems is extensive and complicated.

There are two apparent drawbacks to the electrical subcontractor employing these lower-tier trades. First, the electrical subcontractor will put a markup on the lower-tier subcontractors' quotes. Secondly, the electrical subcontractor will commonly take bids from only one or two companies with whom they have an established working relationship. As a result, they may not receive the lowest bid available. These drawbacks are overshadowed by the importance of the electrical subcontractor working closely with the low-voltage subcontractors. Before they receive bids, the electrical subcontractor will make an effort to ensure that all necessary conduits and other raceways for the design-assist and/or design-build work are properly included in the bids. This close relationship continues throughout construction with the respective crews working side-by-side. When contrasted with the drawbacks, the costs of the coordination and management efforts that are acquired through these arrangements are purchased at a bargain.

Because most low-voltage firms are commonly under a lower-tier agreement with the electrical subcontractor, these trades will also be discussed in this chapter.

---

* MasterFormat Specifications Division 26

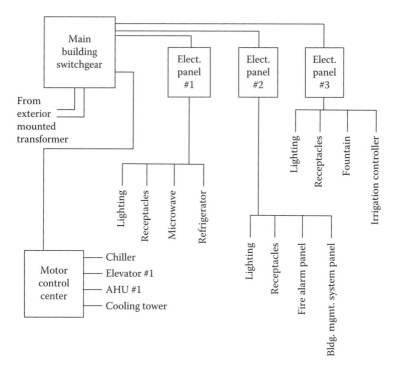

**FIGURE 41.1**   Basic electrical system.

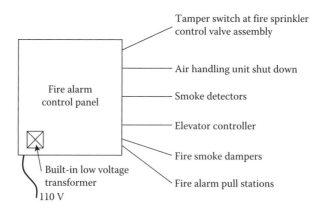

**FIGURE 41.2**   Fire alarm system.

## SCOPE OF WORK ISSUES RELATED TO ELECTRICAL

1. The electrical permit will be handled similarly to the plumbing and HVAC permits. The electrical subcontractor will routinely obtain the permit from the building department. Whether the owner or subcontractor pays for this

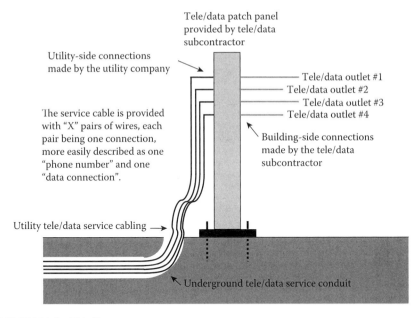

**FIGURE 41.3**   Tele/data system.

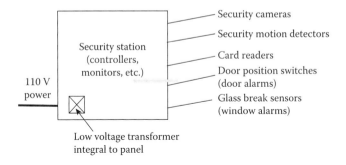

**FIGURE 41.4**   Security system.

permit will be dictated by the contract documents. Although permit costs are a variable, the variation is small and subcontractors commonly include the permit cost in their lump sum bid proposal.

2. While the plumbing and fire sprinkler subcontractors rely on the site utilities subcontractor to extend their utilities to within five feet of the building (from the respective point of connection at the city main line), the electrical subcontractor performs the electrical site work (Figure 41.6). The electrical subcontractor will perform all trenching, backfill, prefabricated in-ground vaults, prefabricated concrete pads, and cabling, and provide any electrical equipment required for the project. There are only a few exceptions:

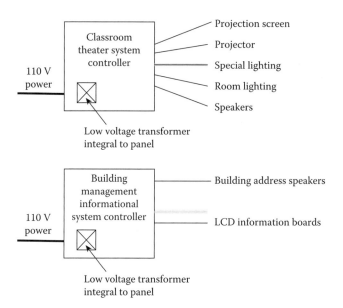

**FIGURE 41.5**   Audio/visual system.

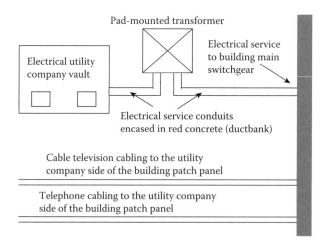

**FIGURE 41.6**   Site electrical.

(a)  While the electrical subcontractor will provide all prefabricated in-
ground vaults and concrete pads, they will not complete any custom
vaults or pads. These custom items will be completed by the site concrete subcontractor. Differentiating prefabricated items from custom
is an obvious task. If the drawings identify a manufacturer and model
number, it is prefabricated. Conversely, if the drawings provide a
structural design, they are custom.

(b)  Demolition of hardscape and stripping of landscaping for the electrical trenching and vaults will usually be performed by the demolition and/ or excavation subcontractors, but can be completed by the electrical subcontractor. This division of work is discussed in greater detail in the chapter on demolition and the chapter on excavation.

(c)  Replacement of man-made items that were demolished to make way for the underground electrical work will be replaced by the respective trade. For example, the installation of underground electrical work might demolish landscaping, sidewalks, gutters, asphalt paving, etc. It is important for the general contractor to identify and allocate this consequential work in their bid instructions. This work will not always be depicted on the construction drawings, especially when it occurs outside the project property lines.

(d)  Removing the spoils generated from the underground electrical work will commonly be excluded by the electrical bidders. The electrical subcontractors often have the expectation that the soil can be used elsewhere on site or that material could be hauled off by the excavation subcontractor in conjunction with their work. The excavated soil is rarely usable elsewhere on site. Furthermore, the excavation subcontractor will normally have demobilized by the time the underground electrical work commences. Therefore, the electrical subcontractor should maintain responsibility for removing their own spoils.

(e)  Tapping into the city electrical grid will be performed by the city's electric company. The electrical subcontractor will typically provide the conduit raceway, including all trenching, etc., from the transformer to the city vault where the tie-in will occur. The electric company will commonly furnish and install the cabling from the point of connection at the city grid to the building transformer. The electrical subcontractor will typically provide all cabling from the transformer to the building. It is important to verify the scope of work provided by the electric company prior to bidding, as the services provided vary slightly among electric companies and from city to city. For instance, in some regions the cabling between the city grid and the transformer may actually be furnished and installed by the electrical subcontractor. The tie-in to the city grid and terminations at the city side of the transformer will always be performed by the electric company. Additionally, the primary transformers for a building are commonly furnished and mounted by the electric company, but at times, these will need to be provided by the electrical subcontractor.

(f)  Telephone and cable television underground conduit will be provided and installed by the electrical subcontractor. This is similar to the conduit for the electrical service. In the case of these low-voltage building utilities, the telephone and cable companies will almost always furnish and install their own cabling. Additionally, each of these companies will terminate their cabling at a predetermined interior location on a patch panel. The telecommunications subcontractor will furnish and install the patch panels. This subcontractor will then commence wiring their system at the building side of the patch panel. (Most commonly, the cable television system will be distributed by the telecommunications subcontractor, not the audio/visual subcontractor as may be intuitively thought by the title).

3. There are many trades requiring a minor amount of low-voltage control wiring. The low-voltage work for sporadic building components is very difficult to locate in the bid documents. This is why low-voltage control wiring is commonly overlooked when bidding a project. When this work is not properly and clearly allocated prior to bidding, the general contractor will bear the financial responsibility for the subsequent scope gaps. There are four key elements of control circuitry that vary in responsibility among the different trades. These four scope elements of a low-voltage controls circuit will be described, followed by a few specific examples.

(a) Furnishing and installing conduit—installation of conduit is performed by the electrical subcontractor. As discussed in the chapter on HVAC, the specific electrical subcontractor performing this work will at times be a lower-tier subcontractor of the mechanical subcontractor. On union projects, the electricians have unwavering jurisdiction over this work.

(b) Furnishing and pulling wire—pulling wire is most often completed by the electrical subcontractor. For this work, there are some exceptions.

(c) Wire terminations—terminating the wire at the control panel or switch and at the device which the control panel or switch is controlling will most often be completed by the electrical subcontractor and is commonly claimed by the electrical trade union. At other times, particularly on non-union projects, this minimal amount of work may actually be performed by the respective system's subcontractor.

(d) Commissioning—start-up of the equipment and testing of the controls will be performed by the subcontractor providing the respective system.

A few examples of miscellaneous building elements requiring coordination with the electrical subcontractor for controls work include the following:

(a) Plumbing pumps with auxiliary control panels, such as booster pumps, will require a controls circuit from the controls panel to the sensor and from the controls panel to the pump. The electrical subcontractor will provide the conduit, wire, and terminations, while the plumber will commission the system.

The circuit between a plumbing pump control panel and the plumbing pump is sometimes a 110V or 220V circuit, not a low-voltage circuit. As a rule, when a controls circuit for any device or equipment runs on medium-voltage power (such as 110V or 220V), the terminations will always be completed by the electrical subcontractor.

(b) Conduit, wiremold (Figure 41.7), cable trays (Figure 41.8), and/or other raceways for the fire alarm, telecommunications, and security low-voltage signal wiring will be furnished and installed by the electrical

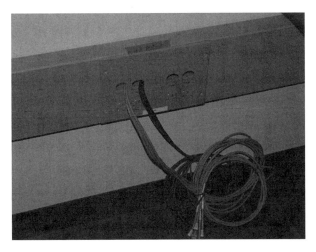

**FIGURE 41.7** Wiremold. (Photo by author, courtesy of Hathaway Dinwiddie Construction Company and The University of Southern California.)

**FIGURE 41.8** Cable tray. (Photo by author, courtesy of Hathaway Dinwiddie Construction Company and The University of Southern California.)

subcontractor. All wiring, terminations, and commissioning of these low-voltage systems will be performed by the respective system's subcontractor.

(c) The circuit between a coiling door motor and the door controller will require conduit, wire, and terminations to be furnished and installed by the electrical subcontractor. This also applies to the interlock circuit between the motor and the manual locking mechanism which is used to ensure that the motor will not operate if the lock is engaged (preventing motor burn-out). Commissioning will always be completed by the coiling door subcontractor.

(d)   A wall-hung plasma screen television has a manufacturer-supplied wiring harness that runs from the television to the control box. This control box is commonly placed in a cabinet near the television. This arrangement requires an in-wall raceway from the cabinet to the back of the television. This raceway will be provided by the electrical sub-contractor and requires a large conduit (as much as 3″) because the prefabricated wiring harness supplied with the television has large plugs on each end that must be accommodated by the conduit. The audio/visual subcontractor who installs the television will also pull the wiring harness through the conduit. Naturally, terminations and commissioning will also be completed by the audio/visual subcontractor.

(e)   As discussed in the chapter on mechanical, the HVAC subcontractor will perform their own controls work via a lower-tier electrical subcontractor when there is an extensive HVAC controls scope of work. When the HVAC system entails few controls, it is more efficient to identify the locations and allocate the associated conduit, wire, and terminations to the electrical subcontractor.

(f)   Dryer duct booster fans, common in residential buildings, have 110V control circuits. For these medium-voltage circuits, the conduit, wire, and terminations must be completed by the electrical subcontractor. The HVAC subcontractor will commission these booster fans, commonly without any assistance from the provider of the dryers.

The above examples provide the basic methodology for controls work allocation among the subcontractors. These examples constitute a sample of the building components requiring controls coordination. Some other examples include the following:

(g)   The controls circuit between the fan coil at a walk-in refrigerator or freezer and the respective roof-mounted condensing unit.

(h)   The controls circuit for an HVAC split system between the wall or ceiling-mounted fan coil and the condensing unit on the roof.

(i)   The controls circuit between a dock leveler and the dock leveler controller.

(j)   The controls circuit between motorized window shades and the respective switch.

(k)   The elevator remote recall (discussed in further detail in the chapter on elevators).

(l)   The controls circuits at door frames for handicap operators, card readers, and security door sensors.

To properly allocate the controls work, two exercises should be performed and the results should be incorporated into the bid instructions. First, the general contractor must review the specifications to identify the building components that require controls work. Secondly, the general contractor must review the mechanical and plumbing equipment schedules on the drawings to identify equipment with auxiliary control panels and/or sensors.

4. Where low-voltage circuits begin, there will be a step-down transformer converting a 110V circuit to a 24V circuit. Most control panels provided

by various manufacturers will have a built-in step-down transformer, few will not have them. Unfortunately, the custom-made control panels, which are common for building management systems, do not routinely include this step-down transformer. In such a case, the ideal approach is for the subcontractor who provides the control panel to also include the step-down transformer.

If the subcontractor providing the control panel does not include the step-down transformer in the panel, they will probably not provide a space for the transformer inside the panel either. This will mean that the step-down transformer will need a separate electrical enclosure mounted adjacent to the control panel.

5. Generators are provided by the electrical subcontractor, but they require considerable coordination with other subcontractors.

    (a)   The formwork, rebar, and placing and finishing subcontractors will construct the equipment pad for an interior generator. A concrete pad for a generator mounted outside the building footprint will be constructed by the site concrete subcontractor.

    (b)   Depending on their size, interior walls may need to be left out until the interior generator is in place to accommodate the path of travel through the building. This means that several trades will need to return to erect and finish the remaining walls. The amount of come-back work and the effected subcontractors will vary by project. It is important to incorporate a project-specific plan for the generator into the bid instructions. Begin procurement as early as possible to obtain and install the generator, as well as all other large interior equipment, before the walls are scheduled for construction.

    (c)   All conduit, wiring, and termination work for the controls circuits will be the responsibility of the electrical subcontractor. The commissioning work will be completed by the generator manufacturer. Note the disparity in the generator scope in that unlike other MEP equipment the installation, start-up, and commissioning work is actually bid and executed by the generator manufacturer, not the electrician or a separate specialty subcontractor.

    (d)   The generator will include a muffler, which is shipped loose with the generator and installed by the HVAC subcontractor. On union projects, the HVAC subcontractor will mount the muffler and complete any exhaust work beyond the muffler. An interior generator will always have an exhaust stack, though an exterior generator may not require an exhaust stack beyond the muffler. All materials beyond the muffler will also be provided by the HVAC subcontractor.

    (e)   Where the exhaust stack penetrates the roof, a curb will be necessary. The curb is to keep the roofing membrane a sufficient distance from the hot exhaust stack. The type of curb will vary from project to project, but, regardless, the curb will be furnished by the same subcontractor(s) furnishing the other roof curbs. The general contractor or framing subcontractor will provide wood curbs. The concrete (formwork, rebar, and placing and finishing) subcontractors will provide concrete curbs and the HVAC subcontractor will provide prefabricated manufacturer's curbs.

The exhaust stack will need to extend a substantial distance (commonly 10 feet) above the roof; therefore the stack will need to be guyed. In addition to the curb around the exhaust stack, three short posts may be needed to anchor this guying, which can easily be provided by the HVAC subcontractor.

(f) The HVAC subcontractor should furnish and install the rain cap (also called a mushroom cap) on top of the stack. This rain cap is sometimes excluded by HVAC bidders that assume the flashing subcontractor will furnish and install it.

(g) Generators commonly have a fuel tank built into the generator unit. Larger generators and generators that may be required to operate for an extended duration will also have auxiliary fuel tanks, which are normally buried underground. These fuel tanks, associated piping, fuel pumps, excavation, and backfill are the responsibility of the plumbing subcontractor.

(h) Controls conduit, wire, and terminations for the fuel tank and pumps will be furnished and installed by the electrical subcontractor. Commissioning of the fuel system controls will be a joint effort of the electrical and plumbing subcontractors.

(i) In-wall backing for the generator control panel will be provided by the framing subcontractor.

(j) The fuel tank may be a significant distance from a roadway. In fact, the hose from a fuel truck may not be long enough to reach the tank. This is when a remote filling port is necessary. This port and associated piping will also be completed by the plumbing subcontractor.

6. Plywood backboards lining the walls of electrical and data rooms will always be shown on the electrical and telecom drawings, but sometimes are not shown on the architectural drawings. These backboards should be furnished and installed by the general contractor's carpenters, as electrical subcontractors cannot perform this work, especially in union jurisdictions. At the general contractor's discretion, the framing subcontractor can also perform this work.

7. In climates subject to freezing, all exterior, above ground, water piping must have heat trace tape. A heat trace tape is also required on a condensate line running from the fan coil inside a walk-in freezer. The heat trace tape is essentially a heated wire running along each water pipe to keep it from freezing. This heat trace tape will be furnished and installed by the electrical subcontractor.

8. The trash chute manufacturer will provide the door interlocks and master control switch FOB jobsite with the chute delivery. These door interlocks allow only one trash chute door to be open at a time. This feature serves several purposes. For example, this is a safety feature to keep someone from having their hand injured in the chute by trash dropping from above.

Also, a maintenance worker can lock all doors while the trash bin at the base of the chute is being serviced. The electrical subcontractor will install these interlocks turnkey, including all conduit, wiring, terminations, commissioning, and furnishing of materials beyond the interlock devices and master switch.

9. Some projects have landscaping on elevated decks, such as plaza decks over basements. Any related irrigation control wiring running through the building should be run in conduit, which is provided by the electrical subcontractor. This is similar to (as discussed in the chapter on plumbing) the plumbing subcontractor providing the copper pipe irrigation lines running through the building. All wiring, terminations, and commissioning of these irrigation controls will still be completed by the landscaping subcontractor.

10. Most mechanical and plumbing equipment will be equipped with factory-mounted motor starters, but there are still exceptions. Similarly, few manufacturers will factory-mount disconnect switches. The electrical subcontractor must provide motor starters and disconnect switches for all equipment that does not have them factory installed. The equipment that requires these items will be indicated on the electrical drawings.

11. A common omission that general contractors make when scheduling a project is the failure to account for burn-in time of a new audio/visual (abbreviated to A/V) system. It will take as long as 60 days for a fully commissioned A/V system to essentially be broken-in. This is not intuitive, but very important. The absence of a fully functioning A/V system could easily cause a project delay in reaching substantial completion.

There are several indirect items of work that are the responsibility of the electrical subcontractor, but are not indicated in the contract documents. This electrical work is required solely for construction purposes and can be very expensive. Items 12, 13, and 14 provide examples of this indirect work.

12. The first and most important indirect electrical work is the temporary construction power. This temporary service is the responsibility of the electrical subcontractor and will not be addressed by the design team. Having temporary power is one of the first concerns a general contractor will worry about on a project. This power is needed on the first day of work, but obtaining this power requires a considerable effort. Even though this power is temporary it will still come from the city power grid. The application for temporary power will undergo the same scrutiny and red tape as the permanent power service. This temporary power will be fully engineered by the electrical subcontractor based upon a criterion provided in the general contractor's bid instructions that quantifies the size of the temporary power service. The electrical subcontractor will also go to the electric company for approval, will interact with the electric company to answer their questions, and will pay for and pull the permit. One difference between permanent and temporary power is that for temporary power the electrical subcontractor will actually provide the transformer in lieu of the electric company. Because it

is smaller and only used for a short period of time, permitting a temporary power service is generally much more expedient than obtaining approval for the permanent power plan. Despite this, it is never as quick as desired and snags in obtaining temporary power are a common cause of project delays.

13. For projects in dense urban environments, or other areas where pedestrians will be sufficiently close to the building to be in danger of being struck by objects falling from the scaffolding or elevated floors, a roofed pedestrian barricade will be required to protect the public. Because the roofs of these barricades block out the overhead street lighting, they must have ceiling-hung lighting. This lighting in the pedestrian barricades will be furnished, installed, maintained, and finally removed by the electrical subcontractor. The maintenance of this lighting includes the replacement of lamps (bulbs) that burn out and repairs necessitated by vandalism. Replacing lamps can be effectively estimated and included in the subcontractor's base bid, but estimating the repair costs due to vandalism is highly variable. Vandalism includes tearing down conduit, punching out lamps, and general destruction in the barricades. Because vandalism is out of the general contractor's control it is most appropriate for the owner to assume this risk. This risk can be minimized by having the electrical subcontractor include an allowance in their base bid for a stipulated number of hours of labor and a stipulated amount for materials to make electrical repairs required by vandalism.

    (a)  As discussed in earlier chapters, it is preferable for subcontractors to include allowances in their base bids in the form of a stipulated number of hours and a stated sum for materials. In this way the bidders will include the hours at their bid rate, which is regularly a lesser cost than their change order rate. Also, the materials included in this allowance will already have the markups included in their estimate. Thus, no additional markups will be added to this allowance. As with the labor rates, the subcontractor's material markups in their estimate are generally lower than the markups used in their change order requests.

14. The electrical subcontractor will provide, maintain, and remove temporary construction lighting throughout the building. The extent of this lighting must be clearly defined by the general contractor and conveyed to all subcontractors, not just the electrical subcontractor. This lighting can range from the code-required illumination of exit corridors and stairwells to illuminating all areas of the project. The actual lighting requirements usually fall between these extremes. All bidders must be informed of the extent to which temporary lighting is provided. All subcontractors will be held responsible for all lighting necessary beyond the temporary lighting plan that is necessary for the performance of their work.

    (a)  For example, the typical extent of temporary lighting for a new condominium building would illuminate the parking garage, but not the mechanical, trash, or utility rooms located at the garage levels. The corridors and stairs would be fully illuminated. The living rooms and bedrooms would be illuminated, but the illumination would not extend to the bathrooms and closets. Therefore, subcontractors working in the utility rooms, bathrooms, and closets must furnish their own lighting, commonly referred to as task lighting.

15. Tower cranes and material hoists, also termed manlifts, draw significant power and require significant electrical services. These services must be incorporated into the temporary electrical service calculations and be provided turnkey by the electrical subcontractor. For safety reasons, running these services underground is recommended to protect the electrical lines from accidental damage. The electrical subcontractor will need to be specifically directed to trench and bury these services in the bid instructions.

16. Exterior concrete light-pole bases are commonly provided by the electrical subcontractor, not always by the site utilities subcontractor as might be an intuitive choice. When pole bases are constructed as per the pole manufacturer's standard footing detail (usually an 18″ or 24″ diameter cylindrical footing) the electrical subcontractor will complete the work. If the pole base is custom or part of a larger site concrete element, such as a seat wall or retaining wall, the site concrete subcontractor will construct the footings and set the anchor bolts.

## FIRE ALARM

A. The permitting process for a fire alarm system is nearly identical to that of a fire sprinkler system. That is, both are design-build life-safety systems under the jurisdiction of the fire marshal. The design documents will include a fire alarm specification section identifying the fire alarm device models and general design criteria. The drawings will also identify the locations of the primary system components, such as the control panel location. The fire alarm subcontractor will design, engineer, submit, interact with the plan reviewer, and pay for/obtain the permit. Fire alarm subcontractors regularly perform this permitting work. Over time, like the fire sprinkler subcontractors, they develop excellent working relationships with the fire marshal. There is generally little need for the direct involvement of the general contractor in this permitting process, but the general contractor will naturally monitor and supervise this process.

B. For ductwork penetrating a fire-rated wall, a fire/smoke damper (Figure 41.9), which will shut off the duct in the event of a fire, is required. The signal to this fire/smoke damper comes from a smoke detector mounted inside the ductwork near the fire/smoke damper. If smoke is detected in the duct, the damper shuts down and prevents the smoke from traveling past the rated wall. The HVAC subcontractor will furnish and install the fire/smoke damper. The fire alarm subcontractor will furnish the duct-mounted smoke detectors FOB jobsite to the HVAC subcontractor. The HVAC subcontractor will mount the detectors in the duct and provide a small access door in the side of the ductwork for access to the detectors. The electrical subcontractor will run a branch line of the fire alarm system conduit to the detectors and from the detectors to the respective fire/smoke dampers. The fire alarm subcontractor will complete

**FIGURE 41.9** Fire smoke damper. (Photo by author, courtesy of Hathaway Dinwiddie Construction Company and The University of Southern California.)

the wiring and terminations. Commissioning of the duct detectors and fire/smoke dampers will be a collaborative effort between the HVAC and fire alarm subcontractors. Although this seems to be an extremely convoluted scope of work, it is an identical division of work from project to project. The subcontractors are all well aware of their specific responsibilities and it is rarely a cause of change order debate.

C. The smoke and heat detectors located within a trash chute are specialty devices that will be furnished FOB jobsite by the chute manufacturer for installation by the fire alarm subcontractor. Aside from having these devices supplied by the chute manufacturer, the remaining responsibilities for these devices are typical of other smoke and heat detectors.

D. Fire alarm subcontractors tend to be quite familiar with code requirements of the fire marshal. Nonetheless, one item of work that is commonly overlooked is smoke detectors in elevator lobbies located a few feet in front of the elevator doors. These smoke detectors are actually a requirement in the elevator code, not the fire code. These smoke detectors are tied to the fire alarm system and are specifically monitored by the elevator control panel. Although these detectors are required on every floor, the most important detector is at the elevator primary recall floor, usually the building lobby at the ground floor. In the event of an alarm, the elevator is programed to return directly (no intermittent stops) to its primary recall floor. If the smoke detector in the elevator lobby at the primary recall floor detects smoke, it will signal the elevator that the fire is on that floor. The elevator controls will then send the elevator to the designated secondary recall floor for safe exiting of the passengers.

E. The fire alarm subcontractor will maintain responsibility for tie-in of devices, provided by others, which are required by code to be monitored by the fire alarm system. These devices include tamper switches and flow

**FIGURE 41.10**    J-hook. (Photo by author, courtesy of Hathaway Dinwiddie Construction Company and The University of Southern California.)

switches for the fire sprinkler system, as well as a connection to the elevator control panel.

(a)    Unlike all other devices, the terminations at the elevator control panel will be made by the elevator subcontractor because only an elevator technician is allowed to touch anything inside an elevator controller. Commissioning of this circuit will be a joint effort of the elevator and fire alarm subcontractors.

## TELECOMMUNICATIONS

A. Conduit raceways and cable trays for the telecommunications work (abbreviated to telecom) are provided by the electrical subcontractor for all communications cabling. Simple j-hook (Figure 41.10) or other individual hangers are provided by the telecom subcontractor. The differentiation between these two scopes is simple:

(a)    The most common division of work for the telecom raceways is as follows: First, cable trays will be specifically identified on the contract drawings. Secondly, when conduit is required it will be clearly identified in the specifications and/or drawings. Generally, conduit is required anywhere telecom cabling is run in a non-accessible space, such as within a wall cavity or above a gypsum board ceiling. Lastly, j-hook or other individual supports will be used anywhere that a cable tray or conduit does not occur, such as above lay-in acoustical ceiling grids.

B. Cable trays and equipment racks regularly require grounding to the building structure. This grounding work is performed by the electrical subcontractor, including grounding bars and final connections. This work will frequently be shown on the telecom drawings, but not on the electrical drawings. Electrical bidders are accustomed to this work and regularly include it.

# Questions—Module Seven (Chapters 38–41)

1. Identify four of the five primary subcontractors dealing with water on a typical building project.
2. Where is the point of connection between the fire sprinkler subcontractor and site utilities subcontractor? Where is the point of connection between the plumbing subcontractor and site utilities subcontractor?
3. Where are backflow preventers for the domestic and fire water services typically located—that is, are they located inside or outside of the building?
4. Which subcontractor will furnish and install a grease interceptor located 15 feet from the face of building?
5. Which subcontractor should provide the fire sprinkler drains?
6. Which subcontractor should furnish and install the low-voltage transformer for a sump pump?
7. Which subcontractor should provide the irrigation piping between a waterproofing membrane and a topping slab?
8. Which subcontractor should provide the drain line from a hydronic boiler system?
9. Where is the best place to look for information on the power connection for a condensate pump when the pump is an integral part of the plumbing engineers design?
10. Which party should perform the grouting of the base of a plumbing pump?
11. Which subcontractor should make the final connection for the natural gas line to a hydronic boiler?
12. On natural gas piping work, where does the plumbing subcontractor's work begin?
13. For which type of rooms do the architectural reflected ceiling plans commonly indicate sprinkler head positions?
14. Identify four items the bid documents will include for a design-build fire sprinkler system.
15. Common terminology refers to the masterformat specification divisions 21 through 28 as the MEP trades. In practice, which of the following trades (electrical, plumbing, fire sprinkler, HVAC) does this terminology not include?
16. Which of the following will not be shown on the electrical drawings? (medium-voltage circuits for the fire pump and control panel or low-voltage circuits for the fire pump and control panel)
17. Which subcontractor will install the exhaust muffler for a diesel generator located in a building basement?
18. Which subcontractor should furnish the flow switches for the fire sprinkler system? Which subcontractor should install them? Which subcontractor should tie the switches to the fire alarm system?

19. Which subcontractor should furnish the fire hose cabinets? Which subcontractor should install them? Which subcontractor should provide a fire extinguisher identified to be within the fire hose cabinet?
20. Are fire sprinklers likely to be required under large exterior canopies?
21. Is fire sprinkler protection required for pallet racks, which commonly store a great deal of combustible materials?
22. Describe the two steps involved in activating a preaction type fire suppression system:
23. Which subcontractor should provide the radiators, and all associated piping?
24. (Fill in the blanks in the following statement) A flow meter in the domestic water supply piping that is tied into the building management system will be furnished by the _____ subcontractor, installed by the _____ subcontractor, wired to the BMS by the _____ subcontractor and commissioned as a joint effort by the _____ and _____ subcontractors.
25. Which subcontractor should provide the low-voltage transformer for a custom-fabricated BMS control panel?
26. For an air handling unit (AHU) with an auxiliary VFD: The _____ subcontractor will bring power to the VFD. The _____ subcontractor will furnish the VFD. The _____ subcontractor will mount the VFD. The _____ subcontractor will provide conduit, wire and terminations for the circuit from the VFD to the AHU.
27. Which subcontractor will furnish the duct smoke detectors? Which subcontractor will mount them? Which subcontractor will provide the conduit? Which subcontractor will wire the duct smoke detectors? Which subcontractor will commission them?
28. Which subcontractor will complete the condensate drain from an AHU?
29. Ideally, how many subcontractors should be responsible for the weatherproof integrity of a duct penetration through the roof?
30. Which subcontractor should be responsible for furnishing and installing the flue for a gas-fired domestic water heater?
31. What is the objective of this book in terms of the reader?
32. Which subcontractor should pull the electrical permit?
33. Discuss the statement that a prefabricated concrete transformer pad located in a landscaped area will be completed by the site concrete subcontractor.
34. For an electrical service trench through asphalt pavement that extends well beyond the project property lines, most commonly, how accurate is the depiction of the asphalt pavement demolition and replacement in the construction drawings?
35. Which subcontractor should be responsible for the removal from the site of excess spoils generated from the underground electrical work?
36. Which party will furnish and install the underground telephone cables from the city point of connection to the building?
37. What are the primary components of a controls circuit?

38. Which subcontractor will complete the control wiring for a plasma television?

39. Identify 10 common building components that require control circuitry?

40. Identify the two exercises a diligent general contractor will perform prior to bidding the project to assure the low-voltage controls work is properly allocated to the subcontractors.

41. Which party will be responsible for performing the start-up and commissioning of a generator?

42. Plywood backboards in telecom rooms are not always shown on the architectural drawings, but will be shown on the electrical drawings. Which party should furnish and install them?

43. Which party is responsible for engineering the temporary power service?

44. Which party bears the financial responsibility for replacing lighting at the pedestrian barricades that has been vandalized?

45. Which party should provide an 18″ diameter light-pole footing constructed as per the light-pole manufacturer's standard footing detail?

46. Name two permits common among projects that will be pulled from the fire marshal?

47. By which code are smoke detectors required in elevator lobbies located a few feet in front of the elevator doors?

48. Which subcontractor is responsible for completing the terminations of the fire alarm wiring at a tamper switch in the fire sprinkler system? Which subcontractor is responsible for terminations of the fire alarm wiring at an elevator controller?

# Module Eight

# 42 Site Utilities*

A number of different utilities are required to make most facilities fully functional, including electrical, natural gas, water, sewer, and communications. In some situations, a different municipal department is associated with each utility service. Some utilities (electrical or telephone) may be brought to the site via overhead lines but most utilities are brought in as buried lines. Site utility work constitutes that portion of the utilities where the lines are brought onto the project site and where they are tied into the facility being built immediately outside the building line.

Site utilities, in concept, are simple and straightforward, but their installation can have a significant impact on the project schedule. Further, because most utility lines are installed in trench excavations in unknown terrain, this work has a high potential for project delays and change order caused by unforeseen conditions unearthed during the progress of work.

Unforeseen or differing site conditions can pose a major hurdle in completing the site utilities work. An example of such is where the soil conditions differ materially from those shown in the construction documents. The information presented in the bid documents is commonly derived from soil borings performed by the geotechnical (also termed soils) engineer. Instances may arise where conditions are materially different simply because the locations where the soil borings were taken were not representative of all subsurface conditions. The time of year when the borings were taken may also result in subsurface water conditions being quite different at the time when the site utilities are installed. Additionally, unforeseen conditions, especially in older urban environments, may be encountered when the drawings fail to show previously installed utilities or other past building features that are buried in the ground. Historical drawings can be a good source for information about a site, but they are not always accurate. It is important to try to anticipate the possibilities of encountering conditions that are not depicted in the project documents. A subcontractor who will be performing trenching work cannot reasonably be expected to include the discovery of unforeseen conditions in a lump sum bid, nor can the general contractor assume this risk. Understandably, the ultimate financial responsibility must reside with the owner of the project. For a private project under a negotiated prime contract it is most appropriate to include an allowance for unforeseen conditions and negotiate (in addition to labor rates and markup percentages) unit prices for any unexpected obstructions with the site utilities subcontractor prior to awarding the site utilities work. Naturally, due to public bidding regulations this negotiation is not commonly feasible on a

---

* MasterFormat Specifications Division 33

**381**

publicly financed project. The following are a few common examples of unforeseen conditions:

1. Old concrete or brick building foundations built before municipalities were keeping detailed record drawings may be encountered. Such old buildings were demolished years ago and what records of their construction that did exist were often lost through the course of time.

2. When old telephone lines that were not identified during the underground utility survey are encountered, in most instances they will be abandoned lines, but this is not known until they have been tested. Site utilities work is often completed during swing or graveyard shifts. If a phone line is encountered, the crew will need to stop work, secure the site, demobilize, and wait until the line is tested. This could result in the loss of a full day of work. That is, they will only lose one day of work if that phone line is verified as being abandoned. If verification of the line cannot be made for several days, the project will be delayed even longer. If the line is found to be active, rerouting the new utility installation can take weeks. Note that live lines should not be discovered if proper procedures had previously been followed by all parties. For example, before any digging is performed, it is important to call the utility companies and have them mark their lines for the contractor. This is now a simple process as one call to the "call before you dig" number (or dial 811) will alert all the utilities that they are to mark their active lines within the next 24 hours or within the next two business days, depending on the regulations of the local region.

3. Unforeseen conditions may include the discovery of abandoned sanitary sewer or storm drain laterals. This can be a blessing in some instances, especially if the pipe has not been grouted full, as is common practice when utility lines are abandoned. Depending on the circumstances, it may be possible to reuse the line and save some money. Unfortunately, in most cases such lines will be grouted full and the pipes may also contain asbestos (transite pipe). If asbestos, or any other, contaminated material is found, the excavated trench will need to be secured and an abatement crew should be brought out as soon as possible. It is recommended that an abatement subcontractor be identified early in the project, especially when there is a high potential of discovering contaminated materials. The early inclusion of an abatement subcontractor can save several days. Otherwise, time will be lost as an abatement subcontractor is found, rates are negotiated, the subcontractor's insurance is verified, a formal agreement is issued, and the subcontractor is scheduled to do the work. Also, by making early arrangements with the abatement subcontractor, there is ample time to negotiate their unit costs. If they are hired only after the need for their services is crucial, there will not be sufficient time to negotiate their rates and a premium cost for securing a subcontract agreement on short notice may be realized.

4. The discovery of artifacts or valued paleontological resources may constitute an unforeseen condition and cause project delays. Such discoveries might include the following:

(a) Old animal bone fragments, which will not be clearly ruled out as human remains until the coroner has been given an opportunity to examine the find, are quite common discoveries.

(b) Ceramic or glass pieces, dishes, bottles, or other artifacts may be found, the significance of which will not be known until an archeologist has had an opportunity to examine the find.

(c) Human remains may be encountered when digging in places such as unmarked cemeteries or in Native American burial grounds. These discoveries will also involve both the archeologist and coroner, and if the discovery is suspected of being Native American, a representative of the Native American nation that formerly occupied the area as well. This representative is appointed from a pool of people certified through a genealogy study confirming that they personally are most likely descendants (termed MLDs) of the people who lived in the respective region hundreds or even thousands of years ago. When working with the MLDs, jobsite productivity for trench excavation may stop or be restricted to only a short length of shallow trenching per day. When digging in an area where significant discoveries can be expected, it is important to plan for these unforeseen conditions both financially and with contingency time built into the project schedule.

If the research of the site indicates that some type of valued items might be unearthed by the excavation work, take appropriate precautions by including time in the schedule not only for a slow excavation, but also for a possible work stoppage resulting from a discovery.

In terms of scheduling considerations, the site utilities work is generally scheduled towards the end of a project. This work is routinely scheduled after the scaffolding has been disassembled around the building and before the hardscape and landscape work is performed. Nonetheless, this is not the best time to perform the underground work, because as the scaffolding is coming down around the building, the landscape and hardscape often become critical path activities in a rush to complete prior to opening the building.

Due to the difficulty in having a timely sequencing of the dismantling of scaffolding, installation of site utilities, and the performance of the landscape/hardscape work, it is advisable to complete the site utilities at the onset of the project, during the excavation and site grading work. Unfortunately, at the onset of a project it is easy to become overwhelmed by getting construction underway and the site utilities may be viewed as something that can be done later. The site utilities work can be done later, but as explained, scheduling problems can abound toward the end of the project. At the onset of construction there are generally only a few other trades on site, making the scheduling of site utilities relatively easy. Two to three months into the project the site will be covered with office trailers, cargo boxes, construction vehicles, materials, and many trades, at which time site utilities work cannot possibly be done because of limited free space. Once the scaffolding goes up around a building the site utility tie-ins to the building are extremely difficult to complete safely, as the trenches will undermine the scaffolding legs and supports. Toward the end of the project, the work will be intense and subcontractors will be busy trying to finish their work so they can demobilize their office trailers, cargo boxes, and other essential items.

## SCOPE OF WORK ISSUES RELATED TO SITE UTILITIES

1. Be sure the scope of work of every subcontractor that will be performing digging of some type includes a provision for hauling off any spoil material that will not be used as backfill. Site utility subcontractors will often exclude hauling off their own spoils if it is not specifically included in the subcontract agreement. Likewise, the excavation subcontractor will commonly exclude hauling off any spoils that are not generated from the mass excavation operation unless specifically instructed otherwise. An economical alternate is to have the excavation subcontractor haul off all of the surplus excavated materials generated from the site utilities work because they can easily perform this work as part of their other operations. If the site utilities subcontractor needs to haul off their own spoils on a daily basis (as is common in dense urban environments) it is not as efficient because they may very well mobilize with two dump trucks, only to find that the second truck will be only half full. Naturally, this can only be successful if the site utility work is performed at the onset of the project.

2. When lane closures are necessary, it is imperative that this responsibility is clearly spelled out. In addition to the responsibility for drafting, submitting, paying for, and pulling the lane closure permits, the work involved with the closures themselves must be defined, including the responsibility for setup, break down, arrow boards, cones, flaggers, etc.

3. Separate taps into the water main will generally be required for domestic water and fire water. There will be one backflow preventer for the domestic water and one for the fire water. This is because the fire system always needs to be a standalone system that is unhampered by any device not directly attributed to the fire system itself.

4. For each utility, assign a clear point of connection between the site utilities subcontractor and the building trade subcontractor (Figure 42.1). For instance:

   (a) For most trades the point of connection will be within five feet of the building perimeter.

   (b) For the domestic water, the site utilities subcontractor will provide the meter and backflow preventer, while the plumbing subcontractor will take the line into the building.

   (c) For the fire water service, the site utilities subcontractor will provide the backflow preventer (meters are not required for fire water systems), while the fire sprinkler subcontractor will run the line from the point of connection into the building.

   (d) For the chilled and steam systems the site utilities subcontractor will furnish the vaults (Figure 42.2) and all work within the vaults and then run the service supply and return lines to within five feet of the building line. The HVAC subcontractor will take the lines through the building wall.

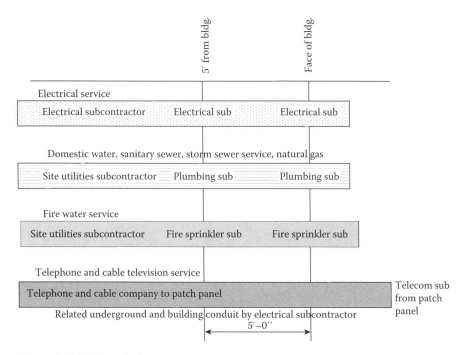

**FIGURE 42.1**   Site utilities.

**FIGURE 42.2**   In-ground vault. (Photo by author, courtesy of Hathaway Dinwiddie Construction Company and Tishman Speyer.)

5. Underground steam piping requires insulation, such as DriTherm (Figure 42.3). While this insulation will be provided by the site utilities subcontractor, it requires formwork provided by either the formwork or framing subcontractor. If the site utilities are completed early in the

**FIGURE 42.3**   DriTherm insulation. (Photo provided by DriTherm International.)

project, the formwork subcontractor would be the appropriate trade to provide these trench forms because they will be on site, while the framing
subcontractor will not yet be on site. The opposite is true for site utilities
completed towards the end of a project when the framing subcontractor
will be on site, but the formwork subcontractor will not.

6. Although it would be intuitively thought that the site utilities subcontractor
will provide the electrical, telephone, and cable television services, these
wired services are actually performed by others, primarily the electrical
subcontractor. The division of responsibility for the underground electrical,
telephone, and cable television services is discussed in greater detail in the
chapter on electrical.

7. Due to tremendous public safety concerns, natural gas service has a unique set
of rules. For the gas service, determine exactly what work the city or public
utility will perform and then have the plumbing and/or site utilities subcontractor perform the remaining work. Most cities will not allow anyone except
their own employees, or their approved subcontractors, to perform the tie-in to
the main line. Due to the safety concerns, this is probably the best route and a
good way of assuring only experienced, qualified, and professional crews will
work on volatile gas lines. Mistakes with this work are unacceptable as they
are potentially catastrophic. Sometimes cities will also perform the trenching
and backfilling work, other times this will be performed by the site utilities
subcontractor. The site concrete subcontractor will provide concrete pads for
above ground gas meters, though the underground gas meters will be placed
in boxes provided by either the city or the site utilities subcontractor. The general contractor must verify the city responsibilities and properly allocate the
remaining work in their bid instructions.

# 43 Asphaltic-Concrete Paving*

Asphaltic-concrete paving (AC paving) work is generally straightforward in scope, but this work can contribute to a major project disruption when closing out a project. The biggest and most common obstacles associated with this work are maintaining access in and out of the jobsite and sustaining the efficiency of other trades while this work is performed, as the area where this work is performed is commonly covered with office trailers, cargo boxes, stored materials, worker parking, and other obstructions that have accumulated throughout the course of the project. This work will be performed at the end of the project, often during the chaotic conditions when other work is frantically being performed as the construction effort comes to a close. It is important to properly schedule and coordinate this work so that access in and out of the jobsite is not hampered to the point that it causes inefficiencies or delays for the other subcontractors who must concurrently perform other work on the project. Phasing this work is a common practice, but the AC paving subcontractor will need to be made aware of any sequenced work prior to bidding the project via either the project schedule or bid instructions.

## SCOPE OF WORK ISSUES RELATED TO ASPHALTIC-CONCRETE PAVING

1. One of the most frequently overlooked aspects of asphaltic-concrete paving work involves milling and patching the pavement that is not specifically noted in the contract drawings. Trucks, cranes, and general construction vehicular traffic will often cause damage to the city streets, but because this damage is a construction issue it is not commonly reflected on the construction drawings. Be sure to address the costs associated with this work. Note that this work is often mandated as part of the city permit requirements.
2. Since pavement striping is a relatively quick and comparatively inexpensive operation the asphaltic-concrete paving subcontractor should ideally perform all striping work, including incidental marking of concrete curbs and underground parking garage work. Some exceptions should be noted. For example, if 90% or more of the striping work is in an underground parking garage, it is more fair and expedient for the general contractor to directly hire a striping subcontractor to do this work.

---

* MasterFormat Specifications Division 32

3. Be sure to properly allocate header boards as necessary and avoid wild paving (without any edge forms) where the paving simply ends on grade. Most asphaltic-concrete paving companies are more than capable of self-performing this work and should include their own headers. Where project conditions allow (and they often do) the asphaltic-concrete paving subcontractor may opt to pave wild, and then come back and saw-cut a clean straight edge. This is generally a cost-effective approach for straight terminations, but where the AC paving makes a radius a header board will be required because saw-cutting machines are large and cannot make clean curves.

4. Asphaltic-concrete paving is an oil-based product and has the potential to leach oils into adjacent porous materials. Be sure to review the compatibility of the asphaltic-concrete paving with adjacent materials such as brick, soft stone, etc. Use a bond breaker when paving comes in contact with porous materials or when deemed necessary. On building projects, architects commonly show the installation of inexpensive bond breakers such as felt or foam in the drawings. At times, caulking joints are used on top of bond breakers to provide for a nicer finished appearance. It is important that the asphaltic-concrete paving subcontractor include this caulking in their bids.

5. The asphaltic-concrete paving subcontractor will invariably include the cost of constructing the asphaltic-concrete speed bumps, as they will only occur at AC paving locations. If prefabricated speed bumps are specified, it is not a certainty that any subcontractor will actually include them in their price quotes. Thus, make sure to properly allocate these items to a designated subcontractor or maintain it as self-performed work by the general contractor. The use of prefabricated speed bumps is common in underground parking garages where it is generally advisable for the general contractor to furnish and install them in the structural concrete decks since the AC paving subcontractor will not have any other work in an underground garage. If these prefabricated speed bumps occur in asphaltic-concrete locations, such as in a parking lot, it is advisable to have the asphaltic-concrete paving subcontractor assume responsibility for supplying and installing them.

   (a)  Also, it is advised that if prefabricated speed bumps are specified an attempt be made to change the construction to, respective to the location, asphaltic-concrete or cast-in-place concrete bumps. The prefabricated speed bumps are more attractive, but they are also less stable and have a tendency to pop out after three to five years.

   (b)  If prefabricated speed bumps are specified in a post-tensioned deck, be sure to place markers indicating tendon locations prior to placing the concrete deck, otherwise the costly and timely effort of x-raying the deck to ensure that the drilling for expansion anchors does not hit and damage a tendon will be required. Some contractors have used adhesive anchoring for prefabricated speed bumps to avoid drilling into a post-tensioned deck, but these adhesives have generally not performed well and have a poor track record for longevity.

**FIGURE 43.1**   Initial AC paving lift for construction purposes. (Photo by author, courtesy of Hathaway Dinwiddie Construction Company and Tishman Speyer.)

6. Be sure protection of adjacent surfaces from splatter of asphaltic-concrete paving operations is covered in the asphaltic-concrete scope of work. Concrete curbs, sign posts, and other items within the asphaltic-concrete paving work commonly receive splatter on the bottom six inches, which can be easily avoided by putting a little extra effort into protecting these items.

7. For a large project with a significant asphalt parking lot surrounding the building(s), it is a good construction practice for the AC paving subcontractor to place the asphaltic-concrete pavement in a minimum of two lifts (layers) and in two different mobilizations. The first pavement lift should be completed early in the construction schedule to provide a clean working area for material storage, staging, and fabrication, particularly during the winter periods (Figure 43.1). The final lift of pavement should not be placed until late in the project, as this lift will suffer construction damage if placed too early. The added efficiency and increased cleanliness of the project far outweighs the cost of the additional mobilization by the AC paving subcontractor.

8. Traditionally the asphaltic-concrete paving subcontractors will be responsible for all parking and traffic-related signage, including signage that is set in landscaped islands or concrete paving. Make sure that convenience signage in the parking lots is also covered. The AC paving subcontractor will only pick up signage required by code, such as stop, handicap, and yield signs. This may result in there being no party with an assumed responsibility for convenience signage, including directional signs such as "Pediatrics This Way" or "Service Deliveries This Way." Depending on the nature and frequency of the convenience signage, this work can be covered by either the asphaltic-concrete paving subcontractor or signage subcontractor. There is commonly a preference to have the signage subcontractor include this signage, but either subcontractor is capable so this just needs to be verified by the general contractor.

# 44 Site Concrete*

Site concrete consists of concrete that is required to complete a project, but lies outside the exterior perimeter of the building being constructed. The general rule of thumb is that a site concrete subcontractor will perform all concrete work outside of the building line, while the formwork, rebar, and place and finish subcontractors will complete all concrete work within the building. Examples of site concrete include concrete streets, approaches, sidewalks, planters, retaining walls, utility pads, and so on. While the site concrete work is fairly straightforward, it is often performed near project completion when tight scheduling constraints are encountered, thus this work can become a critical path activity.

The structural concrete work is completed by a team of three subcontractors (formwork, rebar, and place and finish), each under a direct subcontract with the general contractor. In contrast, the site concrete work is performed by a single subcontractor who will capably comprise all three of the respective structural concrete trades.

## SCOPE OF WORK ISSUES RELATED TO SITE CONCRETE

1. It is important to ensure that the rebar and wire mesh for the site concrete work is appropriately allocated. In the majority of cases, the site concrete subcontractor will complete the rebar and wire mesh work with their own employees, and usually there is no problem with craft union jurisdiction. For projects with project labor agreements (PLAs), the rebar work may need to be performed by a rebar subcontractor. Project labor agreements are special, collaborative, agreements negotiated between owners of high profile projects and the labor unions in which each party to the agreement is provided special incentives. If union jurisdiction is an issue, it is still not a problem as long as the work has been properly allocated in the bid instructions prior to bid time. It will be important to verify whether this work must be done by a rebar subcontractor, be it with the subcontractor employed for the structural work by the general contractor, or a second-tier rebar subcontractor who has contracted with the site concrete subcontractor. Since larger reinforcing steel subcontractors specialize in structural work, they generally have no substantial interest in the project after the structural concrete work is completed. For this reason, it is advisable to have the reinforcing steel work done by the site concrete subcontractor or by a reinforcing steel subcontractor working for the site concrete subcontractor as a sub-subcontractor.
2. Be sure the caulking at expansion joints is included by one of the subcontractors, preferably the site concrete subcontractor. These joints occur where

---

* MasterFormat Specifications Division 32

391

the flatwork meets the building, where the flatwork meets vertical concrete elements, and elsewhere as described by the construction drawings. The caulking subcontractor is also a capable party to perform this work if the site concrete subcontractor is unable to undertake this work with their own crew.

3. Anticipate and include an allowance for work necessitated by utility work and construction damage outside the property lines. Ideally, the architect should coordinate underground utility routing and anticipate where construction damage is likely. This information should be reflected in the drawings, but this is not always addressed in the construction documents. Areas along trucking routes and city sidewalks bordering the site are common locations that may require repair.

4. The site concrete subcontractor must include the labor costs of setting embeds and anchor bolts in the site concrete for tree grates (Figure 44.1), statues, bollards, etc. Embeds and anchor bolts should be provided by the subcontractors providing the final respective product.

5. It is common to exclude the light-pole bases from the scope of work of the site concrete subcontractor, as these tend to be standardized pole bases that are actually completed by the electrical subcontractor. On the other hand, they should be included in the site concrete subcontractor's scope of work if the pole bases are non-standard, architectural in nature, and/or part of a larger concrete element, such as a seat wall.

6. The site concrete subcontractor should include the concrete pads for the gas meter, water meter, transformer, and any other MEP items in the site work on a turnkey basis. Note that these are not always shown on the drawings.

**FIGURE 44.1**  Tree grate. (Photo by author, courtesy of Hathaway Dinwiddie Construction Company and The University of Southern California.)

7. Include the pad for the irrigation controller in the site concrete scope of work. This is usually shown on the drawings, but this information is generally included in the landscaping details. As a result, such pads are often overlooked by the site concrete subcontractor because they will not commonly review the irrigation details when compiling their estimate.

8. The site concrete subcontractor should set bollards, swing gate posts, anchor bolts, and other FOB items that will be provided by the miscellaneous metals subcontractor. Chain link fence posts and parking sign posts, on the other hand, will be by the chain link and asphalt paving subcontractors respectively; all other embed work is typically included in the work of the site concrete subcontractor.

9. It is generally advisable to hold the asphaltic-concrete paving subcontractor responsible for painting the curbs and performing other site concrete markings because they will generally already have a pavement striping subcontractor on board under a lower-tier subcontract and this additional painting work is typically very minor.

10. Coordination between the landscaping and site concrete subcontractor is important. First, be sure all the necessary irrigation sleeves are properly placed (as routed below sidewalks or streets) for both the irrigation water and control wires. Secondly, be sure access for the landscaper is not cut off. Make sure the path of travel from the trucking lane to the point of placement for all large trees (or other large fixtures) is thought through. Trees that must be set with the use of a crane will need to be set before the site concrete work, in the crane's path of travel, is performed. This will ensure that the crane does not damage the new walkways. Also, be careful to check the box size for the trees versus the openings in the sidewalks/plazas where they are scheduled to be set. For large trees the box sizes are often larger than the opening in the pavement, so those trees will need to be set early (Figure 44.2).

11. If stair nosings are required in the site concrete steps, make sure the site concrete subcontractor has included setting them in their base bid. In most cases the miscellaneous metals subcontractor will furnish both the interior and exterior stair nosings for installation by others. There is nothing wrong with this arrangement as long as both furnishing and setting the stair nosings are covered by the subcontractors. At the same time, make sure they are not covered by more than one party (double coverage).

12. It is preferred for the site railings to be field measured after the site concrete work is completed. The concrete can then be core drilled and the railings set and grouted in place. This is a cleaner and simpler installation than trying to accurately set sleeves in the concrete or to set and temporarily brace the railings prior to pouring the concrete. Further, if bike racks, benches, or other site furnishings are shown in the project drawings to have J-bolt anchors, it is recommended that approval be sought to use expansion anchors so they can be set after the concrete work is complete. This will not only be a more efficient construction method, but it will also provide for a better finished product. Small J-bolts and sleeves, which are far less

**FIGURE 44.2**   Large trees must at times be set ahead of the concrete paving the root ball for this tree is wider than the opening in the sidewalk. (Photo by author, courtesy of The University of Southern California.)

critical than the structural steel anchor bolts discussed in the chapter on formwork, are commonly knocked out of place during a site concrete pour. The effort needed to ensure these elements are not knocked out of place is significantly higher than the effort of core drilling the concrete or setting expansion bolts, therefore it is a good construction practice to follow these procedures.

13. The site concrete subcontractor must maintain responsibility for all concrete finishing activities, including sandblasting and/or staining. Concrete staining is a very difficult operation with little room for error; therefore prior to awarding this work, be sure the subcontractor possesses the necessary expertise. The difficulties inherent to a colored concrete finish are discussed in more detail in the chapter on concrete placing and finishing.

# 45 Chain Link Fencing*

Chain link fencing work is not generally a source of large scope changes or expensive change order. Even so, if this scope of work is not properly defined and planned, this work can be a source of small, but irritating, change order requests and problems. Chain link fencing bids are often considered to be no-brainers because of their simplicity. This simplicity often results in bids being received and subcontract agreements being issued without any serious discussions being held about the scope of work. As a result, problems often surface only after the work has begun. This can be avoided with a short discussion with the chain link fencing subcontractor before the subcontract is awarded and construction work begins.

The scope of work for chain link fencing may in fact be quite simple and without any complications, but do not count on this simplicity. On the other hand, this work is more complex if there are security requirements or aesthetic considerations that must be met. Without any unique requirements for security or aesthetics, it may appear to be a simple matter of obtaining bids and executing a subcontract with little need to discuss the work with the chain link subcontractor until it is time for them to erect the fence. Unfortunately, chain link fencing is not always as simple as it appears. Several issues of concern will be described. These tend to apply especially well to interior fencing.

## SCOPE OF WORK ISSUES RELATED TO CHAIN LINK FENCING

1. For conditions where the locking hardware is furnished by others, such as a lockset or a deadbolt, the chain link subcontractor's responsibilities are very similar to those of the miscellaneous metals subcontractor's responsibilities for steel gates. Be sure the chain link subcontractor does the following:
   (a) Provides a plate for attachment of the hardware and properly accommodates the strike plate.
   (b) Performs proper preparations, such as cut-outs and precisely drilled holes, of the plates. This is very important and quite often this work is performed incorrectly. For this reason it is suggested that the locking hardware be furnished by the door subcontractor to the chain link subcontractor for installation. This will help ensure that the preparatory work is properly completed, especially since the tolerances for properly functioning locking hardware are often very small.

---

   (c)   Provides a screen or plate surrounding the locking hardware to prevent someone from reaching through the fence and opening the gate.

   (d)   Provides a plate for the mounting of the card reader. This is often forgotten because the design documents will typically show the card reader only on the security drawings, which may not be examined by the chain link subcontractor.

   (e)   Provides a means of mounting the panic hardware, as well as a plate or screen around the device to keep someone from reaching through and opening the gate.

2. Make sure that posts to be set in concrete footings (not driven) are installed before the asphaltic-concrete or concrete paving work is done so that the paving will run tight to the post and avoid a footing-size patch around each post (Figure 45.1). Even when simple driven posts are specified, the gate posts will still need to be set in concrete because driven posts will not adequately support the cantilevered weight of the gate, so those will need to be set early.

3. For simple inexpensive exterior chain link fencing, the drawings will commonly lack details for the gates and the closing or latching mechanism. For such fencing, be sure the subcontractor provides a locking hasp at all gates and lockable cane bolts at the double gate locations. Providing these items may not be shown on the construction drawings, but because they are staples of a quality chain link fence assembly this should be considered implied by good construction practices.

4. Determine if top and bottom rails (Figure 45.2) are required by the specifications, or if simple tension wires (Figure 45.3) will be deemed sufficient. The rails are often omitted when the design intent is a low-cost fence, but even with inexpensive fencing a top rail in particular is often desired.

5. Ensure that the hinges are tamperproof.

**FIGURE 45.1**   Post set ahead of concrete paving. (Photo by author, courtesy of The University of Southern California.)

**FIGURE 45.2** Chain link fence with top and bottom rails. (Photo by author, courtesy of Hathaway Dinwiddie Construction Company and The California Institute of Technology.)

**FIGURE 45.3** Chain link fence with tension wire.

6. For gates with high security requirements (typically welded chain link panel fencing) that have padlocks, shields will commonly be required over the padlocks so they cannot be cut with bolt cutters.
7. The chain link fencing subcontractor must include mounting plates for security alarms, illuminated exit signage, and any other item mounted to the fencing. For security devices, such as alarms, they must also install plates or screens around the devices and conduit running to the devices to prohibit tampering.

(a)  For both security and aesthetic purposes, it is always best to run the conduit, or low-voltage wiring, up through the post itself as opposed to surface mounting the conduit.

8. For interior fencing it is suggested that welded baseplates and drilled expansion anchors be used to set the posts. In general, avoid drilling and driving posts through slab-on-grade concrete to avoid damaging conduits or piping below the slab and to avoid damaging the waterproofing membrane or vapor barrier.

(a)  For thick elevated decks (10″ or 12″) it is possible to core drill about six inches deep and set the posts, but the structural engineer will assuredly have a problem with coring because of the probability that slab-reinforcing bars will be encountered. There are accommodations that can be made for this core drilling, such as installing additional rebar in the deck to compensate for the rebar that is cut—so plan early.

(b)  For post-tensioned decks, costly x-raying will be required unless markers are properly placed to identify the tendon locations in the deck prior to placing the concrete. This is where early planning is important. In addition to the high cost of the x-raying operation, a premium charge will be inevitable because the x-raying procedure will need to be performed during off hours to avoid a delay impact on other trades. This is because a significant area around the powerful x-ray machine must be cleared of people in order to x-ray safely.

(c)  For drilling into post-tensioned decks for simple 3/8″ or 1/2″ anchor bolts it is a good construction practice to drill 1/8″ pilot holes into the deck to verify that no tendons are present in the drilling location before using a larger bit. It is important to first obtain permission from the structural engineer to drill 1/8″ pilot holes to locate the tendons as an alternate to x-raying, but they rarely have a problem with this. Structural engineers generally agree to this as a 1/8″ drill bit will shatter before it can do any damage to a tendon.

9. For interior fencing, fully communicate the extent of MEP penetrations through the fencing prior to bid time. Chain link fencing subcontractors do not often review MEP drawings, nor are they skilled at such to verify penetrations through their fencing work, and they will request a change order for this trimming work. Further, clearly determine the means of trimming around the MEP systems, as just cutting the chain link fabric will not yield a durable finished product. Depending on the significance of the aesthetic and security considerations, rails may need to be added to frame out each penetration.

10. With multiple elevators sharing a joint pit there is a safety divider between the individual hoistways that is quite often a chain link product, though it will at times be of another construction type, such as concrete. Be sure to verify the specifications of the materials to be used for the pit dividers for this portion of the project and properly allocate the work.

# 46  Landscaping and Irrigation*

There is a common realization about landscaping work and that is that once the project is in the landscaping phase it is also a sign that the project is nearly complete, i.e., most of the hard work has been done by the time landscaping work is scheduled and the project team is eagerly anticipating completion. In general, the landscaping work is done quickly and transforms the exterior of a project in just a couple of weeks from looking like a construction site to looking like a completed facility. As a bonus, there is little that can go wrong with landscaping. It is arguably the most painless scope of work to execute.

## SCOPE OF WORK ISSUES RELATED TO LANDSCAPING

1. Be sure the redwood, plastic, or other material headers are included in the landscaping work. Landscapers often exclude the redwood headers in particular, but it is logical to place the responsibility with them, as their crews are more than capable of handling these items that are integral to their work.
2. Be sure to coordinate planting the large trees with the hardscape work. It is common for the tree well openings in the hardscape to be smaller than the box sizes of the trees (not to mention having to remove the sides, and sometimes full bottoms, of the boxes once the trees are set in place). Also, be sure to plan appropriately for crane access to set the trees. If a crane needs to be set on the hardscape, damage could occur from the point loads of the crane outriggers. Damage caused by the weight of the crane to both hardscape and landscaping in the path of travel from the street to the crane hoisting location is also a common and costly trade coordination error.
3. Any soil specifically scheduled as imported fill at landscaping areas needs to be placed by the landscaper. For project sites with rocky or otherwise poor soil conditions, it is common for the landscape architect to require the native soils to be removed to a designated planting depth and to import new, healthier, soil. In this case, be sure responsibility for removing the native material and importing the new soil is accurately described for both the landscaper and the excavation subcontractor.

---

* MasterFormat Specifications Division 32

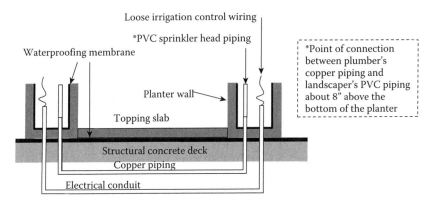

**FIGURE 46.1**   Irrigation piping and controls run through the building.

4. There must be a clear definition of the work to be performed by the landscaping subcontractor and the plumber, especially as it relates to irrigation lines running to balconies, plazas on top of elevated decks, etc. (Figure 46.1). Because PVC piping is not commonly allowed inside a building as per building codes, within the building enclosure the irrigation lines will generally be run in copper, which is the plumber's work. These lines will transition to the landscaping subcontractor's PVC pipe at planters and pots after the plumber has penetrated the waterproofing membrane with copper piping. The division of work between these two trades occurs at the building envelope, therefore irrigation lines running below a topping slab, but above the waterproofing membrane, may also be PVC pipes. The exact transition points of connection must be properly allocated between the plumber and landscaping subcontractor, as the construction drawings may only indicate the size and locations of the irrigation piping, without mention of the type of pipe because landscaping designers commonly deem this coordination effort a contractor's means and methods issue. The unique aspects of each project will result in changing situations that must be dealt with on a project-by-project basis.

   (a)   For instance, if the general contractor opts to route an irrigation line running from one planter to another above the structural podium deck and waterproofing membrane (Figure 46.2), it will be viewed as being above the building enclosure and this will be routed in PVC by the landscaper. Conversely, the line will be in copper if the general contractor opts to route it below the structural deck and the work in this case should be performed by the plumber. This should all be clearly spelled out in the scope of work for the landscaping subcontractor and the plumbing subcontractor to avoid having a gap in the scope of work.

5. If control wiring is required to be run in conduit through the building this must be coordinated with the electrician. To avoid subsequent drilling, plan the placement of sleeves through concrete walls and footings even if the control wiring does not need to be in conduit. The irrigation work performed by the plumber and electrician is often overlooked early in a project.

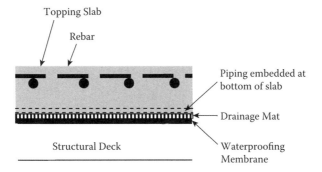

**FIGURE 46.2**   Irrigation piping at the bottom of topping slab.

6. The landscaper is traditionally responsible for providing and installing the site furnishings, such as benches, trash containers, ash urns, and bike racks. It is clearly appropriate for the landscaper to assume the responsibility for the tree grates, tree cages, any lattice work, and similar features, but it is also a regular practice for the landscaper to furnish and install all the other site furnishings to avoid over-complicating this relatively simple scope. In isolated instances, the general contractor may decide to provide and install such items as self-performed work.

7. Another important consideration with the landscape work relates to the maintenance period. Ordinarily, the specifications will require the landscaping subcontractor to perform routine maintenance for a stipulated length of time beyond the project substantial completion date. For most projects this maintenance period is relatively short. The bid quotation submitted by the landscaping subcontractor should be examined carefully and the maintenance period must be coincident with the period stipulated in the specifications. If the landscaping subcontractor's maintenance period is shorter than the period stipulated for the project, an adjustment must be requested from the landscaping subcontractor. In summary, double check that the maintenance period required by the specifications matches the maintenance period qualified in the landscaper's bid. Specifications have a tendency to require somewhat longer maintenance periods (say 120 days) whereas landscapers have a tendency to qualify somewhat shorter durations (say 30 days) in their bids.

# Questions—Module Eight (Chapters 42–46)

1. Name one type of work for which the excavation subcontractor does not need to include hauling off spoils?
2. What kind of reliance for accuracy can be placed on historical drawings?
3. Who bears the ultimate financial responsibility for unforeseen conditions?
4. Give examples of four different unforeseen conditions that might be discovered underground.
5. Should site utilities be completed at the onset of a project or at the end of a project? Explain why.
6. Which subcontractor should provide the trench forms for installing the underground steam piping, which is insulated with DriTherm, at the onset of a project?
7. Which party should perform the tap into the city gas main?
8. Which of the following trades (casework in the building lobby, asphaltic-concrete paving, site concrete surrounding the building, terrazzo in the building lobby) is least likely to perform their work in phases in order to maintain clear access in and out of the project?
9. Assume that a new building is located in a city center with five levels of below-grade parking. There is no asphaltic-concrete paving indicated on the construction drawings, but the general contractor will employ an AC paving subcontractor to mill and patch the street in front of the project because their fenced construction area extends 15 feet into the city street. They expect the paving in this area to become damaged. In this case, which party should employ the pavement striping subcontractor?
10. Discuss the suggestion to use adhesive anchoring for prefabricated speed bumps on top of a post-tensioned concrete deck, to avoid the risk of damaging tendons when drilling for expansion anchors.
11. Which subcontractor should provide the convenience signage in an asphaltic-concrete paved parking lot directing people to the building lobby?
12. Which subcontractor is preferable for completing caulking joints in the site concrete work?
13. Which subcontractor should furnish embedded angle frames for tree grates? Which subcontractor should place them?
14. How well are concrete pads for utility work, such as water meters, gas meters, or transformers, depicted in the construction drawings? Which subcontractor should provide them?
15. Which subcontractor should be responsible for concrete pads for irrigation controllers? Where would this information about concrete pads be located in the construction drawings?

16. Which subcontractor has responsibility for each of the following: exterior chain link fencing posts, exterior pipe bollards, and footings for exterior parking signage?

17. Which party is commonly responsible for furnishing interior stair nosings and which party is responsible for furnishing exterior stair nosings? Which party will commonly set the exterior nosings?

18. Which subcontractor will be responsible for providing the plate for a lockset on a chain link gate and who will be responsible for all preparatory work to the plate, including drilled holes and cut-outs?

19. Discuss the statement that chain link gate posts will always be set in concrete, even when the fencing posts are specified to be driven only.

20. Explain the merits of the construction practice of drilling 1/8″ pilot holes into a post-tensioned concrete deck prior to drilling for 1/2″ expansion anchors. What concerns might the structural engineer have with drilling into a post-tensioned concrete deck?

21. Which subcontractor provides the redwood landscaping headers?

22. When native soil is unsuitable for planting, the landscaping designer may require the native soil to be excavated and removed. The excavated material is then replaced with imported fertile soil. Which subcontractor will perform the excavation work? Which subcontractor will haul the native material from the site? Which subcontractor will be responsible for importing, placing, and compacting the imported soil?

23. Where is the transition between a copper irrigation line installed by the plumber and a PVC irrigation line installed by the landscaping subcontractor?

24. Which subcontractor would generally provide the ash urns outside a building?

25. A landscaping subcontractor will provide routine maintenance for the landscaping work. At what time will their maintenance services no longer be provided?

# Module Nine

# 47 Subcontractor Scope Issues

Each of the building trades possesses unique traits, responsibilities, intricacies, and coordination issues. Despite this, there are many issues that they have in common. In some cases, these commonalities are applicable to selected trades, while in others they are common to all. These commonalities and the best ways to manage them on a construction project will be described.

## COMMON SUBCONTRACTOR SCOPE ISSUES

1. It is important to routinely verify that a subcontractor has adequate liability insurance coverage and that they are current with their payments. A single uninsured accident can cause a significant financial loss to a company and even bankrupt a small or medium-sized company. A project manager may experience many years before being faced with a serious issue concerning insurance. From a personal perspective, one serious accident without insurance coverage can seriously blemish a career. Issues with deficient insurance coverage commonly happen when project managers and superintendents bring small companies onto a project for one or two days' work without thinking about checking their insurance. In most instances, the odds of incurring an accident in this short time are small. Unfortunately, the instance where an accident does occur could be catastrophic to both the company and the project manager's career.

2. Similar to liability insurance, a subcontractor's ability to provide a sufficient bond for a project is generally thought to be a simple matter. A subcontractor's ability to provide a bond is extremely important. This is especially true during economic downturns. Bonding companies are a surprisingly informative resource because they thoroughly review a subcontractor's finances prior to providing a bond. If a subcontractor is unable to provide a sufficient bond for a project, this is an early warning sign that the subcontractor is having financial difficulties. This means that the bonding company does not believe the subcontractor is in a sufficiently stable financial position to successfully complete the project in question. Bonds are a crucial risk management tool that must be purchased immediately after a subcontract is issued and the general contractor must follow up to verify that the bond has been swiftly obtained.

   (a) In addition to verifying that subcontractors have furnished sufficient insurance and bonds, it is important to qualify that the insurance and bonding agencies are reputable. Financial institutions, such as insurance and bonding companies, are given ratings relative to their reliability. A qualification about the insurance and bonding companies should be included in the bid instructions stipulating the minimum rating that the financial institutions must have. The rating should be required to be provided by a specific rating firm, such as A.M. Best Company, as there are many substandard rating companies, just as there are many substandard financial institutions. Subcontractors in dire financial positions may attempt to obtain bonds from offshore bonding companies. Since these firms are located in a different country than the project, they may not face the same legal repercussions as a local bonding company in executing the bond in the event of a default. Having an executable bond in hand does no good if the bonding company declines to make good on the bond.

3. Negotiating, writing, processing, and issuing all the subcontract agreements for a project involves a great quantity of work. Getting the subcontracts issued and signed before the subcontractors commence work is an important task. Several of the reasons timely subcontract execution is important include the following:

   (a)  Some subcontractors take exception to the liability language of the subcontract agreement. It is important to ensure that the subcontractors accept the terms and conditions of the subcontract agreement prior to officially awarding them the project. Even though this language may be provided in the bid instructions, it does not carry much weight until the subcontract agreement has been fully executed.

   (b)  The subcontract agreement will include, or at least refer to, the project schedule. By signing the subcontract agreement, the subcontractor accepts the schedule as prepared. For example, the general contractor may have allowed three months in the schedule for a subcontractor to perform their work. The subcontractor might insist that the work will take six months because they have limited labor resources. This discrepancy must be addressed and effectively resolved prior to the subcontractor being officially awarded the subcontract. After that, the subcontractor can commence their work with certainty that the project schedule is not jeopardized.

   (c)  The scope of work outlined in a subcontract agreement is rarely agreed to as initially written by the general contractor. It is important to confirm the subcontractor's scope of work in writing (in the subcontract agreement) prior to the subcontractor beginning any work. Once the subcontractor has commenced work without a signed agreement, the general contractor is in a defenseless negotiating position and may be viewed as having waived compliance with the contract. This could be costly for the general contractor.

The project manager and superintendent should never permit a subcontractor to work on a project without first providing a proper bond, certificate of insurance, and signed subcontract. To let a subcontractor on site without these documents is irresponsible.

4. Projects located within large cities or other dense urban environments will have many restrictions. These restrictions must be clearly conveyed to all bidders in the design documents and/or bid instructions. Examples of restrictions that might be encountered include the following:

(a) Parking is a costly indirect expense incurred on projects residing within dense urban environments. The cost of parking in a parking garage can easily be $25 per day per vehicle—or even higher in such cities as Los Angeles and Manhattan. The responsibility for employee parking varies among the subcontractors and other project team members. General contractors and architectural firms regularly reimburse their employees for the cost of parking, but this is not a universal practice. On union projects, some unions require that their workers be reimbursed for parking, but other unions do not. All companies (contractors, architects, and subcontractors) must be accountable for their own parking costs unless other arrangements have been made with the owner.

On some urban projects, parking is not a problem. This might be the case involving a new building in a large corporation's complex where the owner opts to provide free use of their parking facilities.

The second preferred option would be to find ample parking within a reasonable distance (a quarter mile is a typical union-agreement-dictated distance) of the project site, as more distant parking areas will be considered remote. On some inner city projects, there may not be adequate parking in the area, necessitating the use of remote lots. This is when bussing will be the only viable alternative to get workers to the job site. When parking is remote, all workers will need to be transported to and from the designated parking location(s) and project site. To simplify the bussing arrangement, the owner or general contractor may rent a parking lot on the outskirts of town and arrange for busses. Multiple busses will make pick-ups at the lot to transport the workers at the beginning and end of each work day, and intermittently during the work day. The rented parking lot may even be a parking garage. Obviously, the cost of parking in the garage will be added to the cost of bussing.

On a union project where bussing is required, it is standard union protocol that a worker's day begins and ends at the parking lot bus pick up location, not at the jobsite. If bussing workers takes 30–45 minutes for each direction of travel, a considerable cost will be incurred for bussing alone. The productive time that workers will be on the project site will be reduced to 6-1/2 or 7 hours. When each of the workers is still paid for 8 hours of work each day, the loss in productivity will be a substantial cost through the life of the project.

For example, an inner city project employed 300 workers who were designated to park one mile from the project site. The arrangements were that these workers would park in a parking garage at a cost of $25 per day. It took an average of 45 minutes for each worker to be transported each direction. The average hourly wage for the workers including all insurance and other markups was $50 per hour. Also, the bus company costs were $1200 per day, equating to $4 per worker per day. The cost computations are as follows:

- $25 per day in the parking garage + ($50 per hour × 1.5 hours) + $4 per worker for bussing = $104 per day per worker for bussing
- 300 workers × $104 per day = $31,200 per day!
- In one year at this rate the total will be $8,112,000 in parking fees and lost productivity (assuming 260 days are worked by all workers)!

If some workers carpool, there will be some reduction in these cost estimates, so it is prudent to develop a carpooling program of some sort on the project. This program may entail entering the carpoolers' names into weekly raffles or providing other preferential treatment.

This is an extreme case on a very large project, but it demonstrates how something as simple as parking can drive up project costs.

The authors are aware of a project in which the owner's representative failed to alert the bidders that parking would be at a remote location. This problem was resolved a short time into the project when the owner cleared a site for worker parking near the construction site. By the time the worker parking problem was resolved, the costs associated with the remote parking had already exceeded half a million dollars.

(b) Noise restrictions, established in allowable decibel levels at different times of the day, are quite common within cities. These restrictions routinely differ among workdays Monday through Friday, Saturdays, and Sundays. Holidays typically have identical restrictions as Sunday. The common impacts these noise restrictions may have on a project include the following:

   (i) The inability to begin work before a stipulated morning hour (commonly 7:00 a.m.) and the inability to continue work beyond a stipulated evening hour (commonly 6:00 p.m.). This restriction would apply Monday through Friday.

   (ii) Saturday work hours might be limited to 9:00 a.m. through 5:00 p.m. and Sunday work might be disallowed altogether.

  (iii) Excessively noisy work, such as pile driving and mass excavation operations, might receive greater work hour restrictions. Such tasks might have to be performed between the hours of 9:00 a.m. and 5:00 p.m. during the work week and might not be allowed at all on weekends or on holidays.

(c) Large cities with heavy traffic congestion may restrict trucking activities during the rush hour commutes. For example, trucking activities might be restricted Monday through Friday to occur outside the hours of 7:00 a.m. to 9:00 a.m. and 4:00 p.m. to 6:00 p.m. It might be possible to get trucks into the site, off-loaded, and out of the city in the early morning hours before 7:00 a.m., as is often preferred. Keep in mind that these night-time activities will not be allowed when they violate the noise restrictions previously discussed.

(d)  Concrete pours requiring lane closures are typically restricted. Lane closures for concrete pours are commonly allowed anytime outside the hours of 7:00 and 9:00 a.m. and 4:00 and 6:00 p.m. to avoid impacting the morning and evening commutes.

(e)  Site utilities work that must occur in traffic lanes is also restricted in many cities to be completed between the hours of 7:00 p.m. and 7:00 a.m. Since it will require shutting down more than one traffic lane, municipalities will weigh the noise impact caused by this night work versus the traffic impact of shutting down multiple lanes during the day. In most instances, the noise impact receives a lower priority. Thus, street utility work is often performed on the late swing shift and during the graveyard shift.

5. Change orders on construction projects are inevitable. To be prepared for these change order requests, it is important to establish subcontractor change order labor rates and markup percentages prior to awarding every subcontract. These quoted rates should also be a factor in determining the low bidder for a project, unless prohibited by public project bidding laws. If the labor rates and markups are not established prior to contract award, they will become known when the first change order request is made. If not previously established, these rates will commonly be much higher than if the subcontractor had provided these rates during the competitive bidding phase knowing their award of the project was partially dependent on the rates.

6. Mock-ups are an important part of the construction process. These trial-runs aid in establishing the acceptable level of craftsmanship for the respective project elements, and like submittals, they also help identify problems that have not been anticipated by the design team, general contractor, or subcontractors. Each mock-up will be clearly indicated in the respective system's specification section, but the support work for these mock-ups will not be coordinated with other specification sections. This support work must be coordinated and clearly allocated among the various subcontractors in the general contractor's bid instructions. Some examples include the following:

(a)  A shotcrete specification may call for one shotcrete test panel for each nozzleman applying shotcrete on the project. These test panels require formwork and rebar ahead of the shotcrete test. This is followed by demolition and removal of the debris once the shotcrete test is complete. It is the general contractor's duty to inform the formwork, rebar, and demolition bidders of this additional work, as well as to inform these bidders of how many test panels will be required. The number of test panels is dependent on the number of nozzlemen that the shotcrete bidders intend to certify, thus adding another level to this coordination effort.

(b) An exterior skin system generally requires a composite mock-up of multiple exterior skin materials, such as plaster, metal panels, and GFRC. This mock-up will be noted in the plaster, metal panel, and GFRC specification sections, but not in the specification sections for the supporting and complimentary work. The framing subcontractor will need to be directed in the bid instructions to construct a structure for these elements to be mocked-up. Other subcontractors, such as the flashing and caulking subcontractors, may be involved and require specific pre-bid directions as well. When the mock-up is deemed no longer useful, the demolition and removal must be accounted for in either the general contractor's or demolition subcontractor's budget.

7. If a subcontractor misses an item of work, they must be held responsible for all consequential costs associated with the missed item. A few common project mistakes include the following:

(a) Suppose the plumbing subcontractor overlooked a grouping of pipe sleeves in a concrete deck pour. The subcontractor will be responsible for coring the sleeve locations, and also for all costs incurred by other subcontractors. Large penetrations through a concrete deck require additional rebar in the deck around the penetrations. In the absence of this rebar in the cured concrete deck, structural angles are commonly bolted to the bottom of the deck to structurally reinforce the demolished section. These angles will be furnished and installed by the miscellaneous metals subcontractor, but all costs will in this case be the responsibility of the plumbing subcontractor. If these sleeves are missed in a post-tensioned deck, the plumber will be responsible for x-raying the deck to ensure that the cores are located such that the concrete core drilling does not sever any tendons.

(b) There is a considerable amount of work that occurs above acoustical grid ceilings and when the tiles are clipped it is most appropriate for the ceiling subcontractor to remove and replace the tiles. Every project will have occasions that the MEP subcontractors need to perform work above the ceiling after the ceiling tiles have been placed. Some of these ceiling grids must be clipped (a description and discussion of ceiling tile clips is presented in the chapter on acoustical ceilings) and when the tiles are clipped, it is most appropriate for the celling subcontractor to remove and replace the tiles. The subcontractor requiring access will be held responsible for all costs associated with the acoustical ceiling subcontractor removing and reinstalling the ceiling tiles, including replacement of damaged tiles. Even with due care, chipping ceiling tiles during removal and replacement is an inevitable occurrence.

(c) If the electrical subcontractor misses a conduit in a wall, they will want to simply surface mount the conduit. Surface mounting is generally not an acceptable practice due to aesthetic concerns. The electrical

**FIGURE 47.1** Surface-mounted steel brackets necessary due to missed embeds.

subcontractor will be responsible for gypsum board removal and replacement, as well as painting. Additionally, to avoid a noticeable patch, the gypsum board and painting cannot simply run the width of two studs; as a good construction practice it will need to be taken to a natural wall break such as a door frame, corner, or reveal.

(d) On occasion, the formwork subcontractor will fail to set a steel embed in a concrete pour. They will be responsible for the remedial repair. Since the concrete demolition is costly and disruptive, the remedial fix is commonly a steel bracket bolted to the concrete (Figure 47.1) by the miscellaneous metals subcontractor. The formwork subcontractor will be held responsible for this premium cost, but they may not perform any of the remedial work themselves.

8. Manufacturers often publish data stating warranties commence the day their products are shipped. Subcontractors often stipulate that their warranties commence the day the respective item of work is complete. Neither of these warranty commencement dates will meet the requirements of the prime contract. The prime contract will state that all warranties are to commence the day the entire project reaches substantial completion and shall be in effect for a stipulated period of time, most commonly for a period of one year. Thus, the issue is not that warranties begin too early, but rather that they will expire too soon, because the subcontractor is still fully responsible for their work in an equivalent or higher manner to the requirements of the warranty during construction until the official warranty period begins. Warranty language contrary to the prime contract will be found in a vast number of bids from a wide variety of subcontractor trades. The warranty terms must be renegotiated in each case. It is important to address these warranty discrepancies early in the bidding phase because only a few subcontractors will request additional compensation for correcting their

warranty to meet the contract requirements while in the competitive bidding environment. More subcontractors will request additional compensation during the subcontract negotiations once they know the job is in hand.

9. The cost of construction documents is a high dollar item (the construction documents for a large project can easily cost over $1000 per set) and a common source of conflict. If the general contractor does not mention whether or not they will provide drawings to the subcontractors in their bid instructions, the subcontractors tend to assume the general contractor is providing them. While general contractors most commonly have the subcontractors purchase their own drawings, it is preferable that the general contractor purchase two sets of documents for each subcontractor, one for their office and one for their field crew. If the subcontractors require additional copies, they should be responsible for purchasing them. Clearly relaying this information in the bid instructions is important to ensure that the drawing costs are either in the subcontractor bids or the general contractor's budget, but not both. By providing two sets of drawings to the subcontractors, experience has shown that they appreciate it and think of this as a collaborative gesture. On the other hand, when the general contractor requires the subcontractors to purchase their own documents, the subcontractors may regard the general contractor as being cheap. This may then be a contributor in making them wary of the general contractor's practices at the onset of the project. Getting the project off to a collaborative start is worth the effort of providing documents to the subcontractors and because these funds will be in the overall project budget in either case (either in the subcontractor bids or the general contractor's budget) it will not cost any more money. Naturally not all subcontractors will be provided full sets of the documents. For instance, a carpeting subcontractor will only be provided the architectural drawings, not the full set, to help keep this cost down. (Keep in mind that this document issuance to the awarded subcontractor does not limit the subcontractor's liability for the project. They will still be held responsible for reviewing the full set of documents prior to bid even if they are only provided limited documents after being awarded the project. Many of the smaller subcontractors will do this pre-bid review in the bid room at the general contractor's office.)

   (a) The subcontractors may be required to purchase their own drawings. If so, it is important that the owner, general contractor, or architect secure pricing with the printer on behalf of the subcontractors. Printing companies often provide low preferential pricing to the architect, owner, and general contractor, but use their standard, elevated, pricing for subcontractors. Since only one printer will have these documents, the subcontractors are forced to pay the inflated rates. Since an astute subcontractor will have the appropriate budget in their bid for these drawings, the owner will ultimately pay the inflated premium. This is another good reason to provide drawings to the subcontractors.

10. Temporary structures, such as a tower crane support (Figure 47.2) and material hoist (Figure 47.3), are often in conflict with the permanent construction.

**FIGURE 47.2**    Tower crane lateral support at building deck.

**FIGURE 47.3**    Material hoist.

Depending on the project and where the team chooses to place these large structures, the exterior skin, stair shafts, roofing, and/or interior finishes may be impacted. Construction elements hampered by a tower crane or material hoist will need to be completed toward the end of the project as return (also termed come-back) work. The general contractor must clearly reflect the locations of these temporary structures on a site logistics drawing in the bid instructions, define the work that will be impacted, and indicate in the project schedule when these temporary structures will be torn down.

11. Scaffolding must be tied off to the building, usually with wire that is routed through sealant joints or plaster. As the scaffolding is dismantled these wires are cut off and simply pushed into the wall. The sealant joint, plaster, or other finish is then patched by the respective subcontractor of the finish system.

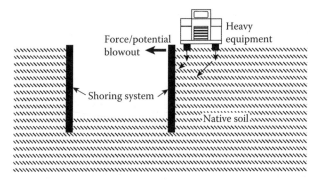

**FIGURE 47.4**   Equipment adjacent to shoring.

**FIGURE 47.5**   Trench plate dispersing load of crane outrigger. (Photo by author, courtesy of Hathaway Dinwiddie Construction Company and Tishman Speyer.)

12. When hoisting adjacent to a shored excavation (Figure 47.4), the weight of the crane must remain a significant distance back from the excavation. The point loads from the outriggers will need to be dispersed across the ground by the use of crane mats or trench plates (Figure 47.5) to avoid overloading the shoring system. Crane mats are commonly constructed of railroad ties fastened together side by side. These ties are heavy and must be transported by a flatbed truck and forklift. Furnishing of these large heavy crane mats or trench plates is the responsibility of the subcontractor supplying the crane. Without clear instructions from the general contractor they will not know that these costly mats are necessary and must be included in their bids. The shoring engineer will verify the criteria for equipment or other heavy items placed next to the excavations (refer to the chapter on shoring for further details). The general contractor must then ensure that all affected subcontractors are aware of this criterion prior to awarding the

subcontracts. This will be based on the maximum weights of various equipment and the respective distance each weight-class must be kept back from the excavation.

13. Subcontractors must be held responsible for their own hoisting, unless there is a tower crane and/or material hoist that they will be permitted to use. Despite the presence of a tower crane, subcontractors will be required to provide their own hoisting in two situations. First, the excavation, shoring, below-grade waterproofing, and rebar subcontractors will need to provide their own hoisting for work performed prior to the erection of the tower crane. Secondly, hoisting of equipment exceeding the capacity of the tower crane, such as large roof-mounted mechanical equipment, will need to be provided by the respective subcontractor. The schedule for the tower crane and its capacity are important items for the general contractor to incorporate into the bidding documents.

    Material hoists are generally available for use by all trades, but the interior dimensions of the material hoist and the weight capacity are important aspects to denote in the bid instructions. Anything that will not fit within the material hoist or is too heavy for the hoist must be loaded into the building by other means. The cost of alternative methods will be borne by the subcontractor.

14. Debris, dirt, dust, and common trash throughout a jobsite are part of the construction process. By some means, this unidentifiable trash must be cleaned up. If the general contractor provides all general cleanup labor for the jobsite, subcontractors tend to take advantage of this arrangement by leaving their trash everywhere. To ensure that the subcontractors maintain direct responsibility for the cleanliness of the jobsite, it is preferable to first hold them responsible for all (100%) of their own cleanup and secondly it is advised that each subcontractor provide laborers for a weekly composite cleanup. They will cleanup unidentifiable trash and lunch litter, sweep floors, and complete other miscellaneous cleanup activities, at the discretion of the general contractor. This cleanup crew is traditionally comprised of one laborer for a stipulated number of total subcontractor personnel on site. For example, suppose the criteria for composite cleanup crew participation is one laborer every Friday from 12:30 p.m. to 2:30 p.m. for every 10 (or fraction thereof) of each subcontractor's total workforce on site. A subcontractor with 35 crew members on site would be required to provide four laborers for composite cleanup. This is a firm requirement a subcontractor can calculate based on their total estimated job labor and include in their base bid. As long as subcontractors are informed of this requirement prior to bidding the project, they typically agree to the participation. (Note: Since it can be difficult to rally these workers in the middle of a workday, it is advised to start the composite cleanup crew either first thing in the morning or immediately after lunch.)

15. Grouting work should be performed by the subcontractor whose work is being grouted. For example, the plumbing subcontractor will grout the pump baseplates, the miscellaneous metals subcontractor will grout their

own post baseplates and the precast subcontractor will grout their own connections. The one exception to subcontractors completing their own grouting work is the structural steel baseplates. This exception occurs on union projects, where the craft jurisdictions dictate that the structural steel baseplates will be grouted by the placing and finishing subcontractor or the general contractor.

16. Finishes and other items are commonly damaged as an inevitable result of construction. The cost of repairing these minor damages must be accounted for. In addition to an allowance in the general contractor's budget, many subcontractors also include allowances. For instance, paint scratches and drywall dings are two of the most common blemishes requiring repair. Accordingly, the painting and framing subcontractors should include allowances in their base bids to remedy such damage. These allowances should be in the form of an estimated number of hours of labor and a stipulated amount for materials to take care of these repairs.

   (a) It is preferred that the subcontractor allowances be provided in terms of the estimated number of labor hours and a fixed amount for materials in order to ensure that this remedial work is completed at the lowest cost. First, the subcontractor's labor rates used for bidding purposes are commonly significantly less than the labor rates used for change order purposes. Secondly, by including the materials allowance in the subcontractors base bid they will have already included all markups in the bottom line of their base estimate. Therefore, by providing the allowances in terms of number of hours and the amount for materials, project costs will be controlled to a better extent. This approach has been discussed in more detail in previous chapters.

      (i) Though this repair work is indirect, it is still a cost of the work and should be treated as such. It is not in any way advised that general contractors make attempts to have change order work completed without paying a subcontractor a fair fee. Since this is an expected cost of work, not a change, these allowances should be treated as such. In this case allowances are used in fairness to the subcontractors simply because repair work is a variable for which they have no direct control. It is not fair to ask subcontractors to include variable costs, and the associated financial risks, in their lump sum proposals.

17. Projects in dense urban environments will have very little open real estate around a project. Subcontractors may not be permitted to bring a large office trailer onto the project, stage significant materials on site, or park their work trucks (which house all of their tools) near the building. Since these restrictions will introduce inefficiencies for the subcontractors it is important for the general contractor to issue a site logistics plan with the bid instructions to clearly convey to the bidders how much of the site they will be allocated. The subcontractors can then reflect any related inefficiencies in their bids.

18. All subcontractors requiring anchor bolts, steel embeds, or other items embedded in the concrete will provide these items FOB jobsite to the

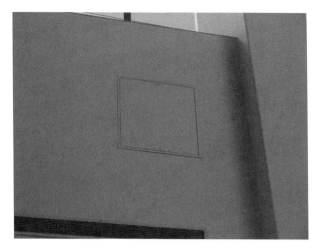

**FIGURE 47.6**   Access door in gypsum board wall.

formwork subcontractor (or the site concrete subcontractor for items outside the building lines). The formwork subcontractor will secure the items
in place prior to concrete placement. The exception to this rule will be MEP
sleeves and conduit embedded in the concrete. These will be furnished and
set by the MEP subcontractors themselves.

19. Access doors (Figure 47.6) in gypsum board walls and ceilings will
be furnished FOB jobsite by the subcontractor whose work is accessed
by the door for installation by the framing subcontractor. Most access
doors are for MEP work and these subcontractors are accustomed to furnishing these items. Other subcontractors are not as accustomed to furnishing access doors because they are rarely necessary for their work.
For example, a coiling door motor might be located above a gypsum
board ceiling. An access door will be required, but the coiling door
subcontractor may not realize this special requirement without having it
pointed out to them. In obscure cases such as this, the general contractor may decide to provide the non-MEP access doors. If so, it will be
imperative for the general contractor to include a budget for these doors
in their estimate and to also inform the affected subcontractor of this
practice to ensure that these access doors are not double covered.

20. Any item penetrating the roofing will require a roof jack, which is a sleeve
for which the roofing wraps up the exterior side. The interior of the roof jack
is then protected by sealants and possibly counterflashing, as dictated by
the architectural drawing details. These roof jacks are typically furnished
and installed by the subcontractor whose work penetrates the roofing. There
are exceptions to this rule. Further, it is advisable that these roof jacks be
placed as one monolithic piece, not split in half and pieced together in the
field. For example, for a steel post the roof jack needs to be slipped over the
top of the post in one piece before anything is mounted to the post.

**FIGURE 47.7**  Roof jack spacing. (Photo by author, courtesy of Hathaway Dinwiddie Construction Company and The California Institute of Technology.)

(a)    Exceptions to this rule are common with PVC and other similar roofing systems. For example, a PVC roof jack (Figure 47.7) will be part of the manufacturer's roofing system and provided by the roofing subcontractor. This exception can be confirmed for any roofing system with a quick phone call to the specified roofing system manufacturer or a review of the standard details on their website.

Subcontractors, through years of experience, become the foremost experts in their fields. This means they are the best resources to brainstorm value engineering ideas, alternative means and methods, scheduling, and all other aspects of their work. During the post-bid subcontractor interviews, the general contractor will be wise to ask each subcontractor if they have any value engineering ideas or if they have identified any aspects of the project that may have been overlooked. The subcontractors should be encouraged to disclose any suggestions that would help the project, including suggestions that would improve the schedule or project coordination. Subcontractors are also generally very willing to provide items they have included in their bids that they think other bidders may have missed because they want to be sure their competitors do not miss items that will otherwise make their bids higher. This same subcontractor input should be sought when allocating subcontractor scopes of work and preparing to issue the subcontracts. This query process may reveal many aspects of the project that can be improved. It is always most beneficial if the suggestions can be incorporated into the project as soon as possible. Otherwise, the problems these suggestions would otherwise solve may be overlooked during the early project planning efforts, only to be first brought up as a change order request.

# 48 General Contractor Responsibilities

Subcontractors all have one common attribute; they are experts in their trade. While subcontractors have an intimate knowledge of their trade, they are less informed about other construction trades. This is a primary reason why general contractors are employed on projects. The general contractor must have a working knowledge of all subcontracted trades, material suppliers, and service agreements. The general contractor effectively understands how these individual experts are brought together to form a cohesive team, then puts this team together and leads them through the project. This cohesive team will often consist of over 100 companies with a sub-contract, purchase order, professional services agreement, or other direct agreement with the general contractor. The scope of a project will be allocated among these 100+ experts ensuring that all aspects of the project are accounted for by one of these companies. It is also imperative that no aspect is accounted for by more than one of these companies. This task is as important as it is immense.

The foremost task of a general contractor entails the project planning effort. This is where the general contractor must identify and incorporate into their estimate all aspects of a project that are not otherwise included by the subcontractors, material suppliers, or professional services firms. Despite the love of the work, each of the general contractor's company representatives has the objective of making a profit for their firms. This is coupled with the ethical responsibility of ensuring all others involved with the project are given a fair opportunity to financially capitalize on their efforts. Overlooking items in the project estimate will ultimately diminish the company profits on a project. This chapter will focus on these potential oversights and how to effectively identify them. In this book, a myriad of activities a general contractor performs to support or supplement the various subcontractors have been described. Many of those issues will be reviewed to exemplify the role of a general contractor.

## GENERAL CONTRACTOR RESPONSIBILITIES

1. For estimating mobilization costs to a project site, general contractors routinely use a boilerplate estimating spreadsheet. This software is used to quantify the costs of office trailers, temporary toilets, site fencing, pedestrian barricades, and the other project staples. Even with this standardized checklist, there are still several items commonly overlooked in a general contractor's estimate for site mobilization. Examples of such items may include the following:
   (a) The general contractor will normally maintain responsibility for con-structing a large structure for the primary project sign and mounting this sign. The owner often purchases the sign. This purchase of the

sign must be verified, as purchasing this sign is sometimes specifically listed in the specifications as the general contractor's responsibility. At times, the responsibility for designing the sign will also be allocated to the general contractor.

(b)   Most inner city projects with a basement have little land available outside the building footprint. By default, the office trailers, dewatering settlement tanks, and other large heavy items, will be placed adjacent to the excavation. While these items are properly covered by the general contractor (or in some cases, subcontractors), expensive items commonly overlooked are the crane mats or trench plates that these items must bear on to disperse the loads along the shoring.

(c)   Site and city street cleanliness, especially during the mass excavation operations, is very important. This will be closely monitored by the city department of public works. The excavation subcontractor will provide street sweepers and wash down the truck tires before they leave the site as a standard operating procedure, but this is not always sufficient. The general contractor should include roll off (also termed rock) plates (Figure 48.1) at the site exit to help knock the mud off the truck tires as a standard practice. Periodic steam cleaning of the street may also be necessary for a block outside the site exit. Work such as this is commonly mandated by city officials.

(d)   The general contractor quite often is required to provide labor to assist the special inspectors. This labor can be quite significant, including cleanup, carrying equipment around the site, and hauling concrete test cylinders on a daily, and sometimes continuous, basis. This work is commonly missed in the general contractor's estimate.

**FIGURE 48.1**   Roll off (rock) plates to knock mud off of truck tires. (Photo by author, courtesy of Hathaway Dinwiddie Construction Company and California State University Northridge.)

2. The most important aspect of a construction project is safety. General contractors will commonly include a line item in their estimate for all-purpose safety. This will address additional guardrails, covering floor openings, purchasing safety harnesses, and other items common to projects. What is not regularly included is a budget for programs to promote safety and to teach safe construction practices. It is recommended that general contractors emphasize the importance of safety on a regular basis. This can be done through safety raffles, incentives for safe work behavior, safety posters, safety slogan contests, safety committees, safety lunches, and a host of other mechanisms that increase safety awareness. For example, a raffle might be implemented whereby each field worker is given one ticket to win weekly prizes through a raffle draw. This draw is held every week that no OSHA recordable injury occurs. The draw might also include only the names of workers who have not been observed performing work unsafely. For safety lunches the general contractor will buy lunch for the entire site to emphasize safety. The company might bring in a speaker from a safety-related company, such as a manufacturer of safety harnesses, to give a short presentation during the lunch. For larger projects, a company that sells safety products may jump at the opportunity to set up a small safety fair where about a dozen salesmen will set up booths to exhibit and sell various safety products. The effective promotion of safety can be achieved through many means and there are many creative solutions that have proven to make a difference in project safety performances.

(a)  When given an opportunity to set up a small safety fair on a large project, a company selling safety equipment will invariably do this for free. They may even sponsor the lunch. There are many companies who sell safety equipment and many would seize an opportunity to exhibit their products to the crews on a large project.

(b)  A collaborative approach for safety lunches is to have six or eight lunches sponsored by different major subcontractors on the project. Subcontractors do not mind doing this, and some really enjoy the opportunity to throw a small party. The subcontractors just need to know about this requirement prior to finalizing their bid for the project.

As important as these programs are, they are not required by law and can be quite expensive. As a result, they are not typically included in project budgets, especially for competitively bid public works projects. On public projects, the general contractors have an incentive to cut every non-essential dollar out of their estimates to win the project. In this case in particular, owner's representatives should be encouraged to add safety programs to the project budget. For instance, if the owner's representative for a public project is willing to pay for safety lunches and safety raffles throughout the course of a project, the general contractor should collaboratively agree to bill the owner at cost, with no markups or management fees, for these safety programs.

3. Safety rails, floor opening cover plates, and other safety items will either be included in the general contractor's estimate or the subcontractors' bids. Sometimes the maintenance of safety items, such as wood guardrails

**FIGURE 48.2**  Wood guardrails. (Photo by author, courtesy of Hathaway Dinwiddie Construction Company and California State University Northridge.)

(Figure 48.2), is overlooked. The general contractor should provide maintenance and removal of all safety items, even the ones installed by the subcontractors. This is because the subcontractors that install the safety items generally will leave these items in place for use by other subcontractors. The subcontractors who installed the items will not always be on site when the safety item is no longer necessary. For example, the formwork subcontractor will provide guard rails at the perimeter of a concrete deck. The formwork subcontractor will then leave the site many months before the permanent construction reaches a point that these temporary railings are no longer necessary.

4. During the period when the earthwork subcontractor is on site, they will maintain responsibility for dust control (Figure 48.3). Once the excavation subcontractor has left the site, dust control will become the responsibility of the general contractor. General contractors commonly include dust control on large sites for the parking lot and the yet-to-be landscaped areas. Dust control is often overlooked in the general contractor's estimate on inner city projects with deep excavations. On these projects the excavation subcontractor will leave the site for short periods of time while the shoring subcontractor installs the tiebacks. The general contractor is

**FIGURE 48.3**   Water truck for dust control. (Photo by author, courtesy of Hathaway Dinwiddie Construction Company and Tishman Speyer.)

responsible for dust control during this time period. This responsibility also extends from the time the excavation is complete until the water-proofing membrane, or vapor barrier, is placed.

5. A mass excavation operation will emit a great deal of airborne dust, even with diligent dust control. This dust in a city center will migrate outside the job to neighboring buildings and cars parked along the street adjacent to the site. The resultant residue on the neighboring building walls and windows must be cleaned by the general contractor near the completion of the project. Depending on circumstances, this cleaning might even be required intermittently during construction. Also, a common good neighbor policy added to a project scope at the owner's discretion is to provide occupants of neighboring buildings with coupons for free car washes once or twice during a project. Providing these coupons is not required by law or the project documents, but the neighbors will need to wash their cars more frequently due to the construction dust. Therefore, it is only fair that the general contractor, or the owner, include money in their budget for this neighborly gesture. Project owners are generally receptive to the cost for these cleaning activities, as the affected people will be their neighbors for years to come.

6. Cleanliness on a project is extremely important, especially lunch trash that will attract pests and rodents. In an effort to keep the building free of lunch trash the general contractor should designate a lunch area outside the building where all workers are required to eat (Figure 48.4), and then strictly prohibit any food inside the building. The general contractor should include a budget for a lunch tent and picnic benches for this purpose. This approach is not always possible. For instance, projects with little clear area outside the building footprint, such as inner city projects, will not have a feasible ability to keep food out of the building. If the project has an underground

**FIGURE 48.4** Lunch tent with picnic benches. (Photo by author, courtesy of Hathaway Dinwiddie Construction Company and California State University Northridge.)

parking garage a lunch area could be set up there to isolate where food will be in the building and to keep the food in an area where few, if any, finishes will be applied. Further, when workers park a significant distance from the site, keeping their lunch in their vehicles is not feasible. Nonetheless, they will need a safe place to keep their lunches on site. To avoid the workers keeping their lunches with them in the building, the general contractor could provide lockers in the eating area large enough for the small ice chests commonly used by trade workers.

7. Pest and rodent control is an important responsibility of the general contractor that begins early in the project. It is important to ensure that rodents are never given an opportunity to become established inside the building, which they may do as early as the structural phase of work.

8. As discussed in greater detail in the chapter on common subcontractor scope issues, the general contractor may incur added worker transportation costs. For example, the general contractor will be required to provide bussing for workers if the closest available parking lot is remote, commonly defined as more than a quarter mile from the project site. This remote lot is most often an empty lot rented by the general contractor. The general contractor will grade the site, lay filter fabric and rock for mud control during rains, fence the site, provide a parking attendant, and arrange for the busses to transport workers to and from the project. In addition to these staples, there are a few items general contractors may miss in their estimate. These include constructing roofed bus stops at the parking lot and the jobsite to provide shade and rain protection for workers waiting for the bus. The general contractor must remember to include parking/bussing signage, a parking attendant booth, and lot lighting. Lighting is not always necessary, especially if work is being completed during the normal daytime working hours. If there is

little or no ambient light for the parking lot from street lighting and the workers are beginning work before dawn or working beyond dusk (remember the sun sets early in the winter time) lighting at the parking lot will be necessary. Even if normal daytime hours are maintained, dusk in the northern states in winter may occur early enough to make lighting necessary. To avoid bringing power to the lot, this level of lighting can be provided by solar-light stands, which are powered by battery packs that are charged by solar panels during the day.

9. Construction operations such as mass excavation, concrete pours, and large crane hoists may require closing of traffic lanes and rerouting traffic. Closing lanes and rerouting traffic may be necessitated by a variety of subcontractors. The lane closure plan itself will commonly be the same every time. Since the lane closure plan is repetitious, it is most efficient for the general contractor to complete the design and to take care of the initial permitting. Then, each time a subcontractor requires a lane closure, they will not have the hassle of drafting, submitting, and pulling a new lane closure permit. Nonetheless, each time the permit is used a permit fee will likely need to be paid by the subcontractor.

   For projects in dense urban environments and other areas where trucks exit project sites onto congested or fast-moving traffic routes, a flagger will be necessary. Flagger expenses are traditionally covered by the general contractor. Since the general contractor is providing a flagger, the subcontractors commonly also assume the general contractor will provide all traffic control for the project, including absorbing all associated costs for lane closures. The general contractor's responsibilities do not extend to all aspects of traffic control. The subcontractors must maintain responsibility for all work associated with their own lane closures. This responsibility includes paying for the permit, setting the cones, renting flashing directional arrows, traffic signage, flaggers, and police officers, when deemed prudent.

10. An example of a simple, obscure and nonetheless important item that is also easy to overlook is the Knox Box. This is a small wall-mounted safe that is permanently mounted near the exterior entrance to a building and locked with a key possessed by the fire department. It contains the building keys that firefighters can retrieve in emergencies for full unobstructed access throughout the building. A Knox Box commonly takes six to eight weeks to procure, and a temporary certificate of occupancy may not be issued without one. If the acquisition of this box is not addressed until the end of the project, this miniscule item could cause a significant project delay. This box is normally recessed in an exterior wall, so if it is overlooked, additional costs could be incurred for cutting into the finished wall.

11. Subcontractors commonly are responsible for access to their own work, an important qualification in all subcontracts. For projects with a wide variety of exterior skin components, scaffolding will be erected for use by multiple subcontractors. Where multiple users are involved, it is suggested that the general contractor furnish this joint-use scaffolding. The alternative is to choose one subcontractor to provide the scaffolding. In some cases, all

subcontractors who will use the scaffolding will compensate the scaffolding provider a proportionate amount. This alternative, while used on occasion, is not recommended. It is difficult to convince one subcontractor to provide scaffolding that will be used by others. It is even more difficult to obtain equitable compensation from the subcontractors who use the scaffolding.

## ETHICS

A general contractor is in a position of considerable power and influence as the key player on a construction project. They have an obligation to exercise that power fairly when dealing with all project participants from the top of the ladder to the bottom. This includes decisions that relate to the owner, the subcontractors, the manufacturers, and the laborers in the field. There are temptations in any business to make decisions that favor one party over another. In many cases, the decisions are made because of the temptation to increase profits. Some of these decisions might deny just compensation to others or result in unfair blame being placed. Such practices do not go ignored. When the general contractor acts in this way, subcontractors are likely to adopt similar practices in an effort to "fight fire with fire." Subcontractors are likely to increase their change order requests and bid future projects differently with general contractors that regularly resort to unscrupulous practices. The entire reputation of the company can become tainted through such practices. Honesty and fairness are not only the cornerstones to being a great general contractor and a great manager, but also to being a great person.

## CONCLUSION

It is common for several different parties to be involved in the construction of most construction projects. The various parties, whether subcontractors or suppliers, will typically enter into an agreement with the general contractor. Each of these agreements is a binding legal arrangement. While these contractual arrangements place an aura of legal seriousness to the various relationships, it is important that all parties recognize that the most successful projects are those in which all the parties work together as a team. The contract should clearly outline the responsibilities of the contracting parties so that each party is fully aware of all their obligations. This prevents subsequent misunderstandings. All parties will have a more profitable project if they all work together as a cohesive unit. When the parties work well together, the parties will be eager to work together again on future projects. With the increase in project procurement by referral in the private sector (not competitive bidding), there is a long-term benefit in establishing good working relationships with other parties.

# Questions—Module Nine (Chapters 47–48)

1. Is it necessary to obtain proof of proper insurance from a scaffolding company when their contractual agreement primarily consists of rental costs and they perform very little labor on the project, just erecting and dismantling the scaffolding?
2. Give three reasons why it is important for a subcontract agreement to be fully executed (signed by all parties) prior to a subcontractor commencing work on a project.
3. When bussing of workers is required for a project, when does the workday begin?
4. Explain why site utilities work is commonly performed at night in lieu of during the typical workday hours?
5. When is it not feasible to require subcontractors to include their change order labor rates with their bid proposals?
6. Assume that the electrical subcontractor forgot to place the sleeves for a group of large conduits in a concrete deck prior to pouring concrete. The rebar subcontractor did not provide additional reinforcing because their work practice is to only add trim bars around deck penetrations physically placed on the deck. In this case the concrete deck will need to be reinforced with steel angles due to the absence of the rebar trim bars, and then the deck will be core drilled for the conduits. Which subcontractor will furnish and install the steel angles? Which subcontractor will bear the additional costs associated with the installation of the steel angles?
7. According to typical prime contract agreements, when does the effectiveness of the warranty for an air handling unit begin?
8. What is the relative cost of a full set of the construction documents? Are they expensive or inexpensive?
9. Give two instances in which subcontractors will be required to provide their own cranes for hoisting despite the fact that there is a tower crane scheduled for the project.
10. Which trade does not perform its own grouting work?
11. When planning a project, discuss the value of utilizing subcontractors as a resource for brainstorming ideas, discussing work sequences, identifying value engineering opportunities, and exploring alternative means and methods.
12. What is the extent of the general contractor's ethical responsibility of ensuring all others involved with the project are given a fair opportunity to financially capitalize on their efforts?
13. Discuss the suggestion that the general contractor should never be required to provide support work for the special inspector, such as carrying concrete test cylinders and cleanup after concrete slump testing, because such assistance is deemed a conflict of interest.

14. Describe four different ways a general contractor can promote safety on a project beyond what is required by law.

15. Lunch trash will attract bugs, flies, rodents, and pests, so it should be kept out of the building. Discuss jobsite protocols that can help mitigate this problem.

16. A comment on the effectiveness of pest and rodent control effort that are suggested to begin when the building finishes are nearing completion and to continue until the project reaches substantial completion.

17. Which party should be responsible for the lane closure plan and initial permitting procedures for a project that will require identical lane closures for a variety of subcontractors? If a separate permit fee is required each time the lane closure is necessitated, who should pay the fee?

18. Explain the purpose of a Knox Box.

# Index